DEATHSCAPES

Deathscapes

Spaces for Death, Dying, Mourning and Remembrance

Edited by

AVRIL MADDRELL
University of the West of England, UK

and

JAMES D. SIDAWAY
University of Amsterdam, The Netherlands

ASHGATE

Published by
Ashgate Publishing Limited
Wey Court East
Union Road
Farnham
Surrey, GU9 7PT
England

Ashgate Publishing Company
Suite 420
101 Cherry Street
Burlington
VT 05401-4405
USA

www.ashgate.com

British Library Cataloguing in Publication Data
Deathscapes : spaces for death, dying, mourning and
   remembrance.
   1. Death--Social aspects. 2. Death--Psychological
   aspects. 3. Burial. 4. Memorialization. 5. Memorials.
   6. Sacred space. 7. Place (Philosophy) 8. Death in art.
   I. Maddrell, Avril, 1964- II. Sidaway, James D.
   306.9-dc22

Library of Congress Cataloging-in-Publication Data
Deathscapes : spaces for death, dying, mourning and remembrance / [edited] by Avril Maddrell and James D. Sidaway.
   p. cm.
   Includes index.
   ISBN 978-0-7546-7975-2 (hardback) -- ISBN 978-0-7546-9935-4 (ebook)
1. Death. 2. Bereavement. I. Maddrell, Avril, 1964- II. Sidaway, James D.
   GT3190.D425 2010
   306.9--dc22

2010028776

ISBN 978 0 7546 7975 2 (hbk)
ISBN 978 0 7546 9935 4 (ebk)

Mixed Sources
Product group from well-managed
forests and other controlled sources
www.fsc.org Cert no. SA-COC-1565
© 1996 Forest Stewardship Council

Printed and bound in Great Britain by
MPG Books Group, UK

Contents

List of Figures and Tables

Figures

Tables

Notes on Contributors

Andy Clayden teaches landscape architecture at the Department of Landscape, University of Sheffield and is also a practising landscape architect who specialises in aspects of sustainable design. His research interests focus on the design and management of cemeteries and specifically natural burial. He has co-authored books on different aspects of sustainable landscape design, has contributed to the development of government guidance on natural burial, and has published refereed articles and book chapters on this subject.

Penelope Davies is Associate Professor in Roman art and architecture at the University of Texas at Austin. Winner of the Vasari Award for *Death and the Emperor: The Funerary Monuments of the Roman Emperors from Augustus to Marcus Aurelius* (Cambridge 2000, UT Press 2004), she is also co-author of *Janson's History of Art* (Prentice Hall 2007 and 2010). Her current research focuses on public art and politics in Republican Rome.

Bel Deering is researching the recreational uses of cemeteries as a part-time doctoral student at the University of Brighton. She has Bachelor and Masters Degrees in Plant Science from the University of Oxford. Prior to her current post she worked at the University of Aberdeen where she carried out a botanical research project in graveyards in Moscow and Scotland. Bel won a Millennium Fellowship in 1999 to carry out environmental research in Western Australia and has also delivered training programmes in Belize, Bermuda, Spain and Greece. She works for the RSPCA, managing an education centre, wildlife rehabilitation centre and nature reserve and is the author of *Animal Welfare* (Heinemann) and numerous articles on animal welfare education for journals such as *Child Education* and *Primary Times*.

Ken Foote is a Professor of Geography at the University of Colorado at Boulder. His work focuses on the social and geographical dynamics of public memory and commemoration, especially the imprint of violence on landscape in the US and Europe. Some of his works in this area are *Shadowed Ground: America's Landscapes of Violence and Tragedy* (2003) and *Shadowed Ground, Sacred Place: Reflections on Violence, Tragedy, Memorials and Public Commemorative Rituals* (2010) as well as articles he has co-authored with Maoz Azaryahu including *Historical Space as Narrative Medium: On the Configuration of Spatial Narratives of Time at Historical Sites* (2008) and *Toward a Geography of Memory: Geographical Dimensions of Public Memory and Commemoration*

(2007). He is currently working with co-author Sylvia Grider on a larger study of school and campus tragedies and with geographer Dominique Vanneste on plans for marking the centenary of the First World War in Belgium.

Clare Gittings is author of *Death, Burial and the Individual in Early Modern England* (Croom Helm 1984), and co-editor of *Death in England: An Illustrated History* (Manchester University Press 1999). A historian, she employs as historical data not only written texts but also paintings and sculptures. Several of her published articles concern visual representations of death and dying. Clare works at the National Portrait Gallery in London.

Paul Gough is Professor of Fine Arts, and Pro Vice Chancellor (Research) at the University of the West of England, Bristol. His interests are the processes and iconography of commemoration, the visual culture of the Great War, and the representation of peace and conflict in the twentieth/twenty-first century. As a painter he has exhibited widely in the UK and abroad, and is represented in the permanent collection of the Imperial War Museum, London; Canadian War Museum, Ottawa, and New Zealand War memorial. Details of his scholarly papers can be accessed on his website www.vortex.uwe.ac.uk as can extracts from his monograph on the painter Stanley Spencer (2006) and his new book on British war artists; *A Terrible Beauty* (2009).

Polly Gould is an artist and writer. She works with telling stories through drawing, print-making, video and audio with the current themes of landscape, representation and death. She exhibits regularly in the UK and elsewhere. She makes hybrids of theory and practice in the form of performance lectures. She has a BA in Fine Art from Central Saint Martins College of Art and Design and MAs in Theory and Fine Art from the Jan van Eyck Academie for Postgraduate Research in Maastricht, The Netherlands. Her writing is included in a collection of short stories by contemporary British artists *The Alpine Fantasy of Victor B. and Other Stories*, edited by Akerman and Daly (Serpents Tale) and *Telling Stories: Countering Narrative in Art, Theory and Film* (Cambridge Scholars Publishing) edited by Tormey and Whiteley. Polly Gould is an AHRC PhD candidate at the Bartlett School of Architecture, UCL, London.

Hilary Grainger is a Dean of the London College of Fashion, and Professor at the University of the Arts, London. A graduate of Leeds University, she was awarded a BA (Hons) History of Art and English and a PhD in Architectural History. A leading authority on the late Victorian architect Ernest George and the architecture of British crematoria, she has published widely in these areas and has lectured both in the UK and America. Her book *Death Redesigned: British Crematoria – History, Architecture and Landscape* was jointly commended in the category of best reference book published in 2006, by the Chartered Institute of Library and

Information Professionals. She is a Council member and Trustee of the Council of the Cremation Society of Great Britain and Chair of The Victorian Society.

Trish Green is a Research Associate in the Department of Landscape, University of Sheffield. She has a doctorate in Gender Studies from the University of Hull. Her main research interests lie in the relational aspects of life course transitions; ageing and gendered subjectivities and the emotional meanings of time, space and place. Her book, *Motherhood, Absence and Transition* (Ashgate) was published in May 2010.

Sylvia Grider is Senior Professor Emerita at Texas A&M University, where she taught in the Department of Anthropology. She has a doctorate in Folklore from the Folklore Institute at Indiana University and specialises in material culture. Following the fatal collapse of the student bonfire at Texas A&M University in 1999, she served as Director of the Bonfire Memorabilia Project. Since then, her research interests have focused almost entirely on spontaneous shrines and roadside memorials. She is particularly interested in the collection, documentation, and archiving of the artefacts which comprise spontaneous shrines. She has published numerous journal articles and book chapters dealing with vernacular mourning practices.

Jenny Hockey trained as an anthropologist and is Emeritus Professor of Sociology at the University of Sheffield. She has published extensively on death, dying, disposal and memorialisation. Her most recent publication is *The Matter of Death. Space, Place and Materiality*, co-edited with Carol Komaromy and Kate Woodthorpe (Palgrave, forthcoming).

Leonie Kellaher is Emeritus Professor at the Cities Institute, London Metropolitan University. As a social anthropologist placing an emphasis on material culture, her research has focused on the places in contemporary western society where the remains of the dead are buried, placed, scattered, strewn or, as ashes, reserved for later disposal. Those who visit and work in cemeteries, as well as bereaved people and professionals responsible for arrangements for ash disposals, have informed her research and publications.

Avril Maddrell is Senior Lecturer in Geography at the University of the West of England, Bristol. Her research includes gender and geographical thought and practice (*Complex Locations. Women's Geographical Work in the UK 1850–1970*, RGS/Wiley-Blackwell 2009); charity shops and volunteering (Horne and Maddrell, *Charity Shops, Retailing, Consumption and Society*, Routledge, 2002); and spatialities of death, mourning, and remembrance (Anderson, Maddrell, McLoughlin and Vincent (eds) *Memory, Mourning, Landscape* (Rodopi 2010).

Mary Murray's main teaching and research interests are the sociology of death and dying; animals and human societies; and aspects of social theory. She is currently Senior Lecturer in Sociology at Massey University, New Zealand. Before that, she taught at Durham University in the UK, and the University of Limerick in Ireland as well as for the Open University. Mary's PhD in Sociology was from the University of Glasgow, and her BA in Sociology with Social Administration, was from the University of Warwick. Her book and journal article publications span feminist theory, historical sociology, death and dying, and the relationship between humans and animals. She is currently writing a book about bereavement and grief, and researching near death experiences in New Zealand.

Anna Petersson is a PhD-student at the Department of Architecture and Built Environment, Lund University, Sweden. Anna has a Master of Fine Arts in Design and a Licentiate degree in Theoretical and Applied Aesthetics. Currently, she is working on her PhD thesis which examines why, how, and in what way material places and things are invested with meaning when in proximity to death, specifically regarding the function of such places and things as links between the space of life and the space of death. Her main publication is *The Presence of the Absent. Memorials and Places of Ritual*, PhLic diss. (Department of Architecture, Lund University, 2004).

Mark Powell is a Research Associate at Durham University, working on an ESRC-funded research project that considers disaster responses to the post-tsunami situation in Sri Lanka. He completed his PhD in Social Anthropology at the Queen's University of Belfast in 2006. His doctoral research considered issues of identity, belonging and community in the context of the Eastern Caribbean. He has since pursued his interest in identity through his work as a researcher on two ESRC-funded projects at the University of Sheffield. The first of these examined the landscape experiences of first generation migrants living in Sheffield, and the second focused on research into the cultural, social and emotional implications of 'natural burial' in the UK.

James D. Sidaway is Professor of Political and Cultural Geography at the University of Amsterdam. He has also taught at several British universities as well as being based for four years at the National University of Singapore and for a year as a visiting Professor at the University of Seville. His main interests are in geopolitics, development and the history and philosophy of geographical thought. He is an associate editor of the journal *Political Geography*.

Tony Walter has written and lectured extensively on funerals, public mourning, afterlife beliefs, representations of death in the media, mediators between the living and the dead, the Body Worlds exhibition, dark tourism, pilgrimage, and the concept of spirituality in palliative care. His books include *Funerals – and How to Improve Them* (1990), *Pilgrimage in Popular Culture* (1993) and *The Revival* *of Death* (1994). A sociologist, he is also interested in the spaces in which people die and encounter the dead, leading to articles in journals such as *Landscape, Landscape Research,* and *Health and Place.* He runs the MSc Death and Society at the University of Bath.

Jacqueline Watts is Senior Lecturer and Staff Tutor in the Faculty of Health and Social Care at the Open University, UK. This role follows a long first career as a training and technology transfer consultant in the civil engineering industry. She has postgraduate degrees from the University of Bath and Middlesex University. Her research and writing interests include feminist theory, gendered labour markets, professions and the social context of death and dying. She has recently completed a four-year role as Chair of the Open University's *Death and Dying* course. Her work has been published in a number of journals including *Qualitative Research, Gender Work and Organization, Work Employment and Society, Feminism and Psychology, European Journal of Palliative Care, Medical Sociology Online* and *Illness Crisis and Loss.* She is the author of *Death, Dying and Bereavement: Issues for Practice* published in 2010 by Dunedin Academic Press.

Eric Venbrux is Professor of the Anthropology of Religion in the Faculty of Religious Studies, and leader of the research group Refiguring Death Rites, funded by the Netherlands Organisation for Scientific Research (NWO), at Radboud University Nijmegen, The Netherlands. Eric has published a monograph on mortuary rites of Australian Aborigines, entitled *A Death in the Tiwi Islands*, with CUP, and he is co-editor of *The Oxford Ritual Studies Series* (OUP).

Joanna Wojtkowiak studied cultural psychology and is currently a PhD student within the Netherlands Research Council project on 'Refiguring Death Rites' in the Faculty of Religious Studies at Radboud University Nijmegen. Her dissertation concerns identity around dying, death and grief rituals in The Netherlands.

Kate Woodthorpe is a lecturer in Sociology at the Centre for Death and Society, at the University of Bath. She is the co-editor of *The Matter of Death: Space, Place and Materiality* with Jenny Hockey and Carol Komaromy, and has published on the experience of undertaking research into spaces associated with death. Her current research is exploring the role of the anatomical pathology technologist in hospital mortuary settings.

Ken Worpole is a writer and broadcaster, and the author of many books on social policy, landscape and architecture. He was a member of the UK government Urban Green Spaces Task Force, and an Adviser to the Commission for Architecture and the Built Environment (CABE) and the Heritage Lottery Fund. His books include, *Here Comes the Sun: Architecture and Public Space in 20th Century European Culture* (2001), *Last Landscapes: The Architecture of the Cemetery in the West* (2003), and most recently, *Modern Hospice Design: The Architecture of Palliative*

Care (2009). He is currently a Senior Professor in The Cities Institute at London Metropolitan University.

Foreword

Death foregrounds the most important social and cultural values that we live our lives by, including those values that we acknowledge and express, but also those that are neither ordinarily recognised nor explicit. Regardless of the ways in which death is managed, marked and memorialised across cultures, fundamental values are revealed through death and its remembrance. Spaces for the dead and dying are a reflection of the changing conditions of the living, as well as shifting meanings and discourses about life, for these spaces have cultural and symbolic meaning invested by the living, representing microcosms of the society within which they are established. Understanding death and dying, and particularly their related practices, rituals and spaces, thus offers insights into life and the living.

This volume reminds us particularly that death and dying are intensely anchored in space and place. Death and dying draw attention to the meanings that we invest in space and place, in as much as spaces and places offer a lens through which to understand death and dying. The ability of spaces and places associated with death and dying to evoke the deepest of memories and to stir an intensity of emotions is evidence of the power of place, and is a reminder that the very nature of our meaningful experience with place is fundamentally anchored in emotions, not functions. At the same time, the process and experience of death and dying is intensely place-based, deeply associated as they are with particular sites and locations. This spatial significance is expressed, for example, through the creation of memorial sites, the choice of which carries deep symbolic meaning. Yet, this locatedness of grief and memory faces increasing challenges as pressures on space for alternative uses erode the ability to anchor meaning in place. In some urban settings, this has led to the virtualisation of memorial sites. The ways in which the emotional relationship between people and place is altered more fundamentally deserves further study.

Deathscapes, as a collection offering multidisciplinary perspectives on the relationships between space/place and death, takes a major step forward in demonstrating to readers the significant relationships between space/place and death and dying. Through the lenses of sociologists, anthropologists, geographers, historians, psychologists and many other scholars of diverse disciplines, the volume offers a spatial lens through which to understand the intensely emotional experiences surrounding death and dying, including mourning, bereavement and remembrance. The chapters in this collection remind us that the experiences of death, dying, mourning and remembering, are mediated through multiple spaces – the body, the site of death, the home, the hospital, the hospice, the mortuary, the cemetery, the crematorium, the memorial, the sites germane to the identity formation of the deceased, and even the virtual spaces of the cyberworld. Through analyses of these various spaces and the associated experiences in specific contexts, we are brought face to face with a range of emotions: grief, fear, disgust, anger, belonging and community, sense of togetherness, camaraderie, and inspiration, for example. These are at once private and personal experiences, as well as public and collective ones. Numerous social and cultural values are also thrown into relief: filial piety, racism, environmentalism, gender (in)equalities, sacred/secular, and the meaning of care to name just a few. The volume thus covers much ground, and if it goes some way, as I believe it does, in helping us understand some of the most intensely emotional experiences in our lives, it will have succeeded in its task.

Lily Kong
Professor of Geography
National University of Singapore

Preface

Each contributor to this collection will have their own thanks and acknowledgements to make. As editors, there are those we would like to take this opportunity to highlight. We are pleased that Lily Kong was able to take time out of a busy schedule to pen the Foreword and appreciate her contribution here. Our thanks also go to Tony Walter for suggesting papers from the Bath 2007 conference for this volume. We should also acknowledge here not only the forbearance of our authors, who accommodated several rounds of our suggestions and editorial requests, but the efficiency of our copy editor and assistant, Claire Kelly. Without her hard work the book would have taken much longer to come to fruition.

In common with other researchers in the field (e.g. Walter 1996, O'Neill 2009), we as editors of this volume, became engaged with questions of death, dying and mourning through our own experiences of bereavement, specifically the death of a child, a son in the case of Avril and a daughter in the case of James. It was the personal experience which brought us into conversation about deathscapes and prompted our collaboration on this project. We hope that we bring insight from our own experience to this volume and within the spirit of materialising the ongoing place of our deceased loved ones in our lives, we dedicate this book to them.

Avril Maddrell and James D. Sidaway
Oxford and Amsterdam

Chapter 1

Introduction: Bringing a Spatial Lens to Death, Dying, Mourning and Remembrance

Avril Maddrell and James D. Sidaway

Deathscapes is an edited collection which focuses on the relationships between space/place and death, bereavement and mourning in western societies (with examples from the Netherlands, New Zealand, Sweden, the UK and USA). Contributors are drawn from a variety of disciplinary backgrounds (geography, sociology, art, history, psychology, health, planning and architecture), reflecting both the 'spatial turn' within the wider social sciences and humanities, as well as the growing engagement with death, mourning and memorialisation within geographical research. The inspiration for the collection came from the Death, Dying and Disposal 8 conference (held at the University of Bath, UK, September 2007), where there was a notable recurrence of papers explicitly addressing spatial themes from varying disciplinary, theoretical and empirical perspectives. A number of those papers have been rendered into chapters here, supplemented by some invited contributions.

The first section of this introduction examines the significance of space and place in relation to death, mourning and remembrance. This is followed by a discussion of 'deathscapes' as a concept, with the final section outlining the content of the chapters, their insights and linkages.

Bringing a Spatial Lens to Death, Dying, Mourning and Remembrance

Death is at once an everyday – universal – occurrence, and an extraordinary experience in the lives of those affected (Maddrell 2009b). Within contemporary western discourse, death itself is often described in spatial terms: a 'final journey', 'crossing to the other side', 'going to a better place'; but the experiences of grief and mourning are typically represented in temporal terms: 'time heals', 'give it time'. Nonetheless, grief and mourning are experienced and marked in space, as well as time. As Maddrell (2010: 123) has argued elsewhere:

> Mourning is an inherently spatial as well as temporal phenomenon, experienced in and expressed in/through corporeal and psychological spaces, virtual communities and physical sites of memorialisation …[these include] individual mappings of bereaved people's experiences of significant spaces/places and how

these change over time, how they are expressed though performance in space, written as corporeal, landscape or literary texts; and how these individual [and collective] emotional maps impact on particular places.

Thus death and bereavement are intensified at certain sites (such as the regulated spaces of the hospital, the cemetery and the mortuary) but affect and unfold in many others: the home, public spaces, places of worship, and sites of accidents, tragedy and violence. They are both intensely private and personal, while often simultaneously experienced and expressed collectively and publicly. Furthermore, these experiences of death, dying and mourning are mediated through the intersections of the body, culture, society and state, and often make a deep impression on *sense of self*, private and public identity, as well as *sense of place* in the built and natural environment.

Work within death studies has long focused on the hospital, the battlefield, the cemetery or crematorium and the undertakers; but research from a variety of disciplinary backgrounds has begun to recognise and articulate the significance of different spaces as spaces *per se*, rather than simply an area or institution – a container – where things happen. This attention to the significance and particularity of space can be seen in Morris and Thomas' (2005) analysis of the negotiations around the desire to die 'at home'; in Hockey et al.'s (2005) nuanced account of domestic space after the death of a partner; as well as in studies of the cemetery as social space (Francis et al. 2005), the crematorium as ritual space (Grainger 2006) and the hospice as palliative space (Worpole 2009).

In parallel, emerging work within geographical research is addressing new perspectives on space/place through interrogating death, loss and remembrance (e.g. Johnson 1995, Hartig and Dunn 1998, Teather 1998, Kong 1999, Foote 2002, Cloke and Pawson 2008, Maddrell 2006, 2009a, 2009b, 2010, Sidaway 2009, Wylie 2009); these are topics which have yet to be explored fully within geographical research, but which clearly relate to existing work on identity, memorialisation, sacred place and geographies of emotion and affect. Here we draw on reflections on space and place, including the material, representational and what Lorimer (2005) describes as the 'more-than-representational', liminal spaces of emotion and affect (Anderson and Smith 2001, Bondi et al. 2005, Smith et al. 2009). We are also mindful of marked social changes within western society and how these have impacted on different ideas and designations of space: the growing role of technology which has impacted on engagements with virtual space; how the characteristics of what can often be described as a simultaneously secular and post-secular society has resulted in a fluidity in defining sacred place discursively and geographically; and, not unrelated, how space becomes politicised and contested.

This edited collection considers many other sites and practices that connect the living and the dead, beyond those of burial or cremation. There is the obvious significance of the cemetery, crash site or war memorial, as spaces creating their own emotional geographies for the bereaved, but, as indicated above, if we probe further there are other types of space pertinent to our understanding of death and mourning conceptually and in terms of practical care of the dying and bereaved. There is the body as space, the corporeal living and dying person. There is the domestic space of the home; as site of dying and death, and locus for private, individual and vernacular remembrance, aspects of which are increasingly seeping into the realms of public space in 'spontaneous' or informal memorials. Other quotidian spaces interpellate us, often unawares, speaking to us of loss and consolation. These can be the everyday micro-spaces of the chair, vehicle, or bathroom shelf, or the more public spaces of the park, social club or school (see Hockey et al. 2005); each having the potential to assault the bereaved with a sense of loss or bring back fond and comforting memories. They can also be the places which were central to the deceased's identity formation such as the seat in the bar, the sports club, place of worship or workplace. The increasing significance of 'virtual' forms of space can be seen most notably in the growing number of online memorial sites on which to log grief and tributes, a new form of narrative space and memorial art, e.g. www.gonetoosoon.co.uk and the Nevis Partnership book of remembrance (see www.nevispartnership.co.uk, Maddrell 2010). Having outlined the *variety* of types of space, the second point to stress is the *significance* of those spaces. Whatever the type of space, it is through this engagement with, or being engaged by, particular spaces that their status shifts, they are transformed from mere physical areas into *places* (see Tuan 2001, Cresswell 2004) through being endowed with *meaning and significance*. Personal events, experiences and relationships, as well as wider historical, cultural and political associations, meld to give individuals a sense of 'place', which may be positive or negative. This may, in part, be shared with others in a collective view of place, but much may be personal, individual. This sense of place is accessed in the present and through memory and because of its significance engenders emotion and affect. For the bereaved various significant places can become *sacred*, sacred to the memory of the deceased, and to understand this we need to look at, but also beyond, the memorials and graves, which often bear that textual epitaph.

While more work needs to be done around the negotiations of memorials within private familial space, the conflict arising from sustained performance and inscription of remembrance in public space is well documented, notably in the case of more informal roadside memorials (e.g. see Hartig and Dunn 1998, Clark and Franzmann 2006) where death is considered by many to be 'out of place', not in its 'proper place' (Petersson, this volume). This idea of 'in/out of place' (Cresswell 1996) is at the heart of many of the individual and collective negotiations around death, dying, mourning and remembrance in contemporary western society. Ultimately it is found in each individual's evolving relation to the absence-presence of the deceased and the places associated with them. It is reflected in debates about expressions and markers of private grief in public spaces and related disputes of what constitutes 'sacred' ground; it is found in the creation of innovative forms of vernacular memorialisation in virtual space, the secular domestic shrine and the memorial bench in a favoured location. Benches, in common with other spaces of remembrance, exemplify the intersection of different narratives and performances of loss and remembrance. They are material, representational and more-than-representational: there is the subtext of their form and location; the explicit, if brief, text found on plaques 'storying the deceased' (Walter 1996, 1999, Neimeyer 1999); and the performances of remembrance, the maintainance of 'continuing bonds' or 'letting go' of the deceased which attending the bench invites. For all their solid materiality they can be seen as spaces of threshold, with liminal qualities (Maddrell 2009a), where each narrative, or more accurately, set of narratives, needs to be read in its individual context. Writing about memorial benches placed 'in loving memory' on the coast of Cornwall at Mullion Cove, leads John Wylie (2009: 279) to see 'absences at the heart of the point of view' and to argue that: 'the entire experience of the memorial benches at Mullion Cove seemed to me to be sensed more in terms of a slipping-away, a letting-go, a failing to grasp or even to touch'. In contrast, Maddrell's (2009a) analysis of the memorial benches in Peel, Isle of Man, demonstrates that many articulate belonging, rootedness in the locality and the continuing attachment of the bereaved to the deceased. These benches can be seen to constitute a sort of 'third emotional space' between home and cemetery/crematorium, which affords a public mapping of private emotion; a place where ongoing negotiation of absence-presence can happen and expressions of mourning and remembrance 'can be located and negotiated in the medium term' (*ibid..*); a threshold space where the work of meaning-making can be pursued – including that of 'letting go'.

As memorial benches illustrate, just as the sense of sacred has become less confined, more fluid, in contemporary society, so too have *forms* of memorial, remembrance *practices,* and their *location*. Thus, in addition to the act of dying and bodily disposal, it is this creation of performative as well as inscribed space/ place of remembrance which transforms everyday landscape into deathscape.

Deathscapes

The idea of a variety of 'scapes' as a means of understanding contemporary social processes was proposed by the anthropologist Arjun Appadurai (1990, 1996). His reference to the interplay of ethnoscapes, technoscapes, finacescapes and ideoscapes has since been supplemented by an edited collection on 'borderscapes' (Rajaram and Grundy-Warr 2007) and taken up in work on 'memoryscapes' (Ballinger 2003). In a similar context of places, spaces, sites, flows, disjunctures and landscapes, we might think of deathscapes. The idea of deathscapes was set out by Kong (1999) and employed by Hartig and Dunn (1998) in relation to informal memorials for road accidents. A search on Google reveals several other texts that have adopted the term, in a range of contexts. We are adapting the broad heading of deathscapes to invoke both the places associated with death and for the dead, and how these are imbued with meanings and associations: the site of a funeral, and the places of final disposition and of remembrance, and representations of all these. Not only are those places often emotionally fraught, they are frequently the subjects of social contest and power; whilst sometimes being deeply personal, they can also often be places where the personal and public intersect. Deathscapes thereby intersect and interact with other moments and topographies, including those of sovereignty (sovereignty-scapes), memory (memory-scapes) and work, life and beauty (landscapes). Thus, as another edited collection that chose to use the 'scapes' suffix noted in its introduction:

> Cultural geography understands landscapes as repositories of contesting interpretations of the meaning of a piece of land and of its appropriate use. Landscapes denote different and contesting technologies of the self...They assert particular moral geographies that denote a hierarchy of land use, and in this way act as an instrument of governmentality, attributing a sense of correct and incorrect behaviour. (Rajaram and Grundy-Warr 2007: xxv)

Of course, western society is no stranger to the mechanisation of death. Yet the focus in *Deathscapes* is not on sites and spatialities of genocide (see O'Lear and Egbert 2009). We are acutely aware that the notion of deathscapes cannot be causally invoked without also conjuring with war, destruction, violence and genocide. Indeed, as one of us has argued elsewhere; ' ... as critical geographers we should therefore seek to remember and reconnect the spaces, places, homes, and, indeed, graves ... that have been thereby forgotten with those that are repeatedly remembered.' (Olds et al. 2005: 478). With time, battlefields (of all kinds) become caught up in other narratives; the meanings of the Somme, or Dresden continue to evolve in the twenty-first century. And as they came into vision and have been mobilised within political discourses since the end of the Second World War, the crimes conducted by Nazi Germany and its allies have shaped wider understandings of genocide. Holocaust became a term applied to both the state-orchestrated process of destruction of life and to the threat of wider destruction via nuclear weapons that accompanied the Cold War. In his intoxicating account of the visible and invisible legacies of the Cold War, Tom Vanderbilt notes how the absence of direct battles between the superpowers, means that this 'war' is not subject to the same memory work that many of its component parts (Korea, Vietnam – each with their own controversies) have received. Vanderbilt (2002: 135) claims that:

> All wars end in tourism. Battlefields are rendered as scenic vistas, war heroes are frozen into gray memorials in urban parks, tanks and other weapons bask outside American Legion posts on suburban strips. That the Cold War, the so-called 'imaginary war' that was never actually fought (apart from proxy conflicts) – its Atlas missiles never launched, its atomic cannons never fired, its massive retaliations never employed – makes its tourism somewhat odd. This tourism curiously combines 'what if' with 'what was'; as one tours never-before-seen secret installations that seem familiar, one is looking at abstract doomsday scenarios poured in hard concrete.

Perhaps it was the difficulty of imagining the scale of death that unleashing nuclear weapons could produce that lead to the discourse of Holocaust. Indeed the genocide that also came to be signified by the term is hard to comprehend in scale. What does the murder of millions of people look like? A pile of discarded shoes or the ruins of a furnace become the symbols of mechanised deathscapes.

Reviewing the literature on the Holocaust (or Shoah as many prefer to term the attempted destruction of European Jews by the Nazi regime and its collaborators) Andrew Charlesworth (1992: 469), remarks on his experience in leading student field-courses to the sites of Europe's former death camps:

> One geographer with a visible sign of distaste has asked me how I can take students to a place like Auschwitz-Birkenau. No such distaste would have been expressed if I had taken them through the monumental imperialist landscapes of Berlin. Yet the heart and anus of European civilization are inextricably linked. To turn away from one is to cheat intellectually.

Charlesworth evokes the language of disgust for a good political purpose. More widely however, the taboos and silences around 'everyday' death and dying are frequently expressed in either euphemisms or with revulsion. Yet this is a geography that touches all of us. And although some of the chapters that follow do consider violent death, multi-victim tragedies and war, more often they are about the more mundane spaces and sites of mortality. Nor are they only about that; for with death come loss and the bitter-sweetness and power of remembrance. The latter is – as we have noted (and some of the chapters that follow investigate) frequently mobilised by (and shapes) individuals, communities and states. Indeed, in Flynn and Laderman's (1994: 51) words:

> Throughout history, human communities have converted the dead into sources of living power by grafting symbolic structures onto them and their places of internment. The impact of these structures on society, however, indicates that the 'dead' are understood as more than physical remains. The dead can be imagined also as memories, spirits, or deities, and the physical or spiritual locations where they reside are essential to the vitality of the symbolism ... When conflict arises and the meaning and handling of the dead are disputed by interested parties, the battle for control can lead to important changes in both identity and the distribution of power.

Although they write of Native American cultures, other landmark studies work from diverse localities in Europe and North America indicate this power was finessed in the twentieth century there and has endured into the twenty-first (Ballinger 2003, Johnson 2003, Young 1993, Verdery 2000). For example, Graham and Whelan (2007: 2) highlight a number of politicised and contested sites of remembrance:

As in the US 'Deep South', where commemoration meshes public memory of the Civil War, Civil Rights, and unequal power ... practises and sites of commemoration in Northern Ireland serve as icons of identity and spatializations of memory that transform neutral spaces into sites of ideology.

In Northern Ireland, as in many other places where sectarian, ethnonationalist and ideological divisions endure, graves, memorials and commemorations retain a capacity to mobilise and reflect those divisions. Dealing with the dead and memory reflects traumas. Thus, as Faust (2008: xvii) notes, the course and aftermath of the Civil War in America:

[...] confronted Americans with an enormous task, one quite different from saving or dividing the nation, ending or maintaining slavery, or winning the military conflict – the demands we customarily understand to have been made of the Civil War generation. American North and South would be compelled to confront and resist – the war's assault on their conceptions of how life should end, an assault that challenged their most fundamental assumptions about life's value and meaning ... Americans had to identify – find, invent, create – the means and mechanisms to manage more than half a million dead: their deaths, their bodies, their loss. How they accomplished this task reshaped their individual lives – and deaths – at the same time that it redefined their nation and their culture. The work of death was Civil War America's most fundamental and most demanding undertaking.

In the post-Civil War period, the means of dealing with dead bodies reflected wider social, political and cultural patterns and cleavages. But cleavages over treatment of and marking the place of the dead not confined to conflict and post-conflict contexts. In the case of the UK, for example, Julie Rugg (1998: 111) notes how cemeteries are a relatively recent phenomenon, dating from the nineteenth century, when 'The disposal of the dead was becoming secularised and municipalised', articulating new forms of community, organisation of the local state, and scientific developments that linked corpses to the transmission of disease. In a subsequent paper, Rugg (2000: 259) examines the distinctions between and characteristics of churchyards, burial grounds, mass graves, war cemeteries and pantheons, noting how:

[...] burial space is essentially mutable: its meaning does not remain static over time; and its significance is not uniform over all cultures. Even at a basic level, the significance of such space alters as time accrues between the living and the dead.

Such work (and much of what follows here) points to the ways that analysis must move beyond what Valentine (2006: 57) describes as; 'a conceptual and disciplinary split in which the grief of modern Westerners has been psychologised and medicalised, while the mourning or ritual behaviour of pre-modern and non-Western others has been exoticised and romanticised.'

The Chapters

Within *Deathscapes* the chapters are divided into four broad, but often overlapping, themes:

- At the threshold – living with death.
- Spaces of burial: taboo, iconoclasm and returning to nature.
- Negotiating space for memorialisation in private and public space.
- Art and design in service of remembrance and mourning.

Each of these chapters marks an intersection of social and cultural changes in death and bereavement practices at a particular time and place and examines how these relate to wider socio-economic and political mores, reflecting Howarth's (2007) call to study death and related issues within specific social and spatial contexts. Furthermore, building on earlier work which focused on the materiality of death (e.g. Hallam and Hockey 2001) and cemeteries as 'last landscapes' (Worpole 2003), chapters in this volume address a range of different spaces at varying scales, whereby the material intersects with geographies of emotion and affect, identity and politics, the 'more-than-representational' and the non-corporeal 'out of body'.

Current debates about deathscapes within academia and wider society are notably addressed by chapters on green and garden burial, negotiating emotion in public/private space, and remembrance of violence and disaster. The actual act of death is only touched on here, notably in Gould's chapter on Art and Mourning and in Petersson's discussion of road traffic accidents. But while dying is not the focus here, it is nonetheless the pivot around which all the chapters in this volume turn. Part I, 'At the threshold' is concerned with aspects of living with death. While we might all be described as 'living with death' as an inevitable part of our lives, Jacqueline Watts focuses on those with life-shortening illness, who consequently live with a heightened awareness of their mortality and its often quite immediate limits. In contrast to the more formal respite and/or end-of-life space of the residential hospice most commonly associated with terminal illness, Watts explores the more informal space-become-place of community care and self-help group support. In this everyday drop-in centre, within a multi-purpose community hall, Watts uses Yi-Fu Tuan's (2001) idea of 'field of care' to account for the practices of 'homemaking' undertaken by members of the group, in order to improve the physical space and thereby the experience of being together and being supported through illness, remission or the approach to death. This illustrates the significance of the built environment and its aesthetics (well known within formal hospice design (see Worpole 2009)), but also the performative space through gift-offering to others in the group, exemplified by bringing home made cakes or throws and flowers to brighten the tired décor, as well as giving and receiving personal care and support. It also goes beyond this articulation of a space of shared community and support, to illustrate how, drawing on Tuan again, it becomes a place of 'pause', a retreat where illness and its concomitant issues are permitted and voiced, but where a silent presence is also allowed. Near Death Experience (NDE) is typically a more sudden encounter with death or a death-like episode, something Mary Murray describes as an 'other worldly' liminal space between the living and the dead, frequently including extra-corporeal experience and some form of 'life review'. Murray demonstrates how NDE has become a political border between life and death, a border which is policed by academics and medical professionals who see NDE as an 'imagined geography' fixed within rather than something beyond corporeal-psychological space. As well as an intellectual battleground, Murray identifies NDE as a transformative space, where many of those who experience a temporary porosity of life-death are inspired to live the rest of their lives differently.

In Part II the focus is on spaces of burial, how this is associated with culturally defined 'taboo', those who are prompted to iconoclasm in the face of that taboo, and how others repudiate or re-negotiate that taboo through the dead body's 'return to nature' or the 'back yard'. Kate Woodthorpe sets the scene with her discussion of the East London Cemetery, where she argues that, despite society's growing readiness to engage with death *per se*, buried bodies in the cemetery continue to have a problematic social status. While domestic euphemisms are employed to describe the location of and relation with the deceased (e.g. the grave as a sort of bedroom), there tends to be a silence around the decaying body itself, the abject nature of the body (see Petersson 2006 and in this volume), something only hinted at in the territorial marking of grave boundaries whereby the integrity of the plot, and thereby the body, are maintained. In contrast, Bel Deering examines cemeteries (one of the largest sets of green space in urban Britain), as social spaces for the *living*. Through interviews and online accounts, Deering demonstrates how cemeteries are writ through with social sensibilities and both written and unwritten rules of normative behaviour, rules which some are ideologically or pragmatically drawn to break through practices which are considered anti-social by others and/or under the law. She traces how these rules are negotiated by those who wish to uphold and those who wish to transgress these norms; how dog walkers, often excluded from cemeteries, can become a form of neighbourhood watch, confounding those who see deserted cemeteries as opportunities for 'irregular' activities ranging from sex on tombstones to stashing stolen goods; how graffiti as memorial might be acceptable, even to heritage conservationists; how sites defined by respect for the dead become thrilling spaces, not only for the fashion/lifestyle/ music subculture known as 'Goths', but also for those on ghost tours or other forms of death-related tourism (see for example Foley and Lennon 2000, Lunn 2007, Sharpley and Stone 2009). Taking an historical perspective on private burial practices in England, Clare Gittings and Tony Walter look at questions of location and the longevity of burials in gardens and other private spaces. Comparing past and present practice, they argue that in addition to motives of choice and control (over the *funeral* as much as the actual *burial site*), ideas of home, family, nature and natural beauty inspire these burials; yet the most telling influence in recent cases is that of spousal love: 'They are not eccentrics, they are simply modern individuals who have invested in the intimacy of spousal love ... taking to the grave values that are central to modern society'. There are obvious links between garden burials and the chapter which follows on natural burial on farmland by Andy Clayden, Trish Green, Jenny Hockey and Mark Powell. The use of farmland for burial is a recent dimension of the now burgeoning number of 'green' burial sites in the UK. While this represents a new form of diversification/home working/ income generation on the farm, this welcoming of the dead in to and the bereaved on to farmland represents a particular intersection of material and social relations. Just as the design of these sites and their underlying economies are evolving, the negotiation between famers' visions of woodland or meadow burial sites are also changing in the face of lived mourning practices, for example when trees are planted or stone markers raised contrary to the original design or management plans. While the choice for green burial may be identity based, it appears this of itself is not sufficient to mark the identity of the deceased for many mourners, who need the individual marking or naming of the plot, a precise materiality which Hallam and Hockey (2001) describe as the 'spatial fix'.

The theme of negotiation continues in Anna Petersson's chapter on memorialisation in private and public space which opens Part III of the book. Influenced by De Certeau's discussion of 'proper places', Petersson draws on her earlier work on 'the proper place for death' (Petersson 2004, 2006) to analyse case studies of memorial-making in public spaces in Malmö, Sweden. Empirical work demonstrates the varying perceptions of site-of-death memorials, temporary and permanent: as site of 'blame' for the death; as sacred site to be inscribed, memorialised and visited; as unwanted reminder of death and threat to ontological security. In the face of these tensions, public bodies such as Malmö's Streets and Parks Department have acknowledged the need to mark the untimely death of some community members, but within a strong regulatory framework (as can also be seen in the case of memorial benches in the UK (Maddrell 2009a). Petersson notes that the number, site and tempo of memorials in public spaces are limited by the requirement that only those considered broadly to have been socially significant can be memorialised, by only allowing 'tasteful' memorials to be situated away from the actual place of death, with a focus on positive places of experience and memory – thereby maximising the physical and ontological distance between death/the abject dead body and public memorial. Again, there are links with the following chapter where Leonie Kellaher and Ken Worpole discuss the growing evidence of modest private memorials in public space in London, UK, which they describe as 'a new wave of patterning and inscription of public space across towns and cities'. Kellaher and Worpole describe this form of memorialisation as 'cenotaphisation': 'the spatial and temporal separation of memorialising practices from the disposition of the remains'. Relaxation of strict social and religious regulation and the 'miniaturisation and portability' of the body as ashes for the near 70 per cent of the UK deceased who are cremated, has resulted in greater freedom for the bereaved to bring their dead or their memorialisation of the dead closer to 'home', or to other localities sometimes far from the place of death or cremation. This mobility of remembrance has also resulted in a new map of memorials located within the everyday social space of their community of origin, in the form of benches, trees and plaques. For Kenneth Foote and Sylvia Grider, the colleges and universities of the USA whose communities have experienced the tragic death of students, become the identity home of the deceased. They argue that:

> [...] public memory is [...] part of the symbolic foundation of collective life and often lies at the heart of a community's sense of identity. The question of 'who we are' becomes an issue of what we share and do together as a community and more often than not, this sharing involves locating history and its representations in space and landscape.

Thus campus-related tragedies are part of what defines those student bodies and their educational environment, which in turn becomes marked by 'new ritual forms and spaces', embedding ongoing remembrance of the deceased in the material landscape and culture of the campus. Tracing the trend to memorialise the anniversaries of civil rights related campus deaths, this political impulse appears to have been extrapolated, by a process of accretion, to other deaths, notably multiple deaths in close knit identity groups such as sports teams killed in transit. More recently, this has extended to include victims of murder. While informal 'spontaneous' memorials are often a rapid and figurative response to tragedy, e.g. including photographs of the deceased as well as flowers and candles etc., Foote and Grider identify a tendency for permanent memorials to be more abstract symbols of loss. They also note the liminal nature of temporal thresholds such as anniversaries and spatial thresholds such as arches or gateways as significant in shaping memorial practices and spaces. In the final chapter in this section, Joanna Wojtkowiak and Eric Venbrux examine processes of remembrance and continuing relationships with the dead through the creation of memorials within the private space of the home in the Netherlands. Within the privacy of domestic space, where memories of the deceased are often vivid, remembrance and a sense of ongoing relationship can be maintained in personal and fluid ways: 'a home memorial serves as a focusing lens for ritual space and continuing bonds with the dead'. Drawing on extensive empirical work they elucidate a fascinating shift to domestic shrines, which are often dedicated to parents, centre on photographs and are kept by as many as a third of the population. After Walter (1996), Wojtkowiak and Venbrux argue these shrines represent an attempt to 'find a place' for the deceased within everyday life and practice, as well as being a space in which to reconfigure the identity of the mourner. Interestingly, these shrine-keepers cut across religious and secular social groups and reinforce the idea of a continuum of belief rather than absence of belief in western societies.

In Part IV, the chapters turn to examining art and design in the service of remembrance and mourning. In her account of tomb design in the Roman Republic, Penelope Davies articulates the relationship between memorialisation, familial lineage and political leadership. Outside of state regulation if located beyond the city walls, tombs afforded the opportunity to flaunt power and address the visitor. Funerary ritual included mourners wearing masks to represent esteemed forefathers and this emphasis on lineage, in both performance and material memorial, allowed the still-living to capitalise on an opportunity for self-promotion, as the tombs of the Scipios and Caecilia Metella demonstrate. Thus the tomb transformed 'the fleeting performance of pedigree into a permanent narrative display', and those who exploited this opportunity undermined the state's control of such representation elsewhere. This ancient historical case study demonstrates the perennial nature of negotiation around permanent memorialisation in public space, as well as a more political slant on the representation of identity – including that of the mourner – through the architecture of remembrance. Architecture is the subject of Hilary Grainger's chapter recounting and analysing the process of designing Coychurch Crematorium at Bridgend in South Wales, UK. In the face of the too frequent mismatch between function and ritual need at other sites, Maxwell Fry attempted to encapsulate the 'anatomy of mourning' within the structure and landscaping of the crematorium at Coychurch. He wanted to create a space for religious and secular ritual, which reflected the need for participation and emotional expression and which was also part of the local community. For Fry, the symbolic importance of the crematorium merited the best of design and materials: 'Fry was convinced that architectural language played a central role in articulating emotion and belief'. He wanted his design to facilitate rather than impede mourning and consequently made a reinstatement of ritual participation by mourners central to his design, seen in the curving access road which affords views of the buildings and landscape, and in the highly appointed flowing interior and exterior spaces. Thus the meshing of the buildings, design features and landscaped park were crafted to facilitate pause for thought and as a space in which to engage in remembrance and the performance of marking respect for the dead. Paul Gough's chapter on the representations in art of the battlefields of the Western Front of the First World War focuses on the absent presence of thousands of men, dead and alive, at different times of day, pre and post battle. Trench warfare with its horrendous casualties resulted in a 'numbing' scale of loss, when the living inhabited underground tunnels and dugouts and the dead covered or were barely covered by the surface: the 'crowded emptiness of no-man's land'. Medical staff dealt with broken bodies and minds and most soldiers had experience of the removal and burial of the fragmented and disintegrating bodies of colleagues and friends. Gough demonstrates how artists such as Stanley Spencer, Will Dyson, Otto Dix, Max Beckmann and Will Longstaff used painting and photography to 'bring the dead, the disappeared and the dying back to figurative life'. Thus the image of the war dead rising from the scarred landscape, such as Spencer's 'Resurrection of the Soldiers' at the Sandham Memorial Chapel in southern England, became iconic, a revisualisation for those at home of the distant dead buried abroad.

In the final chapter Polly Gould gives both a critical account of memorial art and a personal account of bereavement in which she travels afar. While travel and bereavement is normally associated with putting physical distance between the bereaved and place-based memories of the deceased (referred to as the 'geographical cure'), Gould intends the opposite, she travelled to Antarctica in order to engage with her loss, to break the 'melancholic stasis of repetition without progression' when the deceased 'will not pass away'. For Gould this was a 'turning of history into geography ... to find a place in which to encounter the ongoing presence of the past that I was not yet willing to let go', and the work of mourning became, in part, a work of travel, in part, a work of art. As Gould articulates, death is a mirror to life, a life experienced in space and time.

References

Appadurai, Arjun. 1990. Disjuncture and difference in the global cultural economy, in *Global Culture*, edited by M. Featherstone. London: Sage, 295–310.

Appadurai, Arjun. 1996. *Modernity at Large: Cultural Dimensions of Globalization*. Minneapolis MN: University of Minnesota Press.

Anderson, K. and Smith, S. 2001. Editorial: emotional geographies, *Transactions of the Institute of British Geographers*, 26, 7–10.

Ballinger, P. 2003. *History in Exile: Memory and Identity at the Borders of the Balkans*. Princeton NJ: Princeton University Press.

Bondi, L., Davidson, J. and Smith, M. 2005. Introduction: geography's 'emotional turn', in *Emotional Geographies*, edited by J. Davidson, L. Bondi and M. Smith. Aldershot: Ashgate, 1–16.

Charlesworth, A. 1992. Review article. Towards a geography of the Shoah, *Journal of Historical Geography*, 18, 464–69.

Clark, J. and Franzmann M. 2006. Authority from grief, presence and place in the making of roadside memorials, *Death Studies,* 30, 579–99.

Cloke, P. and Pawson, E. 2008. Memorial trees and treescape memories. *Environment and Planning D: Society and Space*, 26, 107–22.

Cresswell, T. 1996. *In/Out of Place. Geography, Ideology and Transgression*. Mannitoba: University of Minneapolis Press.

Cresswell, T. 2004. *Place. A Short Introduction*. Oxford: Blackwell-Wiley.

Davidson, J. 2003. *Phobic Geographies. The Phenomenology and Spatiality of Identity*. Aldershot: Ashgate.

Faust, D.G. 2008. *This Republic of Suffering: Death and the American Civil War*. New York: Alfred A. Knopf.

Flynn, J.P. and Laderman, G. 1994. Purgatory and the powerful dead: a case study of American repatriation. *Religion and American Culture*, 4, 51–75.

Foley, M. and Lennon, J. 2000. *Dark Tourism. The Attraction of Death and Disaster*. London: Continuum.

Foote, K. 2002. *Shadowed Ground. America's Landscapes of Violence and Tragedy*. Austin: Texas University Press.

Francis D., Kellaher, L. and Neophytou, G. 2005. *The Secret Cemetery*. Oxford: Berg.

Graham, B. and Whelan, Y. 2007. The legacies of the dead: commemorating the Troubles in Northern Ireland. *Environment and Planning D: Society and Space*, 25, 476–95.

Grainger, G. 2006. *Death Redesigned: British Crematoria, History, Architecture and Landscape*. London: Spire Books.

Hallam, E. and Hockey, J. 2001. *Death, Memory and Material Culture*. Oxford: Berg.

Hartig, K.V. and Dunn, K.M. 1998. Roadside memorials: interpreting new deathscapes in Newcastle, New South Wales. *Australian Geographical Studies*, 36, 5–20.

Hockey J., Penhale, B. and Sibley, D. 2005. Environments of memory: home, space, later life and grief, in *Emotional Geographies*, edited by J. Davidson, L. Bondi and M. Smith Aldershot: Ashgate, 135–46.

Howarth, G. 2007. *Death and Dying: A Sociological Introduction*. Cambridge: Polity Press.

Johnson, N. 1995. Cast in stone: monuments, geography and nationalism. *Environment and Planning D: Society and Space*, 13, 51–65.

Kong, L. 1999. Cemeteries and columbaria, memorials and mausoleums: narrative and interpretation in the study of deathscapes in geography. *Australian Geographical* Studies, 37, 1–10.

Lormier, H. 2005. Cultural geography: the busyness of being 'more-than-representational'. *Progress in Human Geography*, 29, 83–94.

Lunn, K. 2007. War memorialisation and public heritage in Southeast Asia, battlefield tourism. *International Journal of Heritage Studies*, 13(1), 81–95.

Maddrell, A. 2006. *Mapping Grief. Spatialities of Bereavement*. Second International and Interdisciplinary Emotional Geographies Conference, University of Kingston, Ontario Canada, May 2006.

Maddrell, A. 2009a. Mapping changing shades of grief and consolation in the historic landscape of St Patrick's Isle, Isle of Man in *Emotion, Place and Culture*, edited by M. Smith, J. Davidson, L. Cameron and L. Bondi. Farnham: Ashgate, 35–56.

Maddrell, A. 2009b. A place for grief and belief: the Witness Cairn at the Isle of Whithorn, Galloway, Scotland, *Social and Cultural Geography*, 10, 675–93.

Maddrell, A. 2010. Memory, mourning and landscape in the Scottish mountains: discourses of wilderness, gender and entitlement in online and media debates on mountainside memorials, in *Memory, Mourning and Landscape*, edited by E. Anderson, A. Maddrell, K. McLouglin and E. Vincent. Amsterdam: Rodopi.

Morris, S.M. and Thomas, C. 2005. Placing the dying body: emotional, situational and embodied factors in preferences for place of final care and death in cancer, in *Emotional Geographies*, edited by J. Davidson, L. Bondi and M. Smith. Aldershot: Ashgate, 19–32.

Neimeyer, R. 1999. Narrative strategies in grief therapy. *Journal of Constructivist Psychology*, 12, 65–85.

O'Lear, S. and Egbert, S.L. 2009. Introduction: geographies of genocide. *Space and Polity* 13, 1–8.

O'Neill, M. 2009. Ephemeral art: the art of being lost in *Emotion, Place and Culture*, edited by M. Smith, J. Davidson, L. Cameron and L. Bondi. Farnham: Ashgate, 149–62.

Olds, K., Sidaway, J.D. and Sparke, M. 2005. White death. *Environment and Planning D: Society and Space*, 23(4), 475–9.

Petersson, A. 2004. *The Presence of the Absent. Memorials and Places of Ritual*, PhLic diss. Lund: Department of Architecture, University of Lund.

Petersson, A. 2006. A Proper Place of Death? in *Architects in the Twenty-First Century, Agents of Change?*, edited by K. Rivad. Nordic Association for Architectural Research Annual Symposium, Copenhagen: The Royal Danish Academy of Fine Arts, School of Architecture, 110–17.

Rajaram, P.K. and Grundy-Warr, C. 2007. Introduction, in *Borderscapes: Hidden Geographies and Politics at Territory's Edge* edited by P.K. Rajaram and C. Grundy-Warr. Minneapolis: Minnesota University Press, ix–xi.

Rugg, J. 1998. 'A few remarks on modern sepulture': current trends and new directions in cemetery research. *Mortality*, 3, 111–28.

Rugg, J. 2000. Defining the place of burial: what makes a cemetery a cemetery? *Mortality*, 5, 259–75.

Sharpley, R. and Stone, P.R. (editors) 2009. *The Darker Side of Travel. The Theory and Practice of Dark Tourism*. Bristol: Channel View Publications.

Sidaway, J.D. 2009. Shadows on the path: negotiating geopolitics on an urban section of Britain's South West Coast Path. *Environment and Planning D: Society and Space*, 27, 1091–116.

Smith, M., Davidson, J., Cameron, L. and Bondi, L. (editors) 2009. *Emotion, Place and Culture*, Farnham: Ashgate.

Teather, E.K. 1998. Themes from complex landscapes: Chinese cemeteries and columbaria in Hong Kong. *Australian Geographical Studies*, 36, 21–36.

Tuan, Y.-F. 2001 [1977]. *Space and Place. The Perspective of Experience*. Minneapolis: University of Minnesota Press.

Valentine, C. 2006. Academic constructions of bereavement. *Mortality*, 11, 57–78.

Vanderbilt, T. 2002. *Survival City: Adventures Among the Ruins of Atomic America*. Princeton NJ: Princeton Architectural Press.

Verdery, K. 2000. *The Political Lives of Dead Bodies: Reburial and Postsocialist Change*. New York: Columbia University Press.

Walter, T. 1996. A new model of grief: bereavement and biography. *Mortality,* 1, 7–25.

Walter, T. 1999. *On Bereavement.* Milton Keynes: Open University Press.

Worpole, K. 2003. *Last Landscapes. The Architecture of the Cemetery in the West.* London: Reaktion.

Worpole, K. 2009. *Modern Hospice Design: The Architecture of Palliative Care.* London: Routledge.

Wylie, J. 2009. Landscape, absence and the geographies of love. *Transactions of the Institute of British Geographers,* 34, 275–89.

Young, J.E. 1993. *The Texture of Memory: Holocaust Memorials and Meaning.* New Haven: Yale University Press.

PART I
At the Threshold – Living with Death

Chapter 2

'It's Not Really Like a Hospice' Spaces of Self-help and Community Care for Cancer

Jacqueline H. Watts

Introduction

This chapter discusses the ways in which the material features of a community hospice day care setting influence the approach to caring for cancer patients and those who have been bereaved to cancer who make use of a 'drop-in' service offered by the hospice. It seeks to understand the everyday spaces of terminal illness and the performance of care for people with life-limiting disease. The chapter draws on recently completed participant observation research to offer a critique of professionalised and medicalised hospice settings that have become the institutionalised home of palliative care. In particular, the chapter argues that the culture and practice of care for those with life-threatening illness is, in part, a function of the 'culture' and aesthetic of the built environment itself. As an example, some care, housed in 'non-dedicated' settings, is much more likely to be characterised by a 'make do and mend' operational approach that directly influences both who cares and the components of care.

The chapter comprises six parts. The first serves as an introduction to the key features of hospice day care practice that has become a routine feature of services run by most hospices. This is followed by the second part, which considers innovative developments in relation to cancer day care, focusing on the work of the Maggie's Centres that have expanded over recent years across the United Kingdom (UK) from their original base in Scotland. Discussion of these new spaces for caring serves to contextualise the third part, which discusses conceptual meanings of space and place, highlighting in particular their social construction. A detailed outline of the participant observation study on which this chapter is based comprises the fourth part. It draws particular attention to the aesthetic and environmental aspects of the research setting that constitute the key discussion themes of the chapter. The penultimate fifth part focuses on the material and cultural features of care behaviour in the 'make do and mend' setting that is the research context. Discussion of both the theoretical and observational elements of the research brings the chapter to a close.

Hospice Day Care as Institutionalised Practice

Commentators (Higginson et al. 2000, O'Keefe 2001, Payne 2006, Watts 2009a) have identified the variety of day care services now offered by UK hospices to patients with life-limiting illness, with a majority of users of these services suffering from cancer (Addington-Hall 2004, Randall and Downie 2006). Although hospice day care can take many forms, the conceptualisation of provision often dichotomises between two polarities; highly medicalised with close attention to symptom management and pain control on one side and socially oriented, with stronger emphasis on patients' informational and social needs, on the other. Recently, Fisher et al. (2008) have pointed to a 'middle ground' to characterise day care within the hospice setting as therapeutic community space where patient/carer relationships that traditionally are shaped by hierarchy, are renegotiated and redefined to ones of equality within the day care setting.

Studies of day care provision in the hospice sector have prompted an emerging appreciation that the work of supporting patients who live in the community is complex and is subject to higher levels of discretion than has previously been recognised. Clark and Seymour (1999), commenting on research into the nature and effectiveness of day care, argue that there is a sense that this type of service, with its multiple components, is provided in an *ad hoc* and uncritical way to patients whose needs are not regularly assessed (see also Higginson and Goodwin 2001). More women than men use day care services and patients, both men and women, are drawn from the older age groups, mainly 65 years and over (Myers and Hearn 2001).

In general terms, the support offered within day care can be understood to focus on enhancing the quality of life of patients through the provision of information, rehabilitation, arts and crafts activities, beauty therapy, body care using massage treatments, physiotherapy and occupational therapy with the management and monitoring of symptoms, all central to psychosocial support (Myers and Hearn 2001, Dosser and Nicol 2006, Watts 2009b). Holmes (2001: 168) argues that psychosocial support is directed towards helping people to 'die living not to live dying'. Coping with what O'Connor (2004: 126) describes as 'transitions in status from wellness to illness, illness to wellness' is very much a feature of palliative day care that Clark and Seymour (1999) note can sometimes act as a 'stepping stone' to hospice in-patient admission. Lawton (2000) identifies a further aim of day care as the provision of a safe environment in which patients can accept and share the impact of life-threatening illness. Borrowing from Maslow (1970), the need to belong is a basic characteristic of what it means to be human, and this can often be met through talking with others in a similar situation and who have shared concerns. This can lead to the development of an informal network for the exchange of information in ways similar to in-patient settings (see, for example, McIntosh 1977).

Discussion of types of hospice day care has centred mainly on different models of service provision most of which emphasise a multidisciplinary approach to care, focusing mainly on the roles of clinicians, volunteers and informal carers as well as on resource availability that, increasingly, is concerned with evaluation and audit processes (see Andersson and Ohlen 2005, Barker and Hawkett 2004, Dein and Abbas 2005, Douglas et al. 2003, Hearn 2001, Holmes 2001, Low et al. 2005, White and Johnson 2004). The impacts of the physical setting in which day care takes place, however, has to date, received very little attention in the literature. Recent work by Worpole (2009), that develops ideas from Degremont (1998) and Verderber and Refuerzo (2006), has sought to draw attention to the importance of the aesthetic nature of the built environment as space for death and dying. Degremont's (1998: 127) specific question for the architect community is 'can an architect help improve patient management, particularly in the field of palliative care?'

Drawing on this architectural perspective, Worpole (2009: 9) contends that the hospice 'is a designed and constructed setting where the quality and harmonious sequencing of the spaces and functions matters more than for almost any other building type'. Hospices are dedicated spaces where death is acknowledged and attended and, whilst some are adapted premises for hospice purposes, many are now purpose-built comprising a collection of buildings and facilities, often surrounded by sensitively designed gardens and landscaped grounds, the product of evidence-based design (Worpole 2009). Verderber and Refuerzo (2006) note, however, that, although both aesthetic and function inform contemporary hospice design, there is often a lack of clarity in the commissioning process about who is the principal client: is it the funding body, the hospital trust, the clinicians, the care staff or the patients? Within hospice and palliative care philosophy the needs of the patient are paramount and Verderber and Refuerzo (2006) highlight the provision of single room accommodation within new hospices, built with a sense of domestic scale, as an important architectural design feature that accords with the healing and caring principles of palliative care.

High quality aesthetic that takes account of natural light, colour, privacy, clinical space and areas for community amongst both staff and patients has now become the norm in hospice 'new build' design, with attention to aesthetic understood to contribute to promoting dignity in end-of-life care. This approach builds on recognition that there is a need for more congenial environments more generally within health care settings in the interests of both care workers and clients (see Watts 2009b). Despite this, Worpole (2009: 64) argues that 'by and large, the architectural principles informing the design of modern hospitals and hospices remain largely functional, with the efficient delivery of medical and technical services being given priority over the well being of the patient'. These comments point to the ways in which ideas about health provision within the UK have become increasingly influenced by a business culture that emphasises processes of rationalisation, audit and targets, that includes such matters as space utilisation. The application of standardised metrics-based criteria to measure 'quality' within end-of-life care, however, may not be appropriate as part of a clinical machine that has the concepts of 'cure' and 'through put' at its core.

Responding to the physical, emotional, psychological and spiritual needs of dying people as *individuals* is at the heart of palliative care philosophy and practice that advocates whole person care (Payne and Seymour 2004, MacLeod 2008, Mitchell et al. 2008). In the hospice context, knowing and valuing patients as individuals has resulted in high quality care in both in-patient and day care settings. Aspects of palliative care philosophy, particularly its emphasis on person-centred holistic care, are now being applied in innovative ways in other settings where new models of day care are being developed. Attention to the physical environment in which this care can be accessed is seen as increasingly important as demonstrated by the recently established Maggie's Cancer Caring Centres, first established in Scotland, and now with sites operating across the UK. These centres have been the subject of widespread architectural and public interest, not least because of the cutting-edge design of the buildings and the way in which the quality of the aesthetic is accorded the same priority as the quality of the care service provided to cancer sufferers. Details of these new spaces for caring are considered further below.

New Spaces for Caring

Maggie's Cancer Caring Centres are a network of drop-in centres that aim to help anyone who has been affected by cancer including people with active disease, those who are in remission from the disease as well as people who have been bereaved through cancer. Each centre, built alongside an existing National Health Service (NHS) cancer hospital, is not intended as a replacement for conventional cancer treatment, but as a caring environment that can offer support, information and practical advice for problems associated with cancer. The organisation stresses its close working links with the NHS and specifically its remit to augment NHS provision that increasingly is target driven, reducing the possibility for individual patients to be given 'time' by clinicians with heavy caseloads (Seamark et al. 2008). The service provided by Maggie's is not appointments-based and is not related to a referral system, as are so many cancer services within the NHS. It is a free, small-scale, informal but professionally led provision intended to help cancer sufferers make a healthy adjustment to living with cancer (see www.maggiescentres.org. uk).

Maggie's Centres are run by a charitable trust named after Maggie Keswick Jencks who died of cancer in 1995. During the course of her illness, Maggie Jencks's experience of many clinical spaces (hospital waiting rooms, patient wards, consulting rooms, out-patient clinics), as unwelcoming and stress inducing, shaped her perception of these as essentially negative spaces that undermine rather than enhance an ethos of care and hope. This led her to reflect on how buildings (and the space within them), intended primarily as places for care, can de-humanise those who use them and negatively impact the well being of ill people. The largely disappointing experience of the hospital environment acted as a compelling motivation for Maggie and her husband, the architect, Charles Jencks, to consider different kinds of space or building types that elicit beauty, style and comfort for those confronted by life-limiting illness such as cancer (Keswick Jencks 1995). Ball (2009: 112) makes the point that once a person has been given a diagnosis and prognosis that is life-limiting, intimacy with strangers in the health care workforce is likely to become a familiar feature of life. Maggie Jencks recognised that the space in which that intimacy occurs can profoundly affect the quality of the 'intimacy encounter'.

The result has been the development of a series of architecturally stunning buildings, some of which have been designed by leading architects such as David Page, Richard Murphy, Zaha Hadid, Richard Rogers and Frank Gehry. Their key design features are open plan space, no corridors, windows that are placed to enhance a sense of light and connection to the outside, a central kitchen with a large table that acts as the focal point of the centre, high quality comfortable (and often expensive) furniture and accessories that exude style and a feeling of smart 'unfussy' luxury. Bowls of fresh fruit and vases of flowers contribute positively to the sensory ambience of a place where, first and foremost, people gather. The centres eschew labels, signs and closed doors and are designed to uplift and refresh the human spirit. The organisation's website (see reference above) explains that the 'centres are built to be as enticing and intriguing as possible to help you to take that step to come through the door'. The centres that I have visited resonate a sense of cheerful chic sanctuary. Worpole (2009: 20), writing about the architectural history of buildings as places for both care and control, notes that 'sometimes the ethos of an institution and its staff can overcome the restrictions of the building'. In the case of Maggie's Centres, however, staff in the service they provide, are culturally enjoined to 'live up to' the high expectations created by the aesthetic of the building that sets a tone of 'only the best care will do'.

The extent to which staff working in these environments experience their stylish and pristine features as a pressure to 'over perform', may well form the subject of further enquiry but, for this purpose, it is sufficient to note that the built environment of this care setting both shapes and is central to the programme of care offered which has five core elements provided in a supportive non-institutional setting. The five elements are: emotional and psychological support (including group and one-to-one counselling); relaxation and stress management (using such techniques as visualisation); information (this covers a wide spectrum including clarification of clinical terms to aid understanding of diagnoses); benefits advice; other support such as nutrition workshops and tai chi (Maggie's Centre 2006). Although, as illustrated above, a stylish high quality physical setting is central to creating optimal caring conditions at the Maggie's Centres, the possibilities of the virtual 'e' age have not been lost from view, and the establishment of an online community as 'e' caring space, that acts as a further strand of support, has been incorporated within the programme. The introduction of this support follows the spirit of Brown (2009: 69) who claims that the use of online blogs reduces the scariness of cancer which, he says, 'does not have to be the end till The End'.

The light, airy, uncluttered, but aesthetically rich, spaces of the Maggie's Centres are in stark contrast to the clinical spaces of mainstream NHS settings. They are also different from the spaces within hospices that traditionally have fostered a home-like, almost domestic atmosphere that Worpole (2009) describes as 'cosy'. This emphasis on 'cosiness' within many hospices may be a reflection of the fact that the hospice, as a residential as well as a day care setting, is for many of its patients, their last home. The Maggie's Centres have no residential facility and many of the people they help are younger and highly value life-style elements as a way of maintaining independence and self-confidence (Worpole 2009).

There is no doubt that the residential care element of the majority of hospices is a significant differentiator in terms of the public perception of these institutions that are still seen by many as places where people go to die. They are this but, as the above critique illustrates, they also serve a number of other functions and in many cases can now be seen more accurately as a 'hub' of specialist expertise, with much of their care work conducted in the community in people's homes (see Watts 2010). The person-centred approach of hospice care has become a type of benchmark of excellence, even though this is available to very few people (Randall and Downie 2006, Sinclair 2007). Despite the primacy of person-centredness in palliative care philosophy, patients who are living their last days and weeks in a hospice are cared for *en masse* (even if not on a large scale) and Peace (2003) reminds us that living or being cared for *en masse* can lead to a form of living/care executed and managed for the organisation rather than the person, often as a function of the pressures induced by institutionalisation, with issues of control, safety and surveillance at the fore. Goffman's (1961) ideas about the construction, nature and abstraction of institutional life have made a key contribution to understanding of the ways in which organisations can develop a 'life' of their own that sublimates the needs of those whose interests they have been established to serve. Meanings of space and place, that enhance Goffman's (1961) theoretical perspective, are considered below to provide further context for later empirical discussion.

Understanding 'Meanings' of Space and Place

Space evokes diverse meanings in different situations and under different circumstances. The definition of space is complex but what helps to define it are the boundaries applied to space, with these boundaries principally shaped by the way in which people use space and have access to it. One denominator, for example, is the distinction between public and private space (Sommer 1969), with connotations of home as private space contrasted with public buildings, such as hospitals, as public space. Homes are seen as appropriate spaces for intimacy and provide a base for family life and mutual care giving. Hospitals, on the other hand, are 'official' spaces that are dominated by professional clinical activity and regulated by procedures such as health and safety protocols. They are also, however, places for intimate care with this care given as part of a service encounter. This example points to how fixed definitions of space shaped, for example by utility, may not reflect the reality of ambiguity in relation to experiences within space when considered according to discrete categories. Nevertheless, the way in which space is perceived and primarily defined is reproduced and reinforced in the design of buildings as a function of architectural practice. In recent times the principal shift in understanding issues of space has been towards its transformation as place, which has the core attribute of value accorded it by people.

The nature and meaning of space and place have featured as prominent themes in economic and human geography literature in recent years. Commentators in the field of health and well-being have also begun to show an increased engagement with spatial themes and the work of Jonas-Simpson (2006), which explores how changing the meaning attached to place influences health and lifestyle choices, is one example. More widely, interest in how people both individually and collectively form a meaningful relationship with the locales they occupy by transforming 'space' into 'place' has underpinned extensive academic debate. Cresswell (2004: 12) captures the essence of this debate arguing that 'place is how we make the world meaningful and the way we experience the world'. His critique of place as meaning highlights the dynamics of places that he argues are continually in formation and are 'never finished but always the result of processes and practices' (Cresswell 2004: 37). Places, as culturally determined 'artifacts' (Shields 1997: 187), are shaped by what goes on within them specifically by the things people do within them and, in this sense, are embodied and constantly performed.

Casakin and Kreitler (2008) develop these themes focusing on place as both attachment and connection arguing that 'by interacting with their environments individuals create bonds and links. In the course of this interaction, anonymous spaces are converted into places endowed with meaning, which serve as objects of attachment'. They make the further point that attachment is characterised by an emotional bond to a location, with the topic of spatial emotional engagement now an important area of debate within the human geography literature (see, for example, Davidson et al. 2005) that, in particular, has begun to critically distil the relational aspects of space and emotion in respect of chronic illness and disability. Morris and Thomas (2005), writing about emotional, situational and embodied factors in preferences for end of life care amongst cancer sufferers, highlight how place with its 'normal' or 'regular' meanings (such as in the case of home, for example) can become problematised in the context of care relationships shaped by emotional concerns about dependency, loss and suffering. Concerns centred on the management of the unbounded body of the dying person (see Lawton 2000) point more generally to the ways in which bodily boundaries are vulnerable and permeable. Longhurst (2001), for example, identifies the threat to spatial 'norms' from leaky, messy and awkward zones of the inside/outside of bodies that she argues are not to be trusted in public places. Bodies that are insecure or 'out of control' can also alter the orderliness of domestic spaces with resulting significant emotional impacts for those involved.

These ideas, all to a greater or lesser extent, draw on the work of Tuan (2001) whose ideas about the ways in which people think and feel about space have been influential in developing theoretical insight into the concepts of place and space. At the level of simplicity he suggests that place is security and space is freedom; we are attached to the one and long for the other. He recommends that space be considered as that which allows movement while place be understood as pause. Working from Tuan's (2001) thesis, the interplay between space and place has a temporal dimension with time for transformation implicit in his model. In the case of people with life-limiting illness this is set against a backdrop of consciously time-defined lives (Small 2009).

The concepts of physical space and personal space are two further 'categories' of space that have received considerable attention in the health and social care literature (see, for example, Lawson and Phiri 2003). Writing about care homes for older people, Peace (2003) argues that the personalisation of physical space in this setting is particularly important in helping an older person, who has had to leave their 'own' home, adapt to a new living environment that is likely to be their final home. Twigg (2001), commenting on issues of autonomy in caring relationships, makes the point that people who receive care often have to allow other people into their space that may include the deeply personal space around their body. This suggests that territory, particularly territorial boundary, constitutes another way of looking at space, with ownership of space, especially personal space, a key feature. Longhurst's (2001) contention that human embodiment, as spatial category, is shaped by the materiality and fluidity of the body itself, argues for corporeality to be acknowledged as central to understandings of the constructions of place and space.

The Study: Setting, Participants and Conduct

The research that provides the empirical basis of the chapter was set in a community hospice trust that is housed within a local community centre in an urban area in the South of England. The trust has been constituted as a non-clinical service aimed at providing informational, social and therapeutic support to cancer sufferers. It has no medical facility, no in-patient unit and no employed clinical staff so it feels very different from a traditional hospice. The emphasis is on meeting social rather than clinical need in an informal environment.

The trust operates twice-weekly afternoon cancer 'drop-in' sessions with variable numbers attending. The aim of the research was to explore why users of the drop-in service came to the sessions and what they perceived to be the main benefits of attending. A group of roughly eight to ten regular users of the drop-in constituted a core group who usually were present on both days. There was a very wide age range represented amongst users of the facility with the youngest being 32 and the oldest 88. Proportionately more women than men use the service and nearly all the volunteers are women. The trust has a manager and fundraiser with informational and practical support provided by a team of dedicated volunteers, a now well-documented and familiar feature of the cancer narrative (Lawton 2000, Armstrong-Coster 2004). The voluntary efforts of a range of health and therapy practitioners contribute to a portfolio of different treatments available to users of the drop-in; reflexology, Indian head massage and aromatherapy being the most popular.

Initial contact with the centre was serendipitous through engagement with work of the local voluntary sector in an unrelated area. The opportunity to visit the centre and become an informal volunteer, helping with social aspects of the drop-in sessions, was a pre-cursor to the researcher role. It is this volunteer function that has shaped the participant observer role. The research was conducted over a fifteen-month period, enabling the development of close relationships with many of the participants. This long and close familiarity with the setting and participants framed an ethnographic research approach to data collection, and other writing (Watts 2008 has discussed its particular features in some detail).

Following approval from the management committee for this study I set about planning the ethical framework for the conduct of the research with particular initial attention directed towards the issues of confidentiality and anonymity for participants. Concerns about ethical rigour in the design and conduct of the research initially centred on the issue of informed consent and were far from straight forward. Asking individual participants to sign consent forms, as one way of acknowledging the researcher aspect of my volunteer presence, did not feel very meaningful. Instead, a brief outline statement of research interests was made available at the sessions and, as part of interaction with new users of the service, taking care to refer to research as well as volunteer features of my role, contributed to ethical conduct. The methods used were a mix of participant observation and informal conversations with users of the twice-weekly drop-in sessions. Because these interactions were not interviews in the accepted sense, audio recording of these was neither possible nor appropriate, particularly given the public space in which they occurred and the associated ethical concern of confidentiality.

Some of the data have been drawn, not from conversations between participants and myself, but from listening to talk between group members and from watching their body language. These observational elements of the research were illuminating and this confirms Jones and Somekh's (2005) claim that observation is an important, but often under-rated, method of data collection. The making of detailed handwritten notes in the form of a research journal (Rager 2005) constitutes the documentary data. A thematic approach to data analysis was adopted applying grounded theory principles (Glaser and Strauss 1967).

The Influence of Setting and Aesthetic on Care at the Drop-in

Other work (see Watts 2009a, 2009c) has reported on the findings of the research noting that reducing isolation and fostering a sense of hope for continued survival were the main motivation for attendance. For this chapter, the organisation and utilisation of the space of the care setting is the main focus.

The drop-in sessions were accommodated within two rooms in a small community centre which is shared by two other voluntary groups. These two groups did not hold meetings on the afternoons of the drop-in sessions so the two rooms that acted as meeting/care spaces were for dedicated use. Although volunteers joined the sessions (helping to make tea, join in board games and generally chat to users of the service), it was the service users themselves who organised the space to suit their preference. The furniture at the centre was mainly second-hand and some of it was shabby and not very inviting, giving the space a rather drab feel. Sofas (one that was threadbare on the arms and another with a hole in one of the seat cushions), single chairs and a couple of old coffee tables were available; these were arranged slightly differently each time. Often volunteers would be involved in helping to move furniture, as some of it was heavy and bulky; such help was at the direction of service-users. The dilapidated nature of most of the furniture was only one aspect of the dreariness of the setting; in both rooms used for the sessions paint had started to peel off the walls and the use of low wattage light bulbs made the space feel rather dingy.

As part of setting up the room, some of those attending the session would bring their own throws, shawls or light blankets (particularly in the winter months) to make the setting more comfortable and give the room a more pleasant and 'cosy' feel (Worpole 2009), creating a homelike environment. Sensitivity to light, cold and draughts was experienced by a number in the group and bringing these accessories helped to ameliorate their effects. Degremont (1998) makes the point that clinical settings, such as hospitals and hospices, have to compensate for the lack of structural and sentimental warmth by, wherever possible, creating a homely feel. This 'homely effect' was reproduced in this non-clinical space that temporarily assumed the status of a simple but safe afternoon retreat.

On more than one occasion I observed some attendees bringing small plants or a vase of flowers for the coffee tables, contributing to the aesthetic and ritual life of the drop-in sessions. As a response to a remark from me about how good it was to have the pleasure of flowers, one attendee, Joan,[1] commented: 'it brightens the place up'. One regular attendee, Doris, would bring candles for the window ledge and on one of my earlier visits to the sessions she told me 'You see, it's not really like a hospice', indicating that this community setting was clearly perceived as non-clinical/anti-hospice space which resonates with the way in which naming is one way that place is given meaning and imparts a certain character (Tuan 2001, Cresswell 2004). This differentiation appeared to be important to Doris and to the others who saw the sessions as an opportunity for creating their own kind of small society, attending to aesthetic features as well as to functional aspects as a way of transforming the space into a place that would be less austere and better suited

---

1 Personal names and some other details have been changed to protect confidentiality.

to their needs. This transformation, almost entirely enacted by women coming to the drop-in, was only marginal and it was only possible to achieve a temporary 'masking' of the essential shabbiness of the setting through these superficial adaptive measures, but these appeared to be sufficient for the group's purpose. So, although this community centre could, at one level, be considered a low quality environment where a lot is missing, on another level, it could be made 'good enough' for the sessions, with participants being actively engaged in improving the space, as an act of care to each other.

Another material aspect that was important was the bringing and sharing of cake, sweets and biscuits, with Doris, Molly and Pam providing most of the cake that they had baked at home. This was an act of generosity although they did not see it that way with Molly explaining that all through her 'married' life she had had what she called a 'baking day' most weeks and it was nothing to bring some cakes to the drop-in. For the first part of the session tea and cake was always the initial socialising activity. Sitting around the large coffee table with a hot drink was a way of everyone settling down to begin to exchange news and stories about their week that, for some, included visits to the doctor or hospital. This was very much self-directed talk as there was no professional to guide the discussion. Their talk covered a variety of topics including their families, issues in the news, celebrity gossip, forthcoming trips and holidays, the progress/treatment of their disease and what was happening in their day-to-day lives. I observed varying levels of openness about their illness and there were times when some attendees showed signs of distress; when this occurred, others present were quick to reassure and offer comfort. Given the increasing frailty of some attendees, it was difficult at times to maintain an optimistic and positive atmosphere at the sessions, but a sense of group belonging appeared to shape the way that mutual care was transacted.

Although the space at the community centre was far from pristine, it was just about adequate and had no clinical associations. It was also flexible and attendees had very much taken ownership of the organisation of the space for their purposes over the three hours of the drop-in. In this non-clinical environment, power and control over the use of space were not bounded by the hierarchies that shape clinical settings, and those attending were not recipients of professionalised care, as is the case within traditional hospice day care (Myers and Hearn 2001). Those who came were not in denial about the centre's shabby qualities but had adopted the space, choosing not to be limited by its deficiencies because within it they could have time to be with others in similar circumstances; time to be themselves, time to tell their stories and time to care for each other in small but meaningful ways.

The acts of caring took different forms. Joan, for example, making sure that Rita had remembered to take her tablets; Molly giving Jason some information she had downloaded for him from the internet about DVLA regulations concerning having his car off the road; Rita reassuring Tania about her appearance despite Tania's rapidly deteriorating physical condition; Pam sitting holding Maureen's hand, valuing Maureen's presence among them, even though Maureen was silent most of the time. There appeared to be a collective sense of Maureen's vulnerability that shaped a particular kind of sensitivity toward her, expressed by respecting her withdrawn presence and her need for her emotions not to be intruded upon. Maureen was very aware that she had begun her dying and her wish to be silently present was absolutely respected. On one occasion, Molly, who had had a wide experience of cancer over many years, provided comfort and solace to Reg, who, still in a state of shock having been recently diagnosed, was very frightened about forthcoming chemotherapy treatment. She could be heard patiently explaining to Reg the possible side effects and how he might help himself and that, in her words, she would 'be there for him'.

What I observed in this caring encounter was the quality of 'tenderness' that Thorne (2005), writing in the context of counselling, characterises as a state of being that involves a high degree of contact with one's intuitive self. Such intimate exchanges were not unusual and seemed to be the vehicle that enabled those present to offer each other their own brand of healing through accessing each other's emotions. The developing close relationships between group members at the drop-in were the conduit for this intimacy that was shaped by what Granovetter (1973), writing about the development of social networks, terms 'the strength of weak ties'. Young et al. (2009) make the further point that friends and neighbours, who are in caring relationships with dying people, can have a significant role in enhancing their every day quality of life. Their critique of friendship as kinship points to the potential for the formation of deep relationships outside of professional networks and family at this critical time.

Most of those in the group were in their seventies and eighties and some had become 'veterans' of cancer, with others in the group feeling that the more 'experienced' cancer sufferers among them, were a source of knowledge about the disease. In one sense this created an informal experiential hierarchy within the group and, at times, this was a significant feature of the group dynamic that was shaped by a 'cancer solidarity' derived from mutual support and identification. This model of mutual support groups is now widely in evidence with parallels across a broad interest spectrum including bereavement groups, victim support forums and parent support groups as examples. The shared cancer experience that brought with it a shared sense of responsibility for one another provided the underpinning framework for a safe space to express emotions and opinions in a self-directed way, without fear of professional judgement or intervention and away from 'the clinical gaze' (Stacey 1997: 51). This gave the space a particular meaning, in relation to care giving, and this is explored more fully in the conclusion below.

Conclusion

The interplay of space and place discussed by Tuan (2001) provides a useful framework with which to draw out how the care space described in this chapter is accorded meaning by those who use it, especially in relation to issues of attachment and links to others (Casakin and Kreitler 2008). Unlike the disciplined spaces of clinical health care settings that are ordered by both hierarchy and function, the space at the cancer drop-in was 'transformed' into a place for mutual care and support by those who attend the drop-in sessions. For two afternoons each week this community space, with its limited resources and poor aesthetic, became 'their' place for different types of caring exchanges enacted outside of a pattern of formalised service provision. Creating a sense of homeliness gave a feeling of belonging with care exchanges as central to this 'borrowed' space. Both Tuan (2001) and Cresswell (2004) note that home, more than any other space, is seen as a centre of meaning and a field of care. The space, claimed by attendees, was given value because of the relationships formed and enacted within it, with interactions and bonds shaped by the experience of cancer. Holmes (2001: 167) has argued that 'the relationship that develops between a carer (nurse or volunteer) and patient is extremely important in the day hospice setting'. At the drop-in, however, the status of patient for those who attended was not at all in the ascendant because the time spent there was experienced principally as 'time off' from being a patient. What were important were the care relationships among group members. The 'non-patient status' at the drop-in was only interrupted by interaction with the 'volunteer' professionals who offered a range of complementary therapies with it necessary at these consultations to discuss details of disease status, change in symptoms and current medication.

The approach to care at the drop-in followed the social model (Myers and Hearn 2001), with particular emphasis on non-professional care that was centred on the strengths and capacity of those coming to the centre to offer care to each other as a vehicle for fostering hope. Holmes (2001: 168) contends that this heavily socially oriented provision is in decline with many day hospices becoming more medically orientated that she argues may make them 'more like a hospital', with, for example, features such as discharge criteria and policies in greater evidence. The emergence of 'experts' amongst those making use of the drop-in resulted in an informal hierarchy of care giving that, although outside of the professional/client dyad, signified an empowered status. 'Cancer maturity', in particular, formed the basis for informational care that was a significant feature of this 'cancer subculture'. Expertness has many dimensions and, although referred to most commonly as part of professional jurisdiction, the role of experience in developing knowledge points to the emergence of 'lay' expertise in many areas of health care. Fisher et al. (2008) report similar findings with some participants in their study asserting themselves as 'experts' in identifying and consuming services they require rather than passively accepting services seen as appropriate by 'professional others'.

The discussion above has provided a detailed account of the physical features of the day care setting which, in its raw state, always struck me as disappointing. Whilst attendees may have shared this view, they never voiced these sentiments. What they valued was the space as opportunity to both be and go beyond themselves with others similarly situated. This space can thus be conceptualised as emotional space (Bondi et al. 2005) within which the deeply personal yet patterned emotions felt by those affected by cancer can be authentically shared. The non-clinical environment with 'volunteer' professionals present only to offer complementary therapies (see above), removed the reality of death and dying from the midst of this care setting, this despite the physical frailty of some of those present who clearly were in 'the dying phase of life' (Holloway 2007: 122). The association of hospice with space for dying was made by some individuals but this community space, as contrast, was collective space away from those direct associations that Worpole (2009) contends remain powerful despite hospice patients being regarded as 'honoured guests'. From comments at the sessions it was apparent that some participants were of the view that by entering dedicated clinical hospice space they would become complicit in their own dying. Although professionally led, the success of the Maggie's Centres may in part also be attributed to the absence of clinical features within their 'high value' space.

Most of those who attended the sessions lived alone and many had very modest incomes. This, combined with the physical impacts of cancer, served to socially 'bind' them. As a consequence, the drop-in sessions provided a greatly valued regular opportunity to 'get out' within the locality, other than visits to the doctor or to the hospital that seemed to form part of the routine of their lives. Because, as Fisher et al. (2008) note, modern treatments have resulted in many people living longer with their illness, the social and therapeutic benefits of attending the drop-in sessions were for many ongoing long-term benefits that made a positive difference to their lives becoming what Seamon (1980) terms a 'time-space routine'. Although the setting was undifferentiated space, attendees' experience of the space endowed it with value and meaning through familiarity and community with others that were not shaped by the constraints of many professional medicalised settings where the goals of management, staff and patients are in constant tension.

References

Addington-Hall, J. 2004. Referral patterns and access to specialist palliative care, in *Palliative Care Nursing Principles and Evidence for Practice*, edited by S. Payne, J. Seymour and C. Ingleton. Maidenhead: Open University Press, 90–107.

Andersson, B. and Ohlen, J. 2005. Being a hospice volunteer. *Palliative Medicine,* 19(8), 602–609.

Armstrong-Coster, A. 2004. *Living and Dying with Cancer*. Cambridge: Cambridge University Press.

Ball, P. 2009. Intimacy and relationships: the view from a hospice, in *Making Sense of Death, Dying and Bereavement: an Anthology,* edited by S. Earle, C. Bartholomew and C. Komaromy. London: Sage Publications, 112–16.

Barker, L. and Hawkett, S. 2004. Policy, audit, evaluation and clinical governance, in *Palliative Care Nursing*, edited by S. Payne, J. Seymour and C. Ingleton. Maidenhead: Open University Press, 713–33.

Bondi, L., Davidson, J. and Smith, M. 2005. Introduction: geography's emotional turn, in *Emotional Geographies,* edited by J. Davidson, L. Bondi and M. Smith. Aldershot: Ashgate Publishing Ltd, 1–16.

Brown, C. 2009. Cancergiggles, in *Making Sense of Death, Dying and Bereavement: an Anthology,* edited by S. Earle, C. Bartholomew and C. Komaromy. London: Sage Publications, 69–73.

Casakin, H. and Kreitler, S. 2008. Place attachment as a function of meaning assignment, *The Open Environmental Journal,* 2(2), 93–100.

Clark, D. and Seymour, J. 1999. *Reflections on Palliative Care.* Buckingham: Open University Press.

Cresswell, T. 2004. *Place: A Short Introduction.* Oxford: Blackwell Publishing.

Davidson, J., Bondi, L. and Smith, M. 2005. *Emotional Geographies.* Aldershot: Ashgate Publishing Ltd.

Degremont, N. 1998. Palliative care and architecture: from hospital to people. *European Journal of Palliative Care,* 5(4), 127–9.

Dein, S. and Abbas, S.Q. 2005. The stresses of volunteering in a hospice: a qualitative study, *Palliative Medicine,* 19(1), 58–64.

Dosser, I. and Nicol, J.S. 2006. What does palliative day care mean to you? *European Journal of Palliative Care,* 13(4), 152–5.

Douglas, H.R., Normand, C.E., Higginson, I.J., Goodwin, D.M. and Myers, K. 2003. Palliative day care: what does it cost to run a centre and does attendance affect use of other services? *Palliative Medicine,* 17(7), 628–37.

Fisher, C., O'Connor, M. and Abel, K. 2008. The role of palliative day care in supporting patients: a therapeutic community space. *International Journal of Palliative Nursing,* 14(3), 117–25.

Glaser, B. and Strauss, S. 1967. *The Discovery of Grounded Theory.* Chicago: Aldine.

Goffman, E. 1961. *Asylums.* London: Penguin Books.

Granovetter, M. 1973. The strength of weak ties. *American Journal of Sociology,* 78(6), 1360–80.

Hearn, J. 2001. Audit in palliative day care: what, why, when, how, where and who, in *Palliative Day Care in Practice,* edited by J. Hearn and K. Myers. Oxford: Oxford University Press, 94–115.

Higginson, I.J. and Goodwin, D.M. 2001. Needs assessment in day care, in *Palliative Day Care in Practice,* edited by J. Hearn and K. Myers. Oxford: Oxford University Press, 12–22.

Higginson, I.J., Hearn, J., Myers, K. and Naysmith, A. 2000. Palliative day care: what do services do? *Palliative Medicine,* 14, 277–86.

Holloway, M. 2007. *Negotiating Death in Contemporary Health and Social Care.* Bristol: The Policy Press.

Holmes, J. 2001. The changing face of the day hospice. *European Journal of Palliative Care,* 8(4), 166–8.

Jonas-Simpson, C. 2006. The possibility of changing meaning in light of space and place. *Nursing Science Quarterly,* 19(2), 89–94.

Jones, L. and Somekh, B. 2005. Observation, in *Research Methods in the Social Sciences*, edited by B. Somekh and C. Lewin. London: Sage Publications, 138–45.

Keswick Jencks, M. 1995. *A View from the Front Line*. London: Maggie's Centres UK.

Lawson, B. and Phiri, M. 2003. *The Architectural Healthcare Environment and its Effects on Patient Outcomes*. London: NHS Estates.

Lawton, J. 2000. *The Dying Process*. London: Routledge.

Longhurst, R. 2001. *Bodies: Exploring Fluid Boundaries*. London and New York: Routledge.

Low, J., Perry, R. and Wilkinson, S. 2005. A qualitative evaluation of the impact of palliative care day services: the experiences of patients, informal carers, day unit managers and volunteer staff. *Palliative Medicine*, 19(1), 65–70.

MacLeod, R. 2008. Setting the context: what do we mean by psychosocial care in palliative care?, in *Psychosocial Issues in Palliative Care,* edited by M. Lloyd-Williams. Oxford: Oxford University Press, 1–20.

Maggie's Centre 2006. *Living with Cancer: A Six Week Course for People with Cancer*. Available from the Maggie's Centre, Kirkcaldy, Scotland.

Maslow, A. 1970. *Motivation and Personality*. New York: Harper & Row.

McIntosh, J. 1977. *Communication and Awareness in a Cancer Ward*, New York: PRODIST.

Mitchell, G., Murray, J. and Hynson, J. 2008. Understanding the whole person: life-limiting illness across the life cycle, in *Palliative Care: A Patient-Centered Approach,* edited by G. Mitchell. Oxford: Radcliffe Publishing, 79–108.

Morris, S.M. and Thomas, C. 2005. Placing the dying body: emotional, situational and embodied factors in preferences for place of final care and death in cancer, in *Emotional Geographies,* edited by J. Davidson, L. Bondi and M. Smith. Aldershot: Ashgate Publishing Ltd, 19–31.

Myers, K. and Hearn, J. 2001. An introduction to palliative day care: past and present, in *Palliative Day Care in Practice,* edited by J. Hearn and K. Myers. Oxford: Oxford University Press, 1–11.

O'Connor, M. 2004. Transitions in status from wellness to illness, illness to wellness, in *Palliative Care Nursing Principles and Evidence for Practice*, edited by S. Payne, J. Seymour and C. Ingleton. Maidenhead: Open University Press, 126–41.

O'Keefe, K. 2001. Establishing day care, in *Palliative Day Care in Practice*, edited by J. Hearn and K. Myers. Oxford: Oxford University Press, 43–58.

Payne, S. and Seymour, J. 2004. Overview, in *Palliative Care Nursing,* edited by S. Payne, J. Seymour and C. Ingleton. Maidenhead: Open University Press, 15–38.

Peace, S. 2003. The development of residential and nursing home care in the United Kingdom, in *End of Life in Care Homes,* edited by J.S. Katz and S. Peace. Oxford: Oxford University Press, 15–42.

Rager, K.B. 2005. Compassion stress and the qualitative researcher. *Qualitative Health Research*, 15(3), 423–30.

Randall, F. and Downie, R.S. 2006. *The Philosophy of Palliative Care: Critique and Reconstruction*. Oxford: Oxford University Press.

Seamark, D., Seamark, C. and Hynson, J. 2008. Life-limiting illness: the illness experience, in *Palliative Care: a Patient-Centered Approach,* edited by G. Mitchell. Oxford: Radcliffe Publishing, 47–78.

Seamon, D. 1980. Body-subject, time-space routines and place-ballets, in *The Human Experience of Space and Place,* edited by A. Buttimer and D. Seamon. London: Croom Helm, 148–65.

Sheilds, R. 1997. Spatial stress and resistance: social meanings of spatialization, in *Space and Social Theory,* edited by G. Benko and U. Strohmayer. Oxford: Blackwell Publishers, 186–202.

Sinclair, P. 2007. *Rethinking Palliative Care*. Bristol: The Policy Press.

Small, N. 2009. Theories of grief: a critical review, in *Death and Dying A Reader,* edited by S. Earle, C.M. Bartholomew and C. Komaromy. London: Sage Publications, 153–58.

Sommer, R. 1969. *Personal Space: The Behavioural Basis of Design*. Englewood Cliffs, New Jersey: Prentice-Hall, Inc.

Stacey, J. 1997. *Teratologies: A Cultural Study of Cancer*. London and New York: Routledge.

Thorne, B. 2005. *The Mystical Power of Person-centred Therapy: Hope Beyond Despair*. London: Whurr.

Tuan, Y.-F. 2001. *Space and Place: The Perspective of Experience.* Minnesota: University of Minnesota Press.

Twigg, J. 2001. *Bathing – The Body and Community Care*. London: Routledge.

Verderber, S. and Refuerzo, B.J. 2006. *Innovations in Hospice Architecture*. Abingdon: Taylor & Francis.

Watts, J.H. 2008. Emotion, empathy and exit: reflections on doing ethnographic qualitative research on sensitive topics. *Medical Sociology Online,* 3(2), 3–14.

Watts, J.H. 2009a. I come because I can. Stories of hope from the cancer drop-in. *Illness, Crisis and Loss*, 17(2), 151–68.

Watts, J.H. 2009b. Illness and the creative arts: a critical exploration, in *Death and Dying A Reader*, edited by S. Earle, C.M. Bartholomew and C. Komaromy. London: Sage Publications, 102–108.

Watts, J.H. 2009c. Meanings of spirituality at the cancer drop-in. *International Journal of Qualitative Studies on Health and Well-Being*, 4(2), 86–93.

Watts, J.H. 2010. *Death Dying and Bereavement: Issues for Practice*. Edinburgh: Dunedin Academic Press.

Worpole, K. 2009. *Modern Hospice Design*. Oxford: Taylor & Francis.

White, S. and Johnson, M. 2004. What do doctors actually do in the day hospice? *European Journal of Palliative Care*, 11(3), 107–109.

Young, E., Seale, C. and Bury, M. 2009. It's not like family going is it?, in *Making Sense of Death, Dying and Bereavement: An Anthology*, edited by S. Earle, C. Bartholomew and C. Komaromy. London: Sage Publications, 95–97.

Chapter 3

Laying Lazarus to Rest: The Place and the Space of the Dead in Explanations of Near Death Experiences

Mary Murray

Introduction

Since the publication of Raymond Moody's book *Life After Life* in 1975 there has been considerable popular and professional interest in near death experiences (NDEs). Popular interest in NDEs has been reflected in the publication of books and magazine articles aimed at the general reader, television, radio, documentary and film coverage, and the establishment of internet sites about the phenomena. Professional interest in near death experiences has been expressed through the discourses of western scientific bio-medicine, psychology, sociology and religion. In this chapter I will argue that biomedical, psychological, sociological, and religious arguments about near death experiences tell us something about the way in which relationships between the living and the dead have been spatialised in modern western society. I will argue that professional and academic discourses that have attempted to account for the near death experience articulate an 'imagined geography' of the living and the dead.

In the cosmological cartographies of such 'imagined geographies' the places and spaces that the dead occupy are usually held to be profoundly 'other' to the ones that the living generally occupy. However, near death experience testimonies, as resurrective narratives, connect the world of the living with the world of the dead in a way that both confirms and confounds the boundaries between the living and the dead. Boundaries between the living and the dead, normally held in by rituals of separation, are transgressed as they are upheld in the resurrective narrative of the near death experience. Professional and academic arguments that attempt to explain the near death experience perform the function of 'border control', situating the dead in the place of the dead and the living in the world of the living.

Whilst this chapter draws on existing literature about NDEs, it does so in a way that develops a perspective on place and space. The chapter begins with a description of the NDE, noting some of the circumstances within which it can occur, and signals its occurrence through history and across cultures. I then look at ways in which the language of space and place features in the NDE narrative. This is followed by a summary of biomedical, psychological, religious and sociological explanations of the NDE. Next, I suggest a way in which the sociological account of the NDE may be further developed. In this respect I look at ways in which the relationship between the living and the dead has been spatialised across time and place. My focus here is on changes that are thought to have occurred following the European Reformation, and I link this to ideas about the place and space of the living and the dead in accounts and explanations of the NDE. By way of conclusion, I summarise the main arguments and contribution of the chapter, and situate debates about NDEs within political space and the politics of the supernatural.

The Near Death Experience

Near death experiences can occur in a variety of circumstances. Some NDEs may occur spontaneously when a person does not appear to be close to death. They can also result from pain, fear or extreme stress, as well as during anaesthesia and serious, but not fatal, illness (Fenwick and Fenwick 2008: 204–208). The immediate cause of apparent death in the NDE includes a whole range of circumstances including accident, childbirth, drowning, diabetic crisis, heart attack, hysterectomy, internal bleeding, miscarriage, overdose, pneumonia, poisoning, stroke, and suicide (Sutherland 1992: 58–64).

In a modern western context, regardless of a person's religious faith or lack of it, their sex or age, common elements are clearly identifiable in the NDE and some or all of the following characteristics may be present. Following an accident or serious illness a person's heart stops beating. The person is however resuscitated and can remember an extraordinary experience which often begins with an out-of-body experience (OBE). Those experiencing NDEs frequently report that they witness attempts being made to resuscitate them and many recount an experience of moving through a tunnel, often toward bright lights. Encounters with deceased friends and relatives, as well as beings of light and archetypical religious figures, are also fairly common. A 'life review', in which the person becomes aware of all of their past actions and the effects of those actions on others, is also a feature of NDEs in modern western societies. Whilst many say that their NDE was accompanied by a sense of great peace, some feel distress and fear. Regardless of whether or not the actual experience of the NDE is a positive or a negative one, the NDE can trigger significant life changes for the person concerned (Fenwick and Fenwick 2008: 205–206).

Whilst the publication of Raymond Moody's book *Life After Life* (1975) marked the beginning of contemporary popular and professional interest in NDEs, the phenomenon has traversed time and place. Plato (427–347 BC) tells the story of a soldier called Er who died in battle but came back to life (Corazza 2008: 23). There are descriptions of NDEs in legend and myths from over two thousand years ago (Fenwick and Fenwick 2008). In her book *Other World Journeys*, Zaleski (1987) provides an account of NDEs across the places and spaces of different cultures and societies, from biblical and medieval times through to the modern. Sogyal Rinpoche in *The Tibetan Book of the Living and the Dying* (1992) points to the similarities between the Tibetan *delok* experience in which people say they have returned from the dead, and the NDE. Kellehear (1996) and Corazza (2008) provide accounts of NDEs in China, India, Japan, Guam, Western New Britain, Melanesia, and amongst Native American peoples, Aboriginal Australians and Maori New Zealanders.

Space and Place in the Near Death Experience

Although the space and place of the OBE and the tunnel do not appear in all cross cultural accounts of the NDE, encounters with other beings in other realms appears to be common across cultures (Kellehear 1996). Indeed, the language of place and space is embedded in the resurrective narrative of the near death experience across cultures.

The near death experience seems to take place in liminal 'other worldly' space. It is a place and space different to the 'this worldly' place and space that the person usually inhabits. It is also one that appears to lie somewhat betwixt and between the land of the living and the land of the dead. People often report that their NDE begins with an out of body experience in which they find themselves looking down at their own body from the vantage point of perhaps a hospital ceiling or the scene of an accident. From this apparently extra corporeal place and space, people may witness attempts being made to resuscitate their body, and they are frequently able to describe the place, space and circumstances surrounding their resuscitation in great detail. Reports of travelling through a tunnel like space are also fairly common. The tunnel appears to lead to a point of light towards which the traveller is moving, and which grows larger and brighter as the traveller is drawn closer to it.

At a significant point in their journey, the near death traveller comes to a border beyond which they may perceive a cosmic landscape of great beauty. Moody (1977) reports descriptions of concrete and detailed archetypical images of celestial landscapes, with cities of light, radiant palatial mansions, beautiful and exotic gardens and magnificent rivers. On the negative side, Moody also reports accounts of experiences in astral realms with perplexed spirits and confused discarnate entities unable to detach themselves fully from the physical world.

This liminal border like place can also be a meeting point where the individual experiencing the NDE encounters archetypical beings of light, such as angels, and figures such as Jesus or the Buddha. The individual may also meet up with dead friends and relatives who are on the other side of the border. Often, people undergo a 'life review' akin to a Day of Judgment. In what seems to involve a rapid condensation of space and time, the near death traveller reviews their own past actions and the impact that those actions have had on others. Many near death travellers experience a deep sense of cosmic interconnectedness, peace and unity with all that is in the space of the universe during their celestial journey. Others though, who have encountered 'hell' like places have found the experience very distressing (Satori 2008: 17–27). The border is commonly perceived as a point of no return. Here, the near death traveller may be told, or decide for themselves, that they cannot proceed any further – effectively they do not have the required passport – and must return to their earthly space and place.

Once back in the world of the living, many, if not most, of these other worldly travellers report profound psychological or spiritual changes affecting their sense of themselves and their personal inner space. Those experiencing NDEs often say that they have lost their fear of death (Moody 1975: 94–6, Sutherland 1990: 86–9). The inner psychological and spiritual transformations that emerge from the journey of the near death experience also encourage people to value their lives in this world and can give a renewed sense of purpose (Fenwick and Fenwick 2008: 205). The near death experience, then, appears to have transformative effects in terms of inner psychological space and such effects may be mirrored in outer social/societal space.

The following quotes from Cherie Sutherland's (1992) *Transformed by the Light* illustrate different aspects of the NDE:

I sort of sat up on the ceiling. I remember watching all the panic, total panic ... the midwife couldn't find a pulse, she couldn't get a needle in. I was bleeding. I couldn't work out why everyone was so fussed. (1992: 6)

I was up above everything. I could still hear them talking ... I was out of my body and I thought to myself "I must be dead". (1992: 6)

... it's like falling down a big tunnel. (1992: 7)

... And then I experienced a replay of all of my life ... from my birth to the actual operation ... And it was like it was on fast-forward video ... I re-experienced my whole life ... I could see a light – it was like a light I could never describe ... like a silver white light ... I felt myself, just my being, move toward that ... (1992: 8)

I feel I was making contact with people I loved that had died. I think my father was there. I think my great friend and mentor ... was there encouraging me not to be afraid ... (1992: 9)

... And I felt this extreme presence of love, just absolute love ... It shocked me somewhat but there was no problem with accepting that in essence I was being confronted with my creator ... (1992: 10)

... I heard a voice distinctly sing out, a very strong voice: 'Go back!' really loud, but not scary, not angry ... (1992: 9)

... now I know you don't die – you just pass onto another plane. You leave the physical body and the spirit body passes on to another plane. (1992: 81)

Now I have no fear of death! ... (1992:87)

... Now I am more spiritual ... (1992: 99)

Love, love ... Love is the most important feeling in the universe ... (1992: 107)

I came back with a sense of purpose ... (1992: 138)

Since my NDE I've wanted to help others and I've had a real compassion ... (1992: 142)

... changed my relationship with my family and to the planet ... (1992: 145)

... life's more meaningful ... (1992:165).

Accounting for the Near Death Experience

The adventures of these celestial cosmonauts have seized the space of the popular spiritual imagination in modern and late modern society. Much of the contemporary coverage of the NDE in popular culture, though sometimes sceptical and critical, lends support to the idea that death of the physical body is not the end of the human journey, and that we may be destined to travel to other realms when we die. Coverage of the NDE in popular culture also celebrates the profound inner journey and process of transformation and new social values that the NDE can provoke (Kellehear 1996: 76–9).

'Professional' explanations of the NDE have included western bio-medicine, psychology, religion, and sociology. Western bio-medicinal explanations of the NDE focus on physical and chemical changes within the body of the person undergoing the experience. These include anoxia or hypoxia, such as a lack of oxygen to the brain. The effects of anaesthetics as well as temporal lobe epilepsy have also been used to account for aspects of the NDE. So too has the production of endorphins which are naturally released in the brain during particularly stressful times and can cause pain reduction and pleasant sensations (Satori 2008: 59–103, Kellehear 1996: 120–23, Greyson, Kelly and Kelly 2009: 217–28, Blackmore 1993).

Psychological accounts link certain elements of the near death experience such as the OBE, meeting with dead relatives and the life review to psychological defence mechanisms against the threat of impending death. The trauma of imminent death and the possibility of annihilation are thought to produce the NDE as a way of defending against overwhelming anxiety, fear and dread.

It has also been suggested that NDEs may be linked to psycho-pathological conditions such as autoscopy in which people say that they can see a phantom of themselves. OBEs have also been linked to schizophrenia, depersonalisation and psychosis. Similarities have been suggested too between NDEs and multiple personality disorder, where beings of light, hell like experiences, helping spirits, time distortion, floating feelings, telepathic communication and psychic abilities are not uncommon. The possibility that NDEs are connected to memories of the birth process has also been aired (Satori 2008: 103–20).

There are of course striking parallels between descriptions of transcendent 'inner' and 'outer' spaces and places in the NDE and the afterlife teachings of major world religions. Features of typical NDEs that find support in the eschatology of major religions are the OBE, the tunnel experience, life reviews, a blissful existence in a 'realm of light' inhabited by the dead, a Being of Light, and the association of the individual experiencing the NDE with the deceased in that realm (Masumian 2009). There are also parallels between the positive after effects of NDEs and ethical and moral teachings of major religions (Masumian 2009). In this respect, many of those who have experienced NDEs report that their extraordinary experience has encouraged them to lead lives that uphold values such as compassion, love, kindness, humility, charity, service to others, non violence, and respect for life.

According to one study, the majority of clergy interviewed, more than half of whom were Methodist and Catholic, felt that the NDE did provide a fleeting look at the afterlife (Royse 1985). Even so, religious interpretations of the NDE do not necessarily agree that it is evidence of an afterlife. Seventh Day Adventists and some strands of Judaism, for example, teach that the final resurrection of the body will not happen until the second coming of Christ and the end of the world. Interestingly, The Worldwide Church appears to endorse the medical explanation of the NDE, maintaining that a near death experience is simply a state of unconsciousness. Yet others think that the NDE is a trick masterminded by Satan and his demons (Kellehear 1996: 72–3). Meanwhile Catholic theologian Hans Kung (1984) takes the view that the NDE is not necessarily evidence of life after death. As Kung sees it, because the individual returned to the land of the living, they didn't actually die. There is also an emphasis on faith rather than evidence in Christianity.

Nor does religious expert Carol Zaleski view the NDE as a return from the dead. Zaleski sees the other world visions contained in the NDE as:

> ... products of the same imaginative power that is active in our ordinary ways of visualizing death; our capacity to portray ideas in concrete, embodied, and dramatic forms; the capacity of our inner states to transfigure our perception of outer landscapes; our need to internalize the cultural map of the physical universe, and our drive to experience that universe as a moral and spiritual cosmos in which we belong and have a purpose ... we are able to grant the validity of the

near-death testimony as one way in which the religious imagination mediates the
search for ultimate truth. (Zaleski 1987: 205)

In his book *Experiences Near Death,* Allan Kellehear (1996) provides a
sociological perspective on NDEs. This perspective includes explanations of
features of the NDE found in western societies, such as the tunnel and life review,
to culturally specific factors. Kellehear thinks that the sociological concept of
status passage can be applied to NDEs. Status passages involve a social process
of transition from one part of the social system to another. Such passages or
transitions often involve a changed identity or sense of self, and changed behaviour.
Significant status passages for the individual in society affecting behaviour and
identity include those of adult, worker, spouse and parent. Kellehear sees that the
NDE can involve a change in identity and behaviour. Kellehear contends too that
NDE narratives ' ... may be read and be interpreted for their assumptions about
and allusions to the ideal society ... ' (Kellehear 1996: 115). He links positive
transcendental visions of the NDE to discourses about social utopias. He says ' ...
NDEs, whatever else they may be, are social images that belong to the historical
and social discourse about the ideal society' (Kellehear 1996: 101). He thinks that
'If visions of this otherworldly society are prompting people to change their values
and lifestyles, then it is important to understand why ... as a utopian form, the
transcendent society reawakens the pursuit of the ideal society'(Kellehear 1996:
101).

Reformation Protestantism and Relationships Between, and the Places and Spaces of, the Living and the Dead

I would argue that one way in which the sociological perspective on NDEs could
be further developed is by thinking about ways in which the relationship between
the living and the dead has been spatialised across time and place, and ways in
which ideas about the place and space of the living and the dead are to be found in
accounts and explanations of the NDE.

In *The Place of the Dead,* edited by Gordon and Marshall (2000), it is argued
that for the societies of late medieval and early modern Europe, 'placing' the dead
in spiritual, social and physical terms was an important political exercise that
involved conflict and complex negotiation. In the spaces and places of pre-modern
Europe, the boundary between life and death was thought to be more porous
than it is today. It was believed that those in the land of the living could help
those in the land of the dead by means of prayer. Offering masses or good works
for souls in purgatory (a place for the cleansing of sins) were other important
means of effecting such assistance. Most medieval and Renaissance Christians
also believed that the dead could interfere in the affairs and world of the living.
Whilst Saints in heavenly spaces and places might help souls in the land of the
living, the displaced dead wandered restlessly through the imaginative space of

medieval rural and urban communities. Spirits haunted the medieval landscape and popular imagination, and ghosts were thought to be searching for physical places in which to lodge. There was also a widespread belief that corpses could rise from the contained spaces of graves, and that spirits of the dead could enter and posses the bodily space of the living. We might say, then, that the demarcation between the land of the dead and the land of the living was perhaps much thinner and more fluid than it is today.

It is commonly argued that the Reformation fractured this intimate relationship between the living and the dead. With the power of Catholicism challenged in England by the new official state religion of Protestantism, the belief in ongoing relationships between the living and the dead was no longer officially sanctioned. The new view was that the living couldn't alter the fate of the deceased, and the dead couldn't return to the world of the living in any shape or form. The Catholic places of limbo (for the souls of dead babies), and purgatory were ousted from the religious cartographies of the heavens. According to Protestantism, there were only two possible destinies for disembodied souls; either heaven or hell. In more recent times, partly because of the influence of science, heaven has now arguably become a place of vague geographical location. In the late 1990s Pope John Paul II announced that heaven was a blessed community rather than a physical place (Stanford 2002: 11). Indeed, heaven and hell have been postulated as being spiritual states of inner personal space, rather than places existing in the outer physical universe.

One of the supposed consequences of the changed relationship between the living and the dead brought about by the Reformation was that the dead were, according to Protestantism at least, marginalised and ousted from their space in the cultural imaginary. It certainly is the case that in modern times the dying and the dead have, to a significant extent, become institutionally and spatially separated from the living in rest homes, hospitals, hospices and cemeteries. Meanwhile significant rituals associated with death, such as funerals, burial, and cremation, continue to act as rites of separation between the world of the living and the world of the dead.

This rough sketch of the relationship between the living and the dead is somewhat simplistic, however. The physical and symbolic separation of the living and the dead in modern western societies may actually be more apparent than real in both public and private space. Belief in ghosts and other worldly beings, such as angels, is not uncommon. Many people also consult mediums in an attempt to connect with deceased loved ones (Hallam, Hockey and Howarth 1999). The popularity of television programs about ghosts and practicing mediums, as well as the plethora of 'New Age' books about communicating with the dead may in part reflect a belief and hope that the channels of communication between the living and the dead remain open. It would seem that these hopes and beliefs have traversed the spaces of historical time and place.

Contemporary understandings of grief recognise the 'continuing bonds' that the living maintain with the dead (Klass, Silverman and Nickman 1996). The dead also live on in the space of private memory. In this respect, Hallam and Hockey (2001) have also explored relationships between the internal spaces of memory and external material lived environments. Meanwhile, as psychologists may refer to a sense of the presence of deceased family members as 'positive introjects', the 'in memoriam' columns of newspapers continue to reflect the bonds that the living have with the dead. The dead are also alive and more than kicking in the space of public memory. As Benedict Anderson (1991) has argued, Nation States have been constructed in and through resurrective narratives, creating communities of the dead, the living, and those yet to be born. Indeed I would argue that personal and national identities in the land of the living are constructed in and through ongoing relationships with those in the land of the dead.

I would also argue that the relationship between the living and the dead in some explanations of the near death experience can be viewed as a form of 'border control/patrol', the purpose of which is to separate the world of the living from the world of the dead. Whether or not dying and death became taboo subjects in modern society, the spatial separation of the dying and the dead from the living in modern society, as well as contemporary moral panics and anxieties about fatally infectious diseases such as AIDS and bird and swine 'flu, indicates that there *is* a sense in which the dying and the dead are regarded as contagious, dangerous and polluting. Anthropologist Mary Douglas has written a good deal on societal ideas about pollution (see for example Douglas 1978). She points out that 'dirt', once pathogenicity and hygiene are abstracted from it, is simply 'matter out of place'. Matter out of place implies a set of ordered relations and a violation of those relations. The emergent product, dirt, links to symbolic ideas about purity and pollution that are embedded in relationships normally held in by rituals of separation.

The idea of pollution and non conformity to class is, it can be argued, inextricably linked to our ideas about being 'in place' and 'out of place'. Relating this to the relationship between the living and the dead in modern and late modern societies, I would argue that the NDE may destabilise the boundary that was supposedly put into place with the Reformation. The NDE also destabilises binary oppositions between life and death, and in so doing calls into question modes of identity that have privileged, solidity and fixity within the contained spaces of individual bodies and psyches. NDEs, especially where they are accompanied by OBEs, challenge the idea of a closed bodily integrity and with it the modernist view that identity is located within an individual bounded self with a unified identity based on particular states of being including those of alive or dead. Those experiencing NDEs seem to be saying that their sense of themselves and their identity has been profoundly altered by their encounter with death and what they perceive as their journey to the land of the dead.

Where the dead are positioned as 'other' in relation to the living, disembodied deceased beings can be perceived as troubling and troubled. The NDE, as a liminal experience located perhaps at the border and on the margins of life and death, is suggestive of a hybrid identity, even if only transitory in nature, composed of living, and if not dead, then almost certainly dying being/s. There have of course been numerous hybrid beings and creatures that have lingered at the boundary of life and death and haunted the popular imagination across time and place. Such figures include ghosts, vampires, zombies, and those believed to be in the grip of spirit possession. Such hauntings have been regarded as dysfunctional, disorderly, dangerous and polluting. When the dead become disorderly, disruptive and out of place, some kind of reordering is required. One way in which the unruly dead have been disciplined is through religious blessings and exorcism. These special means of ritual separation are designed to caste out the dead from the haunted outer and inner worlds of the living.

I would argue that the dead can also be disciplined and exorcised from the haunted places of the psyche by psychoanalysts, psychiatrists and psychologists. In Foucault's (1978) view, such professionals have taken over the confessional functions of the clergy in western society. In this sense we might say that the soul has become an object of the medical rather than the religious gaze. In a similar vein, in his book *Governing the Soul*, Nicolas Rose (1989) talks about the ways in which subjective private space has become intensely governed in modern and late modern societies. The management of subjectivity is a central task for modern government, and we have become profoundly psychological beings. I would argue that ways in which citizens and persons can and should legitimately relate to the dead has become part of this management of subjectivity.

In modern western societies the relationship between the dying and the dead has been spatialised in both inner and outer worlds. In the outer world of the living, the dying and the dead have been institutionally and geographically separated off. Meanwhile Freud (1917), with his views about so called 'normal' and 'pathological' grief, began the modern psychological management of our personal, intimate, subjective relationships with the dead. Though challenged in recent years by the 'continuing bonds' thesis (Klass, Silverman and Nickman 1996), Freud's view that the bereaved who continue to relate to the deceased for an overly long period of time are manifesting symptoms of a 'pathological' grief response, has influenced generations of psychoanalysts, psychiatrists, psychologists and counsellors. As Freud saw it, to facilitate full and healthy living in this world, the bereaved needed to sever ongoing 'pathological' emotional ties with those in the land of the dead. Meanwhile though, the state and religious institutions legitimise ongoing relationships with the dead amongst subjects and community, through, for example, forms of nationalism, remembrance services, and, for Catholics, 'All Souls Day'. Adherence to such rituals, which can contain and harness the power of the dead to institutional forms, spaces and places, are regarded as signs of good citizenship and moral character rather than symptoms of inner pathology and pollution.

Space and Place in Explanations of the Near Death Experience

The experiences of ordinary people as recounted in near death narratives seem to both transgress and uphold boundaries between the living and the dead. Near death narratives contain numerous references of travelling to, and returning from, the places and spaces of the dead. Individuals undergoing NDEs appear to travel to cosmic places and spaces thought to be occupied by the dead. With only a limited stay visa though, the trip is a short one, and these individuals soon find themselves back in the more prosaic places and spaces of the living. Similar perhaps to illegal migrants in the places and spaces of the living, NDE travellers might be regarded as transgressors of borders in cosmic geo-political place and space. At the same time, the return of the NDE traveller to the land of the living serves to uphold the very delineations of geo-political place and space between the living and the dead that their experience seems to transgress. A quote from Stanford alludes to this complex of transgression and maintenance of boundaries: 'Those who claim to have had a NDE today no longer sit up in coffins, but the effect is much the same' (Stanford 2003: 316).

From the perspective of western bio-medicine, the NDE is to be explained in terms of the inner workings and mechanisms of corporeal space. Though many medical practitioners may of course privately subscribe to beliefs that uphold the view that there are lands of the dead, the biomedical explanation of the near death experience maintains that NDE travellers never actually got there, because they never actually died. According to this perspective, rather than being engaged in an 'other worldly' expedition, near death travellers don't actually leave base camp; the 'this worldly' space of planet Earth.

Similarly, from the perspective of modern western psychology, although the person *may* have been on the way to the lands of the dead, they never actually arrived in those spaces and places because the NDE experience takes place within the inner psychic space of the individual in the land of the living. From this perspective, because the NDE occurs within the place of the individual human body and psyche, the person was, as it were, 'confined to barracks'.

The psychological interpretation advanced by Blackmore (1993) is one that draws on Buddhist views about the empty nature of the self: '... the NDE breaks down, if only for a brief moment, the self-model which was the root of all our greed, confusion and suffering. There never was any real persistent self ... any permanent self ... There was only a mental model that said there was one' (Blackmore 1993: 253–4). As Blackmore sees it, we have no self, and we are simply 'here' (Blackmore 1993: 264).

Meanwhile, religious views contending that the NDE traveller didn't actually die lend support to the view that the NDE takes place within individual corporeal space. The view that NDEs are an expression of the religious imagination (Zaleski 1987) also lends support to the view that the phenomenon occurs within inner psychic space, albeit a space that the cultural and historical variation associated with NDE visions suggest is culturally mediated.

Different religious traditions have, of course, provided accounts of worlds of the dead. As Masumian (2009) observes, these worlds appear to contain many of the features reported in the other world journeys of NDE travellers. According to Hindu beliefs, for example, the deceased are said to enter *Rig-Veda,* which is a place of light. In *The Tibetan Book of Living and Dying,* Sogyal Rinpoche explains that according to Tibetan Buddhism, in the NDE '... the mind is momentarily released from the body, and goes through a number of experiences akin to those of the mental body in the bardo of becoming' (Rinpoche 1992: 325). These include the OBE, mobility, knowledge of the past, meeting others including beings of light, perception of different realms including cities of light, and hellish visions. According to the Pure Land Buddhist tradition, the deceased may travel to the 'Pure Land', which is a realm of lakes, pavilions and gardens laden with multi-coloured flowers and jewels. Christianity and Islam have also upheld beliefs in the spaces and places of heaven and hell.

It could be argued, of course, that religious accounts of lands of the dead, and religious perspectives that take the view that NDE travellers don't actually die, paradoxically perhaps leave the binary opposition between the world of the living and the world of the dead – if there is one – intact. To this extent they may share something in common with bio-medical and psychological explanations of the NDE. However, many world religions teach that the living can assist the dead. Tibetan Buddhists for example have practices and rituals designed to help the deceased on their spiritual journey (Rinpoche 1992). Catholics, too, are encouraged to pray for the souls of the dead. If the living can assist the dead, the relationship between the worlds of the dead and that of the living may not, after all, be one of binary opposition. Vernette (1997) illustrates this view:

> Near death experiences ... seem to reveal the existence of places and times in which communication between our world and the next is possible ... boundary separating one from the other is not an insurmountable barrier, like the former Berlin Wall ... long string of mystics who related a set of authentic experiences of communication with the inhabitants of the other world ... signs of life ... they are like comets streaking across the night sky; they come from beyond our world and cannot be summoned at will ... like guideposts ... they indicate a direction ... they are a privileged form of communication in this borderland where both coincide ... (Vernette 1997: 281–83)

Sociology is frequently associated with critical social inquiry in the land of the living. There is a long tradition of such inquiry within the discipline (see for example Marx 1997, Weber 1930). Indeed, sociologists have often turned this critical gaze onto their own discipline, unpacking the extent to which the discipline may be complicit in maintaining the status-quo in the land of the living. Dorothy Smith (1990, 1996) is one such sociologist. She has observed that the conceptual imperialism of sociology, through which the lived experience of people has often been made to fit into conceptual categories that reflect the concerns of state policy makers, may have more to do with the relations of ruling than the everyday experiential world of the sociological subjects being researched.

Although Durkheim as one of the founders of the discipline wrote about suicide (1951), the macro conceptual framework he adopted, linking rates of suicide to degrees of social integration and regulation was certainly not phenomenological in orientation. Moreover, until relatively recently, sociology didn't concern itself very much with the dying and the dead in modernity. To this extent sociologists didn't admit the dead into the world of the living. Sociologist Zigmunt Bauman (1992) contends that death has been the skeleton in the cupboard of the neat, orderly and functional home that modernity attempted to construct. Death and dying has also been something of a skeleton in the cupboard of sociology, a discipline which emerged with the development of modern society. In recent years, however, sociological interest in dying and death has grown and sociologists are increasingly recognising that social relationships and personal identity in this world are constructed in and through relationships with the dead (Hallam, Hockey and Howarth 1999).

Moreover, Kellehear's (1996) sociological account of the NDE is one that perceives the socially disruptive yet socially progressive potential of the dying and perhaps dead. Unlike status passages such as adulthood, marriage and parenthood, Kellehear interprets the status passage of the NDE as one that is not normative and is an exception to the rule of social regulation. He sees that '... the transcendent society as a utopian image is critical of some modern values (e.g. competition, materialism) ... ' (Kellehear 1996: 112). Kellehear (1996) views the status passage of the NDE as a critical, or potentially critical, influence for social change in this world, though he does not foreclose on neuroscience or rule out the possibility of life after death 'elsewhere' (Fenwick and Fenwick 2008: vii).

Conclusion

The lens of place and space has been used to review literature about NDEs in this chapter. That lens has enabled the development of fresh perspectives on near death narratives and explanations of the NDE. Cultural ideas about appropriate relationships between the living and the dead, and the rightful places and spaces of the living and the dead in modern western societies, can be discerned in near death narratives and explanations of the NDE.

Officially at least, the Reformation undermined intimate relationships between the living and the dead. According to Protestantism, the dead went either to heaven or hell and couldn't return to the world of the living in any shape or form. From this perspective the spaces and places of the dead are profoundly 'other' to the spaces and places of the living. Similar ideas can be discerned in biomedical, psychological and even some religious perspectives reviewed in this chapter. According to these perspectives, the NDE traveller doesn't actually die. Consequently, communications between those who have undergone NDE and the living are not regarded as coming from the grave or a world beyond. From this viewpoint, the NDE traveller doesn't take a trip to heavenly or hell-like lands of the dead, because that particular trip doesn't include a return ticket.

However, Protestant Reformation beliefs about relationships or lack of them, between the living and the dead, and ideas about the living and the dead being located in radically different places and spaces have been far from hegemonic. In popular culture, for example, belief in angels, ghosts and spirit guides is not uncommon. Nor are consultations with mediums uncommon or covert. Meanwhile, the 'helping professions' have come to acknowledge the significance of ongoing and psychologically healthy relationships that the living have with the dead. Religions and cultural traditions, such as Catholicism and Buddhism, also maintain that relationships (albeit ones that are sanctioned by the respective religions) can, and often should, exist between those in the lands of the living and those in the lands of the dead. The relationship between the dead and the living is not necessarily one of binary opposition.

It is perhaps unsurprising, then, that near death testimonies have had such popular appeal. Even so, whilst near death narratives suggest transgression of binary oppositions between the living and the dead and their places and spaces, near death narratives also uphold the idea that there are particular places and spaces for the living and particular places and spaces for the dead. But the popularity of near death testimonies also suggests that the 'border control' between the lands of the living and the lands of the dead effected by biomedical, psychological and some religious interpretations of the NDE, is by no means entirely effective. 'Governing the soul' in this respect may be more difficult than governing the citizen.

On the basis of the material reviewed, it can be argued that professional and lay interpretations of the NDE articulate a politics of the supernatural conducted in the spaces and places of the living. Max Weber (1930) considered aspects of the politics of the supernatural in his analysis of disenchantment and the marginalisation of magical beliefs and imagination that accompanied the European Enlightenment and Protestant Reformation. More recently, Jeremy Northcote (2007) has provided a sociological account of the paranormal and the politics of truth. He links debates about the paranormal to knowledge or 'truth' construction in modern western society. However, neither Weber nor Northcote provide an account of the politics of the supernatural in relation to NDEs. The literature reviewed for this chapter, though, does point to such politics.

Debates about NDEs are linked to knowledge or truth construction in modern western society. Whilst Weber pointed to the eclipse of magical beliefs with the privileging of rationality in the West with the Enlightenment, Kellehear links the popularity of the NDE narrative to the contemporary disenchantment with medicine and science. Meanwhile the professional discourses through which NDEs have been interpreted, for example western bio-medicine, psychology, sociology, religion and science, have been significant in the discursive and ideological formation of modern society. As they have been used to interpret the NDE, ontological positions have been adopted with respect to the nature of consciousness and the meaning and existence of transcendental spaces and places. Where bio-medicine, psychology and even religion foreclose on these questions, they are engaged in a kind of exorcism intent on the expulsion of the dead from any theatre of transgression. Discourses that lend themselves to disciplining the potentially disruptive and dangerous dying and dead return the dead, dying and living to what are deemed to be their rightful spaces and places.

The problem of consciousness, though, remains unsolved. Using the language of place, Fenwick and Fenwick (2008: 185) see this problem as the 'last frontier'. They explain that the current mainstream scientific view is one that limits consciousness and subjective experience to the space and places of the individual brain and body. This is, of course, similar to biomedical and psychological accounts of the NDE that confine the experience to the spaces and places of the individual body and psyche. However, other contemporary scientific approaches to consciousness outlined by Fenwick and Fenwick (2008) suggest that consciousness may exist in places and spaces outside of the brain and body, and that consciousness may be inter-connected across space. This approach to consciousness could help shed light on the experiences of near, or perhaps temporary, death travellers.

The jury may still be out on the nature of consciousness and how this might shed light on the apparently transcendental spaces and places contained in the NDE narrative. However, the accounts of ordinary people, western bio-medicine, psychology, sociology, religion and the science of consciousness studies, may yet contribute additional maps and compasses to help track the archetypical journey of NDE travellers.

References

Anderson, B. 1991. *Imagined Communities: Reflections on the Origin and Spread of Nationalism.* 2nd Edition. London: Verso.

Bauman, Z. 1992. *Mortality, Immortality and Other Life Strategies.* Cambridge: Polity Press.

Blackmore, S. 1993. *Dying to Live: Near Death Experiences.* London: Prometheus Books.

Corozza, O. 2008. *Near Death Experiences: Exploring the Mind Body Connection.* London and New York: Routledge.

Douglas, M. 1978. *Purity and Danger: An Analysis of Concepts of Pollution and Taboo.* London: Routledge and Kegan Paul.

Durkheim, E. 1951. *Suicide: A Study in Sociology.* Glencoe: Free Press.

Fenwick, P. and Fenwick, E. 2008. *The Art of Dying.* London and New York: Continuum.

Foucault, M. 1978. *The History of Sexuality.* Harmondsworth: Penguin.

Freud, S. 1917. Mourning and Melancholia, in *On Metapsychology.* Vol. 11. London: Freud Pelican Library, Penguin, 245–69.

Gordon, B. and Marshall, P. (editors) 2000. *The Place of the Dead.* Cambridge: Cambridge University Press.

Greyson, B., Williams Kelly, E. and Kelly, E.F. 2009. Explanatory models for Near-Death Experiences, in *The Handbook of Near-Death Experiences*, edited by J. Holden, B. Greyson and D. James. Santa Barbara and Oxford: Praeger. ABC-CLIO, 213–34.

Hallam, E., Hockey, J. and Howarth, G. 1999. *Beyond the Body: Death and Social Identity.* London and New York: Routledge.

Hallam, E. and Hockey, J. 2001. *Death, Memory and Material Culture.* Oxford and New York: Berg.

Kellehear, A. 1996. *Experiences Near Death.* New York and Oxford: Oxford University Press.

Klass, D., Silverman, P.R. and Nickman, S.L. 1996. *Continuing Bonds, New Understandings of Grief.* Washington, DC: Taylor and Francis.

Kung, H. 1984. *Eternal Life?* London: Collins.

Marx, K. 1997. *Selected Writings,* edited by D. McLellan. Oxford: Oxford University Press.

Masumian, F. 2009. World religions and Near-Death Experiences, in *The Handbook of Near-Death Experiences*, edited by J. Holden, B. Greyson and D. James. Santa Barbara and Oxford: Praeger. ABC-CLIO, 159–83.

Moody, R. 1975. *Life After Life.* Atlanta: Mockingbird Books.

Moody, R. 1977. *Reflections on Life After Life.* New York: Bantam Books.

Northcote, J. 2007. *The Paranormal and the Politics of Truth: A Sociological Account.* Exeter and Charlottesville: Imprint Academic.

Rinpoche, S. 1992. *The Tibetan Book of Living and Dying.* London: Rider.

Rose, N. 1989. *Governing the Soul: The Shaping of the Private Self.* London: Routledge.

Royse, D. 1985. The Near Death Experience: a survey of Clergy's attitudes and knowledge. *The Journal of Pastoral Care*, 39, 31–42.

Satori, P. 2008. *The Near-Death Experiences of Hospitalized Intensive Care Patients: A Five Year Clinical Study.* New York and Lampeter: Edwin Mellen.

Smith, D. 1990. *The Conceptual Practices of Power: A Feminist Sociology of Knowledge.* Toronto: University of Toronto Press.

Smith, D. 1996. Women's perspective as a radical critique of sociology, in *Feminism and Science*, edited by E. Fox Keller and H.E. Longino. New York and Oxford: Oxford University Press, 17–27.

Stanford, P. 2002. *Heaven: A Traveller's Guide to the Undiscovered Country.* London: Harper Collins.

Sutherland, C. 1990. Changes in religious beliefs, attitudes and practices following Near-Death Experiences: an Australian study. *Journal of Near-Death Studies*, 9, 21–31.

Sutherland, C. 1992. *Transformed by the Light.* Sydney and London: Bantam Books.

Weber, M. 1930. *The Protestant Ethic and the Spirit of Capitalism.* London: Allen and Unwin.

Vernette, J. 1997. Dialogue with Monsignor Jean Vernette, in *On the Other Side of Life*, edited by E. Valarino. New York and London: Insight Books.

Zaleski, C. 1987. *Otherworld Journeys: Accounts of Near-Death Experiences in Medieval and Modern Times.* New York: Oxford University Press.

PART II
Spaces of Burial: Taboo, Iconoclasm and Returning to Nature

Chapter 4

Buried Bodies in an East London Cemetery: Re-visiting Taboo

Kate Woodthorpe

Introduction

Multi-disciplinary academic debates about death in western societies are now well established, frequently articulated in discussion related to its apparent contradictory and complex presence/absence in society (see Kellehear 2007, Klass 1996, Mellor 1993, Smith 2006, Walter 1994). Framed by distinctions made between death 'behind closed doors' and death in the public domain (that is, in communal, shared and accessible spaces) sociological discussion in particular has branched out into expositions of mourning in public (see Grider 2006) and death's depiction in the modern media (see Hanusch 2008) to consider what these might reveal about a societal acceptance/rejection of death. Hypothesising such as this has generated an enduring theoretical debate about the status of death in society and its relation to discourses and institutions such as medicine, science and religion (see Walter 2008 for a useful overview).

As a part of this debate, the concepts of denial and taboo have retained a central position, owing in part to their relative and respective frequent usage across disciplines such as anthropology, sociology and psychology. Towards the end of the twentieth century an argument emerged however that challenged their applicability in contemporary western societies where particular facets of dying and bereavement are now more public (see Walter 1994). Some contributors to this line of reasoning have gone so far as to claim that debates about denial and taboo are not only out-dated, but perhaps even no longer relevant (Mellor 1993, Mellor and Shilling 1993).

This chapter suggests differently. Using data generated from an ethnography of a large cemetery in East London, it considers a particular component of the material culture of the cemetery – the buried body – to suggest that a conflation of societal and individual perceptions of death as taboo has lacked a material and empirical depth (see Willmott 2000). Rather than theorising death as a monolithic or homogenous entity, this chapter thus agrees with Howarth (2007) that death needs to be considered and examined within the social and spatial contexts in which it is encountered.

Drawing on data from staff and visitors to the cemetery in question the chapter first summarises academic discussion about taboo before moving on to consider the problematic status of buried bodies in the cemetery landscape. To support the assertion that these buried bodies are taboo – in the cemetery at least – the chapter explores the gaps and silences in conversation in the cemetery when it came to talk about the materiality of the buried body, and the alternative use of domestic euphemisms when talking about its location. After examining the further problem of the abject nature of the buried body as it decays, the chapter argues that the problematic status of the unbounded buried body is revealed above ground in contestation over the ownership and integrity of plots, and emphasis placed on maintaining plot boundaries.

It is via these contradictions and contestations that the chapter illustrates the ways in which bereaved visitors and staff maintain a social silence regarding the reality of the dissolution of buried bodies in the cemetery. In other words, I argue here, talk of the materiality of buried bodies in the cemetery is 'taboo'. This suggests that the concept still has relevance in sociological discussion and that future dialogue needs to move away from grand narratives of death (Seale 2001) towards an empirically situated debate that considers taboo in relation to particular places and spaces.

Background

This chapter originates from a four year ethnographic case study of the City of London Cemetery and Crematorium (CLCC), co-funded by the Economic and Social Research Council (ESRC), the City of London Corporation (CLC) and the Institute of Cemetery and Crematorium Management (ICCM). Located in Newham, East London (approximately two miles from the location of the London Olympics site in Stratford), the CLCC was opened in 1856 to provide a service for all those living in the square mile of the City of London (see Lambert 2006, Mellor and Parsons 2008). Expanding in the intervening years to over 200 acres, the CLCC is currently one of the largest cemeteries in the UK (Brooks 1989), containing seven miles of roads and a 32 acre memorial garden (CLCC Heritage Brochure 2004). Dealing with around 3,000 cremations and 1,000 burials every year, it had a staff of 90 – far more than most other cemeteries around the country.

The ethnography intended to extend previously published and unpublished work on this cemetery (Francis 1997, Francis et al. 2005) to examine the everyday functioning of the cemetery landscape from the perspective of visitors, staff and the local community. The primary methods of data generation were semi-structured interviews and participant observation, covered in turn briefly here.

Over 100 people were interviewed for the project, divided into three groups of visitors, staff and the local community. In total approximately 70 visitors were interviewed, as well as 20 staff members and 15 local community members. Most of the interviews took place within the cemetery, with interviews with visitors either taking place in the cemetery café or, more commonly, out in the grounds. Visitors to the cemetery were approached on site to ask if they would like to take part in the study, at which time they would be offered a leaflet outlining the research. Sampling visitors was thus opportunistic, with consent sought verbally. Subsequent interviews were either recorded on tape, written in note form at the time of the interview, or involved me writing them up afterwards. Their recording depended very much on the interview context, such as whether or not we were walking, whether the visitor was tending to a grave, if we were sitting in the café, the weather, the time of day and so on.

Staff members were primarily recruited through a snowballing technique (see Fink 2003). With collaboration from the funding partners, key members of staff were identified and subsequently interviewed (and recorded). Often they would then suggest other staff members to interview. For example, in early interviews it was recommended that I speak with the staff member who organises grave selections, and as a result I interviewed them and observed grave selection meetings with bereaved clients, accompanying them into the grounds when they considered their options.

In terms of the local community, cemetery staff responsible for communicating with the public suggested a range of organisations, individuals, and groups that might fall within the 'local community' rubric based primarily on their proximity to the CLCC. Initial contact was sought with all of them via a telephone call or a visit, however most of these attempts proved to be unsuccessful; indeed, communicating with the local community has been a persistent problem for the staff at the cemetery themselves. In an attempt to involve more respondents from the local area, I placed a generic letter in several local newspapers and left an information leaflet on a stand in the CLCC café and distributed the same at the bi-annual Open Day. From these efforts I received seven responses, which led to two interviews with local community members, one conducted in the café and one over the telephone.

At the same time as interviews were taking place, I undertook participant observation over approximately 60 days in a six-month period, where most time was spent observing people in the site and occasionally participating in what they were doing, for example, in helping to carry watering cans for people tending to graves and directing people around the grounds.

The data utilised here is predominantly from speaking to visitors at the grave side or from speaking to the members of staff who have a lot of dealings with visitors in the cemetery (for example, members of the grounds staff team). Like Francis et al. (2005) before me, participants in this chapter are not identified beyond whether they were a visitor or staff member. Not providing any detail about participants or coding beyond whether they are a visitor or staff member is intended to protect their identities according to the usual standards of anonymising participants in sociological research (see British Sociological Association 2002). It is also particularly important as anonymity did not extend to the field site itself, as the venue for this case study is explicitly named as the research location. Hence, whilst identifying features of participants have been kept to a minimum, data regarding the setting has not been obscured so that the reader can garner a better understanding of the *place* as much as the people.

Interpreting Contemporary Death

As noted in the introduction to this chapter, theoretical discussion about the status of death in society has been a fertile area for debate. Within the UK, the origins of sociological discussion about death are often attributed to Geoffrey Gorer's (1965) analysis of death, grief and mourning in contemporary Britain (see Lee 2008 and Walter 1991). In this publication Gorer declared death to be 'pornographic', in that it had replaced sex as *the* taboo of modern society. Proposing that throughout the first half of the twentieth century there had been a gradual loss of shared rules and rituals of how to deal with death in British society; Gorer argued that death had been gradually moved out of the public sphere and become increasingly restricted to the medical and scientific realms (Howarth 2007). This was associated with the industrialisation of society, and the development of medicine and a funeral industry, meaning that dying and the disposal of bodies became events that were increasingly presided over by professional specialists, rather than the local community.

The argument that death in contemporary western society was being concealed from public view was reinforced by French historian Philippe Ariès in his analysis of death across the ages in European societies (1974, 1981). In the sixteenth and seventeenth centuries, Ariès argued, death was 'tame'. That is, it was expected, public and visible (Howarth 2007); people were in regular contact with death. Comforting the dying and disposing of their bodies, they were therefore accustomed to its inevitability. In contrast, Ariès suggested, through the decline of organised religion, rapid changes in the demographic profile of societies as a result of industralisation, and driven by an increased political emphasis on individuals rather than communities, death in modern societies was 'wild'. Comparable to Gorer's argument, Ariès understood a wild death as death hidden from public view, controlled by medicine and governed by experts, constructed as a 'risk' that required management. By controlling that risk, death came to be known as an event that was threatening, something from which people needed to be protected. This line of reasoning was taken further by some sociologists who suggested that in contemporary western society death was sequestered from everyday life (Mellor and Shilling 1993), protecting individuals from the reality of death in order to ensure their continuing sense of self, so that their 'ontological security' was not compromised (Giddens 1984).

So what is the relevance of these arguments to this chapter on taboo? Gorer and Ariès' contributions are significant because they both went some way to instigating the development of academic discussion about whether or not death in modern societies was taboo (Walter 1991). A sociological literature evolved from this question throughout the latter stages of the twentieth century, and into the twenty-first, taking place alongside (and in conjunction with) psychological discussion about death denial (see Becker 1973). Both taboo and denial were heavily utilised to examine whether modern society sought to conceal death; often with the terms being conflated: 'In everyday usage, the word "taboo" refers to something prohibited, forbidden, by custom rather than by law. It may be something too terrible even to think of, its reality denied' (Walter 1991: 295). In the same article from which this quote comes, Walter provided a well articulated overview of death and taboo, where he made a number of cases for why a death taboo may either never have been the case, or was no longer relevant. In this overview, the evidence he used to support his assertions included the establishment of palliative care, the way in which professional support groups have grown to deal with body disposal, and the manner in which death had entered public discourse.

Summarised well by Howarth (2007), debates about taboo and death denial have particularly revolved around theoretical distinctions between the public and private domains, and the development of modern institutional care and expert knowledge. These conversations have necessarily involved a re-examination of Gorer's assertion that death is *the* taboo of modern society (see Walter 1991) and a critique of Ariès as a somewhat romanticised conceptualisation of the past (Hockey 1996).

Within this evolving debate there has been a growing assertion that a view of death as taboo or denied is either misguided or irrelevant in contemporary society due to its growing inclusion in public discourse and its visibility in the modern media (Kellehear 1984, Mellor and Shilling 1993, Walter 1991). In recent years, Noys (2005) has gone so far as to claim that the idea of death as taboo is a cliché; others have suggested that the issue of taboo has been 'overcome' (Lee 2008). Within these arguments however, as Howarth (2007) notes, there has been insufficient attention paid to the social context(s) of death, with sociologists instead often choosing to draw on convenient psychological concepts that may not be suitable for a social analysis (see also Kellehear 1984). Howarth's point echoes Seale's (1998) claim that sociological arguments about denial have become confused with ideas about psychological repression.

A confusion regarding individual psychological repression and societal analysis can be seen in a further critique that has emerged about the way in which denial and taboo have become a means through which sociologists can problematise modernity and self-identity (Lee 2008). In terms of taboo specifically, this has led to a conflation of taboo as both a societal and individual construct, often over-generalised in the production of narratives about death that present it as a monolithic all-encompassing concept (see also Willmott 2000). Of equal concern, in this chapter at least, is how it is unclear as to the ways in which academic theorising about taboo and denial translate into the everyday world (Seale 1998), whether there is sufficient empirical evidence used to support them, and how these debates relate to the spaces and places associated with death.

Is it then accurate to claim that taboo are clichéd or outdated? This chapter argues not, via an empirically-based contribution to the discussion that situates the concept of taboo within a particular death-related setting. In developing a case for the way in which particular facets of death in the cemetery (here, the buried body) are specially worthy of the label 'taboo' (see Walter 1991), it takes its starting point from evidence from this research project that indicated a disparity between academic theorising about death and the everyday reality of individuals operating and visiting a space associated with death. Many participants in the research – visitors, staff and local community members – repeatedly stated that they considered death to be taboo, as the following two quotes indicate:

> I think the problem is we need to get over this taboo problem with death that we have. (staff member)

> It's something you don't really talk about really, death, is it? It's a bit taboo really. (visitor)

These types of comments were echoed by the then Director of the CLCC, Dr Ian Hussein, in his paper to the Memorial Awareness Board at the Houses of Parliament in May 2006 (reproduced in the ICCM's Trade Journal, 2006), where he argued that for cemeteries to be considered a viable topic for political discussion, the problem of the 'taboo status of death' needed to be addressed. Comments such as these indicated that participants found talking about death to be unpopular and also indicated that at a societal level, death – or at least death in the context of the cemetery – was something that was ostracised from public discourse, contradicting the aforementioned assertions that a growing engagement with death via, for example, the modern media indicates that a debate around death and taboo is no longer pertinent.

As data accumulated, suggesting that a number of participants felt that death was 'taboo', I thus began to query whether academic rejections of taboo were somewhat premature. Perhaps, as Walter (1991) pointed out, whether or not taboo is relevant depends on where you look. In this case, where I was looking – the cemetery – evidence suggested that there was something taboo, as it became increasingly apparent that the buried body was excluded from conversation. Building on Bradbury's (1999: 138) assertion that 'the symbolic value of the decomposing corpse holds the key to understanding why we behave as we do when someone dies', data being generated (or rather, a *lack* of data being generated) indicated that in the context of the cemetery the buried body, the 'body-that-once-was' (Hockey and Draper 2005: 48), was hugely problematic. Throughout fieldwork the materiality of the buried body consistently went unspoken. This was in stark contrast to the activity at the surface of the grave that indicated it was highly resisted, as bereaved visitors strove to uphold the individuality and integrity of the plot through boundary maintenance and contestation over the plot's ownership. It was this tension between activity above ground and a simultaneous reluctance to discuss the reality of what was happening below ground that is the focus of this chapter.

Taboo Revisited

So what is meant when we talk about taboo? Aside from the overview in the introductory section of this chapter, it is worthwhile unpacking the concept of taboo to examine its applicability to places, people and talk associated with death. Often the focus of anthropological study of ritual, taboo has been a profuse concept for understanding social behaviour in the context of the knowledge and history of the people in question (Hendry 2008). It is not a static concept, however, and has been taken to mean many different things; mutable and utilised in varying ways, a wide range of definitions have been offered as to what is actually meant by the term 'taboo'. For example, it has been articulated as something that is forbidden, such as incest (see Twitchell 1987); or something that is unmentionable (see Houppert 1999). Elsewhere, Yang and Southwell (2004) have perceived taboo in relation to something rotten or diseased, such as those people infected with sexually transmitted diseases. It has also been considered in relation to dirt (see Douglas 2002 [1966]); as something sacred or sacrificial (see Aho 2002), magical (see Frazer 1954 [1922]), or disordered (see Willoughby 1932 cited in Zuesse 1984). The pervasive link throughout all these varying ideas about taboo however has been the cultural significance and danger of something that is associated with taboo.

In terms of death, taboo has most commonly been associated with the precariousness of the relationship between the living and the dead. Indeed, it has been argued elsewhere that 'one of the most important cultural functions of taboo is the effective regulation of our relationship with the dead' (Koch and Magshamhrain 2003: 80). Possibly the most well known analysis of taboo in relation to death has came from anthropologist Mary Douglas (2002 [1966]), who examined ritual symbolism to consider the importance of bodily boundaries and the way in which they were invested with power and danger throughout various cultures and people. The significance of the breach of those bodily boundaries, for example through excretion, menstruation, saliva, or leakage after death, and the way in which these breaches were managed she saw as acting as a highly symbolic way through which to maintain social order. To illustrate this, she explored the Coorgs, and argued that the impurities they associated with bodily excretions were a means through which social hierarchies could be maintained. In terms of death, those lowest in the social structure (the lowest in the caste system) were required to tend to the polluting bodies of their deceased, thus confirming and reinforcing their low status. Their tending also served to underline the polluting danger of the dead body. Similar efforts to get rid of the 'taboo' from the corpse can be found in traditional Maori rites, where after a burial 'the mourners perform a rite of washing to remove the tapu ('taboo') from themselves' (D. Douglas 2002: 108).

From Mary Douglas's work has developed a small, but very interesting, body of social scientific literature on the management of the corpse in western societies (see Hallam et al. 1999, Harper 2010, Quigley 1996, Troyer 2007 for example). Within anthropology the focus of work on the corpse has particularly focused on non-western societies, such as in Bloch's (1971) analysis of the dissolution of the buried body and the symbolism of the tomb above it as the deceased person inherits their ancestral identity. However, not much has come from sociologists on the topic of buried bodies and the places in which they are located, with more attention having been given to the grief of survivors or the place of death, such as death at the roadside (see Breen and O'Connor 2007). This chapter hopes to redress this focus in part, by making a case for the social significance of the buried body in the cemetery.

Buried Bodies in the Cemetery

So why, in what ways, and to whom, might buried bodies in a cemetery matter? One principal reason is the way in which the dead body – before burial even – challenges the boundaries of the ideal contained and controlled body (echoing Douglas 2002 [1966]). In a society such as the UK where bodies are ordered and controlled (Evans 2002), the uncontrollable, decomposing person-that-once-was (Hallam et al. 1999) presents an absolute rejection of efforts to order and control the body. Consequently:

> ... it is hardly surprising that ageing and death are viewed so negatively – they are unwelcome reminders of the inevitable decay and defeat that are in store, even for the most vigilant of individuals. (Featherstone 1991: 186)

A central tenet of this chapter is that the problem of the dead body is *amplified* in a cemetery upon burial, owing to the fact that in this landscape these problematic bodies quite literally underpin – both metaphorically and materially – everything that happens in the site. Without these buried bodies, a cemetery is not a cemetery.

Bearing this in mind, within the ethnography being drawn upon in this chapter it might have been fair to assume that these bodies would have come up in interviews with staff and visitors. Yet, similar to Bradbury's experience in her doctoral research (1999), there were very few occasions in fieldwork (one or two, out of approximately 100 interviews) when anyone mentioned the buried bodies in the cemetery. Sometimes this was when we were standing literally on top of them. Thus although undeniably present, they were markedly absent in conversation; they were, I argue, *unmentionable*. This paradox of the unspoken – yet implicitly known – existence of buried bodies in the cemetery was intensified when participants used familiar, domestic euphemisms to talk about the dead person being in the cemetery as if they were still animate and the focus of a 'social visit':

> My husband [is in this grave now and], it will be a matter of time before I join him. I like knowing where he is waiting. (visitor)

I like visiting here – it's like knowing where my bedroom is! (visitor)

Argued by Francis et al. (2005: 85) to be a means through which to domesticate the grave space and 'mask the stark truth of death's finality', the materiality of the buried body was thus concealed by euphemistic language that (re)presented the buried body as being in a comfortable, safe space, as it lay waiting to be joined. What is more, visitors would often 'pay a social call' to them:

I think for some people it's not just to do this, lay the flowers and that sort of thing. They find comfort, they do, they talk to them, say things; they ask their advice. It sounds silly don't it?

I always talk to people I come to, tell them different things. I see loved ones as being here.

This intersection of the simultaneous physical presence and social silence of the buried body became a pivotal issue within fieldwork, made more ambiguous through comments such as those above that suggested that the deceased person had a social presence in the cemetery. Indeed, the possibility of the dead having a social presence has become an increasing focus for discussion in recent years (Hallam et al. 1999), explored elsewhere in studies of the role and destination of ashes in families (Prendergast et al. 2006), ongoing relationships with spouses (Bennett and Bennett 2000) and the rituals used to maintain a social presence (Klass 1996). In the cemetery however, this social presence was problematic as it coincided with the location of the deteriorating buried body, and in sustaining a social presence for the deceased person in the cemetery by visiting the location of their body, visitors were implicitly acknowledging their physical presence in the site. This recognition existed in marked contrast to the absence of the same bodies in conversation. Thus, the buried bodies in the cemetery were therefore both *there* and *not there* (Meyer and Woodthorpe 2008), bodily present yet persistently absent from conversation. This paradox was illustrated by a visitor who remarked on how:

I suppose it's like you're still visiting your family – although they're not here are they?

The importance for maintaining a social presence can in part be interpreted using Howarth's (1996) study of the activity of funeral directors in restoring the body of the deceased in preparation for viewing. This they did, she argued, to prevent the loss of the deceased person's social identity, meaning that when viewed it could remain as a site for a meaningful relationship between the living and the dead (see Hallam et al. 1999). In comparison, by deliberately not mentioning the materiality of the buried body in the cemetery, participants were separating the social identity of the person-that-once-was, who they visited and talked to (if they felt like it), from the materiality of the body buried below. The use of domestic euphemisms to account for the physical presence of the dead person in the cemetery reinforced this separation.

To further complicate this is the problem of the mutability of the buried bodies themselves, which can be very difficult to talk about (Walter 1990). Constantly shifting, buried bodies are an unstable entity that are constantly changing as their fluids seep into the surrounding soil and their flesh breaks down to become part of the earth. They do, over time, quite literally become part of the cemetery.[1]

The problem of decomposition – and why it is difficult to talk about – needs to be understood however as something more than the bacterial process of bodily decay. As an inhabitant of a single grave plot, the decay of the body is a transformative loss of the individual's bodily integrity (Hallam et al. 1999), and the creation of a body that is 'neither "self" nor "other", "inside" nor "outside". Transgressive in nature, it respects no borders, rules or positions' (Williams and Bendelow 1998: 124). It is:

> ... the body without boundaries, the permeable body, the liminal body, the leaking, fluid body [that] has become a site of horror, dread and fear for its transgressive nature. (Lupton 1998: 97)

No longer a subject body (a living, breathing, bounded, autonomous and independent person), nor an object body (skeletal remains), the transforming buried body's status is precarious as something that is, to borrow Victor Turner's (1967) well-used expression, 'betwixt and between'. It is perhaps the ultimate source of abjection[2] (see Kristeva 1982), completely uncontrollable in its putrefaction. Boundary-less and un-boundable, the buried body is thus a highly dangerous body, illustrated in how:

> we ... shut them away in coffins, which have been nailed down or (even more securely) screwed down, and which are then enclosed in sealed, concrete burial vaults, under a very heavy stone. And all this is further enclosed in a cemetery surrounded by high walls, the gates of which are kept locked. (Ragon 1983: 16 cited in Quigley 1996: 19)

---

1   This reality of the decomposition of the human body (see Rodriguez and Bass 1985, Rogers 2005, Wilson et al. 2007) depends on a range of factors, including the body itself; whether or not it has been embalmed; the weather and season; and the position of the body in the grave plot. Decomposition is particularly contingent on the conditions of the soil and the body, and can take anything from six to a hundred years or more to decompose entirely. Indeed, archaeological evidence shows us that if the conditions are ambient, bodies can remain preserved for thousands of years, (see Adams 1988, Turner and Scaife 1995).

2   Abjection here is taken to mean the rejection and revulsion associated with the breaching of bodily boundaries through bodily waste (see Cregan 2006).

Further evidence for concealing the danger the buried body presents could be seen in how staff at the CLCC were keen to backfill plots once the grave started to collapse (typically after the first six months after burial). The sinking soil offered very visible evidence of the breakdown of the coffin and its contents below.

Beyond backfilling graves, then, what other activity above ground revealed the potency of the uncontrollable, unbounded buried body below? So far in this chapter I have argued that the buried body in the CLCC was subject to a social silence, concurrently concealed by the use of domestic euphemisms to talk about the presence of the body in the site. I want to suggest here that this concealment was further exacerbated by the way in which bereaved visitors related to the space in which the body was located, played out in uncertainty over ownership of the land and the boundaries of the grave plot. It is to these two points that the chapter now turns.

The Implications Above Ground

As you have seen so far, the data generated in this ethnography suggested that the buried body in the cemetery is problematic; somewhat surprising considering its presence is what makes the cemetery a cemetery rather than simply a garden of remembrance or a memorial park. Beyond its concealment in talk, this problematic status, I want to suggest here, was revealed in tension above ground via contestation over ownership of the plot and its contents, and the boundaries of the plots themselves. These are dealt with in turn here.

The first tension above ground revealed in fieldwork was uncertainty regarding the ownership of the plot and its contents. The data generated from this project suggested that, in the first few years after a burial had taken place, newly bereaved visitors often had difficulty establishing their claim to the land in which the burial had occurred. This was confounded by confusion over their 'rights' to do what they wanted in terms of memorialisation and their 'stakeholder' status in the cemetery (Francis et al. 2005: 83). This uncertainty could loosely be understood as conflicts arising from two (at times competing) ideas about claims to the land in which the buried body 'resided'.

Data suggested that there were two aspects of plot ownership at play within the cemetery: *legal* ownership that was agreed between the plot leaseholder and the City of London Corporation, and *moral* ownership of the land of the plot and its contents. Evidence from staff and visitors suggested a pervasive misunderstanding amongst bereaved visitors between these two aspects of ownership of grave space and associated 'rights' to the use of land:

> ... you're not going to stop people putting things on the grave, what they want on the grave. You can give them rules and regulations, what you can have and what you can't have, but people don't want to listen to that. That's their little memory

to that person isn't it? *They want to put on it what they want.* (staff member, emphasis added)

The analogy of renting a residence is useful here in explaining the consequences of this conflict in the cemetery landscape. Although commonly referred to in fieldwork by visitors as 'purchasing' a grave, in legal terms an individual would actually buy the Grant of Exclusive Rights of Burial to the grave plot for a specific number of years.[3] At the CLCC, the City of London Corporation remained the legal owner of the land and therefore in essence the customer *leased* the interment plot for a number of years. Currently, at the CLCC, this lease was usually for 75 years, after which time, unless the lease was renewed, ownership of the Rights of Burial reverted back to the City.

Although legally the cemetery land continued to be owned by the City of London Corporation, until June 2007,[4] English Law[5] stated that it was an offence to disturb human remains without obtaining a licence from the Secretary of State of the government department that dealt with burial – at the time of writing, the Ministry of Justice (formed following the merger of the Department of Constitutional Affairs and part of the Home Office in May 2007). This law meant that it was illegal to move human remains once buried.[6] As a result, the plot lease agreement at the CLCC equated to more of a 'permanent residency' contract between the landowners and the leaseholders. Unlike a conventional rental agreement though, the leaseholder or resident of the grave plot could not be 'evicted' if they were found to be in breach of their contract. This put the landowners (or their representatives – the staff at the CLCC) in a difficult position as they lacked the ultimate means through which to govern their 'tenants'. It is in this way that the rights of the leaseholder of a plot extended beyond the usual leaseholder/landlord relationship, in that visitors could deliberately break rules and regulations without having to face any sanctions. This had the potential to lead to a greater feeling of autonomy for leaseholders and an expectation that they could do what they wanted at the surface of that plot.

> You've got to remember you've got thousands of people that own, as far as they're concerned, they *own* the grave. They don't temporarily lease it, the rights to it, they *own* that spot. (staff member)

This sense of owning the plot served as a source of entitlement within the cemetery, and reinforced the domestic euphemism of the dead having a residence

---

3   This is referring specifically to private graves.

4   When a London Local Authorities Bill was passed, in which Clause 9 permits the disturbance of human remains to create more burial space in the capital.

5   Burial Act 1857, s.25.

6   Owing to the lack of new space for burial, the decision by the Ministry of Justice to make legal changes that would enable human remains to be disturbed was taken in 2007.

(Francis et al. 2005). Asserting 'rights' and claims of ownership over the individual plot also went some way to maintaining the individuality and integrity of the person within that plot, as their exclusive abode, further concealing what was actually happening in the ground below. It also reinforced the sense that the person within the plot was intact and 'residing' in their plot as they waited to be joined by others.

A very visible outcome of this at the surface of the grave could be seen in tension when it came to marking and maintaining grave boundaries, which served to remind and:

> ... inform the stranger, about the body of the deceased as it was in life. They challenge the observer to recognise and endorse the portrait above the tomb *rather than to dwell on the decaying matter below*. (Hallam et al. 1999: 149, emphasis added)

The making and maintenance of individualised grave boundaries above ground emerged as a central concern within fieldwork, where for many visitors it was apparent that a key issue was that the boundaries of graves were controlled and did not encroach on neighbouring plots:

> It's when they go over the boundary. If you're within the boundaries of the grave and guidelines, it's personalisation for the person buried there.

> It was just round here, a little boundary. And I used to fit the flowers to the ground to make it look like it was growing.

> As long as it doesn't encroach on [to my] area ... when people go over the top, I don't like that.

Hallam and Hockey (2001) have suggested that a focus on the demarcation of space is a key element of maintaining individuality. This was confirmed in the cemetery, where boundaries between plots were vital to many participants, whether these were marked out or not. Those that chose to mark the boundaries were an ongoing cause of concern for staff, whose job required that boundaries did not interfere with their maintenance of the site. Thus, staff members would remove boundaries that extended the full length of the grave, and restrict activity to that around the headstone. This was an ongoing battle however, as it was not uncommon for grave marking activity to creep outwards after being cleared, meaning it was not unusual to see varying boundary sizes along the length of the grave.

This emphasis on grave boundaries existed in stark contrast to the ever-diminishing integrity of the body below ground, revealing the juxtaposition of the lack of control over what is going on in the ground and the primary importance for many visitors of the establishment of boundaries surrounding the plot at the surface. This further served to reflect the ambiguity of the status of the buried body in the cemetery. On the one hand, the body was undeniably present, buried in a plot. On the other hand, as seen earlier in this chapter, the buried body was absent in conversation. Further hidden by the usage of familiar domesticated phrases such as the deceased person 'resting' and 'laying down', the physical presence and social identity of the buried body was separated by those who visited the cemetery to spend time with the deceased.

It is the intermingling of all this activity – the backfilling of graves to hide the presence and reality of the decomposition of the buried body, the omission of the buried body from conversation and its continued concealment through euphemistic language, alongside efforts to maintain individuality at the surface – that leads me to conclude that talk about the materiality of the buried body, in the context of the cemetery, is taboo.

Conclusion

Using data from an ethnographic approach, this chapter has illustrated the way in which buried bodies can occupy a tenuous and uncertain position in the cemetery setting. Simultaneously present and absent, their unbounded and uncontrollable nature contrasts with the intense activity that takes place at the surface in contestation over ownership and the importance attached to grave plot boundaries. Data in this ethnography – or rather, a *lack* of data – suggests that talking about the existence of buried body in the cemetery setting is typically proscribed. Complicated by the way in which their precise location is visited, the physical presence of these buried bodies is often cloaked in language that suggests they are sleeping or resting. Their tenuous presence is further reinforced by activity such as grave backfilling, which conceals the reality of the decaying, transgressive body below ground. It is this contradictory status that leads me to conclude that in the context of this piece of research at the CLCC buried bodies could be considered 'taboo'. Not *denied* – evidenced by visitations and contestation over the boundaries of their 'residence', rather the bodies are concealed and figuratively managed via euphemistic talk and emphasis on grave plot integrity at the surface. Any discernible threat they pose that may serve to remind visitors and staff of their uncontrollable-ness is further neutralised, or compensated for, through the practice of grave back-filling and the (at times intense) management of individual plot boundaries.

Moreover there is much scope to develop this argument, not least in empirically examining in more detail *why* visitors and staff typically do not talk about the materiality of the buried body when in the cemetery setting. Would they be more willing to discuss it when in not such close proximity to the bodies themselves? And why is talking about the buried body so problematic at all? These types of questions were beyond the scope of the study from which this chapter originates and could be examined in future research. What is more, the relationship between the visitor and the buried body they are visiting could be further explored in relation to the heterogeneity of visitors to the cemetery. There is much potential to refine the arguments presented in this chapter along class, ethnicity and gender lines, or situated alongside spiritual beliefs about what happens after death, providing greater insight into how people relate to the buried body as the earthly remnants of a life-once-lived.

Theoretically speaking, in light of the ambiguous status of the buried body in the cemetery and the way its dissolving materiality is excluded from conversation, the chapter also indicates that there may yet be much mileage in discussion about taboo and death in contemporary society. However, this discussion needs to move from the abstract realm towards a quite literally grounded debate that relates to the places and spaces in which death is encountered. Future debate about taboo could benefit greatly moving away from grand narrative explanations towards an empirically focused discussion that considers the materiality of death. Moving beyond talking about taboo in relation to 'death' as a homogenous discursive concept could enable more nuanced and insightful investigation into the environments in which death, and all its different facets (such as the buried body), is encountered and experienced.

References

Adams, B. 1988. *Egyptian Mummies.* Aylesbury: Shire.

Aho, J.A. 2002. *The Orifice as Sacrificial Site: Culture, Organization and the Body.* New York: Aldine Publishers.

Ariès, P. 1974. *Western Attitudes Towards Death: From the Middle Ages to the Present.* Trs. P.M. Ranum. London: Baltimore.

Ariès, P. 1981. *The Hour of Our Death.* Trs. H. Weaver. London: Allen Lane.

Becker, E. 1973. *The Denial of Death.* New York: Free Press.

Bennett, G. and Bennett, K.M. 2000. The presence of the dead: an empirical study. *Mortality.* 5(2), 139–57.

Bloch, M. 1971. *Placing the Dead: Tombs, Ancestral Villages and Kinship Organization in Madagascar.* London: Seminar Press.

Bradbury, M. 1999. *Representations of Death: A Social Psychological Perspective.* London: Routledge.

Breen, L. and O'Connor, M. 2007. Family disputes, dysfunction, and division: case studies of road traffic deaths, in *Remember Me: Constructing Immortality: Beliefs on Immortality, Life and Death,* edited by M. Mitchell. London: Routledge, 159–66.

British Sociological Association 2002. *Statement of Ethical Practice* [online]. Available at: http://www.britsoc.co.uk/equality/Statement+Ethical+Practice. htm [accessed: 28 September 2009].

Brooks, C. 1989. *Mortal Remains: The History and Present State of the Victorian and Edwardian Cemetery.* Exeter, Devon: Wheaton Publishers Ltd.

City of London Cemetery and Crematorium. 2004. *Heritage Brochure.* Updated edition, first published 2003. London: City of London.

Cregan, K. 2006. *The Sociology of the Body: Mapping the Abstraction of Embodiment*. London: Sage.

Douglas, D. 2002. *Death, Ritual and Belief: The Rhetoric of Funerary Rites*, 2nd edition. London: Continuum.

Douglas, M. 2002 [1966]. *Purity and Danger*. Abingdon: Routledge Classics.

Evans, M. 2002. Real bodies: an introduction, in *Real Bodies: A Sociological Introduction*, edited by M. Evans and E. Lee. Basingstoke: Palgrave, 1–13.

Featherstone, M. 1991. The body in consumer culture, in *The Body: Social Processes and Cultural Theory*, edited by M. Featherstone, M. Hepworth and B.S. Turner. London: Sage, 170–96.

Fink, A. 2003. *How to Sample in Surveys*. London: Sage.

Francis, D. 1997. *A Cemetery for Posterity: The Conservation of the Landscape of the City of London Cemetery*, unpublished report. London: Architectural Association.

Francis, D., Kellaher, L. and Neophytou, G. 2005. *The Secret Cemetery*. London: Routledge.

Frazer, J.G. 1954 [1922]. *The Golden Bough: A Study in Magic and Religion*, abridged version. London: Macmillan.

Giddens, A. 1984. *The Constitution of Society: Outline of the Theory of Structuration*. Cambridge: Polity.

Gorer, G. 1965. *Death, Grief and Mourning in Contemporary Britain*. London: Cresset Press.

Grider, S. 2006. Spontaneous shrines and public memorialization, in *Death and Religion in a Changing World*, edited by K. Garces-Foley. London: M.E. Sharpe, 246–64.

Hallam, E. and Hockey, J. 2001. *Death, Memory and Material Culture*. Oxford: Berg.

Hallam, E., Hockey, J. and Howarth, G. 1999. *Beyond The Body: Death and Social Identity*. London: Routledge.

Hanusch, F. 2008. Graphic death in the news media: present or absent? *Mortality*. 13(4), 301–317.

Harper, S. 2010. Behind closed doors? In *The Matter of Death: Space, Place and Materiality*, edited by J. Hockey, C. Komaromy and K. Woodthorpe. Basingstoke: Palgrave Macmillan.

Hendry, J. 2008. *An Introduction to Social Anthropology: Sharing our Worlds*, 2nd edition. London: Palgrave Macmillan.

Hockey, J. 1996. The view from the west: reading the anthropology of non-western death ritual, in *Contemporary Issues in the Sociology of Death, Dying and Disposal*, edited by G. Howarth and P.C. Jupp. Basingstoke: Macmillan, 3–17.

Hockey, J. and Draper, J. 2005. Beyond the womb and the tomb: identity, (dis)embodiment and the life Course. *Body and Society*, 11(2), 41–57.

Houppert, K. 1999. *The Curse: Confronting the Last Unmentionable Taboo: Menstruation*. New York: Farrar, Strauss and Giroux.

Howarth, G. 1996. *Last Rites: The Work of the Modern Funeral Director*. Amityville, New York: Baywood Publishing.

Howarth, G. 2007. *Death and Dying: A Sociological Introduction*. Cambridge: Polity Press.

Hussein, I. 2006. The cemetery as a garden: why a new understanding of the word 'cemetery' is necessary. *The Journal of the Institute of Cemetery and Crematorium Management* (Inc), Autumn. London: ICCM, 13–16.

Kellehear, A. 1984. Are we a death-denying society? A sociological review. *Social Science and Medicine*, 18(9), 713–23.

Kellehear, A. 2007. *A Social History of Dying*. Cambridge: Cambridge University Press.

Klass, D. 1996. Grief in an eastern culture: Japanese ancestor worship, in *Continuing Bonds: New Understandings of Grief*, edited by D. Klass, P.R. Silverman, and S.L. Nickman. London: Routledge, 59–70.

Koch, G. and Magshamhrain, R.L. 2003. Between fear of contact and self-preservation: taboo and its relation to the dead. *New German Critique*. 90(Autumn), 71–83.

Kristeva, J. 1982. *Powers of Horror: An Essay on Abjection*. New York: Columbia University Press.

Lambert, D. 2006. *The Cemetery in a Garden: 150 years of the City of London Cemetery and Crematorium*. London: City of London.

Lee, R.L.M. 2008. Modernity, mortality and re-enchantment: the death taboo revisited. *Sociology*. 42(4), 745–59.

Lupton, D. 1998. Going with the flow: some central discourses in conceptualising and articulating the embodiment of emotional states, in *The Body in Everyday Life,* edited by S. Nettleton and J. Watson. London: Routledge, 82–100.

Mellor, P. 1993. Death in high modernity: the contemporary presence and absence of death, in *The Sociology of Death: Theory, Culture, Practice,* edited by D. Clark. Oxford: Blackwell, 11–30.

Mellor, H. and Parsons, B. 2008. *London Cemeteries: An Illustrated Guide and Gazetteer*, 4th edition. Stroud: Sutton Publishing Limited.

Mellor, P. and Shilling, C. 1993. Modernity, self-identity and the sequestration of death. *Sociology*, 27(3), 411–31.

Meyer, M. and Woodthorpe, K. 2008. The material presence of absence: a dialogue between museums and cemeteries. *Sociological Research Online*. 13(5). Available online at: http://www.socresonline.org.uk/13/5/1.html [accessed: 28 September 2009].

Noys, B. 2005. *The Culture of Death*. Oxford: Berg.

Prendergast, D., Hockey, J. and Kellaher, L. 2006. Blowing in the wind? Identity, materiality, and the destinations of human ashes. *The Journal of the Royal Anthropological Institute*. 12(4), 881–98.

Quigley, S. 1996. *The Corpse: A History*. London: McFarland and Company Inc.

Rodriguez, W.C. and Bass, W.M. 1985. Decomposition of buried bodies and methods that may aid in their location. *Journal of Forensic Science*. 30(3), 836–52.

Rogers, T.L. 2005. Recognition of cemetery remains in a forensic context. *Journal of Forensic Science*, 50(1), 1–7.

Seale, C. 1998. *Constructing Death: The Sociology of Dying and Bereavement.* Cambridge: Cambridge University Press.

Seale, C. 2001. The body and death, in *Exploring the Body*, edited by S. Cunningham-Burley and K. Backett-Milburn. London: Palgrave, 98–116.

Smith, W. 2006. Organizing death: remembrance and recollection. *Organization*, 13(2), 225–44.

Troyer, J. 2007. Embalmed vision. *Mortality*, 12(1), 22–47.

Turner, R.C. and Scaife, R.G. 1995. *Bog Bodies: New Discoveries and New Perspectives.* London: British Museum Press.

Turner, V. 1967. *The Forest of Symbols.* London: Cornell University Press.

Twitchell, J.B. 1987 *Forbidden Partners: The Incest Taboo in Modern Culture.* New York: Columbia University Press.

Walter, T. 1990. *Funerals and How to Improve Them.* London: Hodder and Stoughton.

Walter, T. 1991. Modern death: taboo or not taboo? *Sociology*, 25(2), 293–310.

Walter, T. 1994. *The Revival of Death.* London: Routledge.

Walter, T. 2008. The sociology of death. *Sociology Compass.* 2/1, 317–36.

Williams, S. and Bendelow, G. 1998. *The Lived Body: Sociological Themes, Embodied Issues.* London: Routledge.

Willmott, H. 2000. Death. so what? Sociology, sequestration and emancipation. *The Sociological Review.* 48(4), 649–65.

Wilson, A.S., Janaway, R.C., Holland, A.D., Dodson, H.L., Baran, E., Pollard, A.M. and Tobin, D.J. 2007. Modelling the buried human body environment in upland climes using three contrasting field sites. *Forensic Science International.* 169(1), 6–18.

Yang, C. and Southwell, B.G. 2004. Dangerous disease, dangerous women: health, anxiety and advertising in Shanghai from 1928–1937. *Critical Public Health*, 14(2), 149–56.

Zuesse, E.M. 1974. Taboo and the divine order. *Journal of the American Academy of Religion.* 42(3), 482–502.

Chapter 5

From Anti-social Behaviour to X-rated: Exploring Social Diversity and Conflict in the Cemetery

Bel Deering

Introduction

This chapter examines what goes on in the cemetery alongside the 'official' business of burial, mourning and land management. The discussion is centred on narratives derived from an interrogation of the modern day recreational use of cemeteries and churchyards in England and France using both interview and focus group techniques. Internet sources such as websites and blogs were also used to provide data that added a more international perspective. I found that cemeteries and churchyards have a rich and diverse role in hosting leisure and recreational pursuits, and that these activities in themselves can cast the sites into positions of disagreement and conflict. Such activities can play an important role in the conservation or celebration of the cemetery, but may equally be of significance in the social, physical and mental health and development of young people and adults (Seymour 2003, Harrison et al. 1995).

By the tenth century and up until the nineteenth century, churchyards – the burial land found around a church – were the main burial places in much of Western Europe. From the nineteenth century onwards this role was increasingly filled by purpose-built cemeteries, able to offer burial for people of all backgrounds and beliefs, and with the advantage of having more space available to inter the increasing urban populations of the industrial revolution (Dunk and Rugg 1994). Despite both churchyards and cemeteries having the primary purpose of being for the disposal of the dead, there is a wealth of evidence to suggest that, in the past, they also played an important role in the provision of recreation space. By the thirteenth century it seems that churchyards already had a thriving social role, perhaps due in part to their being located in village and town centres. Ariès (1994) describes an intimate relationship between burial grounds and leisure pursuits with churchyards apparently playing host to activities as diverse as gambling, theatrical performances, music and dancing.

Authors including Worpole (1997), Dunk and Rugg (1994) and Sloane (1991) describe how fairs, markets, walks and carriage driving were all commonplace activities in churchyards up to Victorian times, suggesting that these sites were embedded in the community and its daily life. However, change began to occur during the Victorian era, when many churchyards reached capacity and the novel concept of the garden cemetery was born (Francis et al. 2005). Planners and designers of new cemeteries had a vision of them as parks or gardens, where people might walk and indulge in 'rational recreation' (Dunk and Rugg 1994, Worpole 2003a). Loudon articulates this Victorian agenda: 'The *main object* of a burial-ground is, the disposal of the remains of the dead ... A *secondary object* is, or ought to be, the improvement of the moral sentiments and general taste of all classes, and more especially of the great masses of society' (1981: 1).

Alongside moral education, artistic sensibilities, too, were fostered by the ever more elaborate sculptures and monuments on display at a time when public art was very limited (Ciregna 2004). The supposed transition from general community space to morally instructional space may have been well intended and may have increased visitor numbers enormously but in reality did not change recreational uses of burial ground sites completely. Linden-Ward (1989: 317) describes how visitors still came to seek 'present pleasures' rather than moral guidance, causing cemeteries to suffer vandalism and destruction of property, as well as playing host to activities from horse racing to picnicking. However, as purpose built parks increased in number, the need for cemeteries as recreational space decreased (Cross 1990, Linden-Ward 1989); the huge volumes of visitor traffic may have declined, but research suggests that recreational use is still part of the cemetery's physical and emotional landscape.

Whilst mass recreation in cemeteries may have declined, these spaces are far from redundant today. Dunk and Rugg (1994) argue that disused cemetery sites are a valuable addition to the leisure amenities of the locale and in their research identify four types of visitor: those with an interest in history; nature lovers; people making an educational visit; and walkers. In addition they suggest that improving the provision of facilities such as toilets and seating in cemeteries would increase the throughput of visitors. My research, then, is embedded in a historical picture rich in cemetery usage and where cemeteries and churchyards are replete with life as well as death. This chapter explores the details of what people think and feel about these burial sites today, what motivates their visits and what they do when they are there.

People make choices every day about where to go and what to do in their leisure time. Towns and cities offer amenities such as parks, gardens, swimming pools and museums for the explicit purpose of recreation (Urban Green Spaces Taskforce (UGST) 2002, Harrison et al. 1995). Green spaces are a popular part of this recreation matrix, with all of England's green spaces combined attracting annual visit numbers exceeding two billion (UGST 2002). Green space is defined by Harrison et al. (1995) as land, water and geological features which support plant and animal populations, and which are located within walking distance of the houses of local residents. The classification employed by both Harrison et al. and the UGST categorises cemeteries and churchyards as green spaces alongside more obvious examples such as parks and riverbanks, by virtue of their relatively natural appearance – and indeed they form the second largest reserve of urban green space in England after parks (Welch 1991). But beyond this factual accounting of the extent of burial ground there is no firm agreement as to the exact role cemetery land plays, or could play, within the urban framework of green spaces.

Methodology and Positioning

The first journey in this chapter is a physical one, which documents experiences and observations of interviewees about the activities common to graveyards. This discussion draws on research carried out in a number of different burial grounds; some in Paris, France and the rest in the south of England, and comprises a mixture of traditional town churchyards and Victorian cemeteries. The sites were chosen for their location – being near to residential or commercial areas – and for the fact that they are largely unused for current burials, meaning that research was less likely to impact on ongoing mourning practices.

The second research journey takes place in the virtual dimension, travelling through websites and blogs to gain an insight into the international opinions and attitudes held about graveyards. Taken together, these voyages form the starting point for an exploration of alternative graveyard behaviours ranging from vandalism to drug taking and dog walking. With regards to cemetery recreation I would position myself as an 'insider' in terms of experience, whilst being an outsider at the sites that I visited. Having played in village churchyards as a child, hung out in cemeteries as a teenager, and carried out botanical research on tombstones as an adult, I have many different experiences and impressions of recreation in burial grounds. The fact that my recreation has not always been at the moral education end of the spectrum of cemetery users added to my insight and empathy around the issues raised. I did not, however, make my 'insider' status explicitly known to interviewees unless it arose in our dialogue, and tended to then use it as a prompt or questioning device where it did arise. At all of the sites chosen for my research I was an outsider. This gave me the chance to ask basic questions that would have been unusual for an 'insider', but did mean that I lacked contextual knowledge of local culture and geography.

Setting the Scene – The Language of the Cemetery

Pertinent to the exploration of how people use cemeteries and graveyards is a consideration of how these spaces are perceived by their visitors. The popular culture impression of a cemetery is that of a scary place (Dickens 1994, Harness 1999, Athkins 1993, Stine 1999, Gascone, 1997) and this resonates with some academic opinion that burial sites can inspire fear (Worpole 1997). Hallam et al. (1999) add to this picture and bring the argument into the modern day context with the additional suggestion that with contemporary culture focused on youth, beauty and vitality, society is ill equipped to cope with the ugly reality of death. Corpses are dressed, made up and presented as sleepers, to protect us from the truth of age and decay. The fear we have of aging and dying carries over to the home of the dead, the cemetery.

Although cemeteries and graveyards carry society's baggage of fear and are strongly associated in the public mindset with ghosts, skeletons and vampires, there is also considerable evidence that people are enjoying visits to the cemetery. This finds expression in both fact and fiction across the globe. From children playing hide and seek (Amparo Escandon 1999, Chevalier 2002) to picnics and walks (DeBartolo Carmack 2002), the cemetery is subtly offering an alternative venue for recreation and hobbies. Lynch (1997) even offers a humorous glimpse of the future with his concept of the 'Golfatorium'; a cemetery and golf course combined.

Returning to the question of how cemeteries are perceived in the present day offers an interesting contrast to the horror stereotypes. In focus group and interview work I found that participants rarely described graveyards with horror story language:

> We used to take a short cut through the cemetery. You know the large tombs which have cracks in? I did run past those, because I did wonder if somebody might come out of them … you might have bodies coming up! So that is a question of going through (a stage …) And that is the nearest I can think of me being scared- a kind of childish, reasonable thing. (Interviewee 1, site A, retired female)

> [...] to me it is a peaceful, friendly place where I can forget my own problems. (Interviewee 3, site A, middle-aged female)

Whilst some interviewees expressed a fear or dislike of burial grounds, the majority distanced themselves from emotion by recognising fear as a potential reaction but asserting that they did not specifically feel it themselves in this context. Many described fear that they had felt in the past as a child, but expressed how the site now held a different interpretation for them. When speaking specifically to children the language used held many similarities, suggesting that the sites were interpreted in congruence with the adult perspective, but children were more likely to mention both scary themes and natural themes. Adults tended to focus more on the historical and ancestral themes, picking out the atmosphere as melancholy or tranquil, while the children saw and heard the natural world.

In children's literature the cemetery is persistently portrayed as a place to be feared, or that is home to fearful beings (Harness 1999, Athkins 1993, Stine 1999, Gascone 1997). Children were able to conjure up many of these influences in focus group work, describing graveyards with reference to vampires, ghouls and ghosts and using adjectives such as creepy and scary. On balance though, they used an equal amount of language that reflected other dimensions of the landscape including peaceful, holy, quiet and flowery. Many of the participants in my research were excited by the idea of the scary aspects of the cemetery and took great delight in recounting horror stories, but professed not to believe in them. They enjoyed a certain thrill from feeling scared but at the same time had control over their feelings and used them to foster a momentum of storytelling and competitive anecdote recounting in the group. When considering Worpole's viewpoint that 'the cemetery exerts a continuing influence upon the urban imagination, especially for children, for whom this walled world...is often a source of unease and superstition' (2003a: 22) it appears that the children in my research may not have been forthcoming about the full picture, or perhaps they saw the different layers of meaning and were able to separate reality from the power of imagination. Terry Pratchett eloquently encapsulates the sentiments expressed by some focus group participants:

> It was one of those old cemeteries you got owls and foxes in and sometimes, in the Sunday papers, people going on about Our Victorian Heritage, although they didn't go on about this one because it was the wrong kind of heritage, being too far from London.

> Wobbler said it was spooky and sometimes went home the long way, but Johnny was disappointed that it wasn't spookier. Once you sort of put out of your mind what it was – once you forgot about all the skeletons underground, grinning away in the dark – it was quite friendly. Birds sang. All the traffic sounded a long way off. It was peaceful. (Pratchett 2004: 11)

The children's focus group work articulated something of the duality of the cemetery. Whilst it is a place for the dead, it matters more to the living. Invested in the creation of these sites are our hopes for our own lives and perhaps our thoughts, fears and beliefs about what may or may not happen next. The fact that there is much we do not know – even if we believe in any sort of afterlife – opens the door to imagination and possibility. And in the perhaps clichéd narratives and artwork of the children involved these possibilities are explored in story, language and drawing. The young participants tap unknowingly into the concept of the burial ground as a liminal place – a place with qualities that are not fully understood and where normal rules do not apply.

In the Scene – Visiting Cemeteries

Interviewees in my research consistently expressed an understanding of how a graveyard might be seen as a scary place, but most did not feel that way themselves. The general consensus amongst adults seemed to be that cemeteries were peaceful, beautiful, and thought provoking places – if inclined towards the melancholy. Some used the word 'sacred' or 'holy' to describe churchyards, but on the whole the language used tended towards the secular. Such an impression might lead one to the conclusion that the activities carried out in cemeteries would be compatible with this atmosphere of reflection and tranquillity. However the palette of uses ascribed to graveyards and cemeteries definitely transcends the supposed tranquil feeling of the place. Table 5.1 shows the most commonly cited recreational activities from my interviews.

Table 5.1   Recreational activities in cemeteries

Recreational activities in cemeteries as identified by interviewees
Reading
Picnicking
Jogging
Sunbathing
History/genealogy studies
Nature study
Photography
Walking the dog
Drinking
Taking drugs
Enjoying romantic pursuits

There was a strong association between the age of the user and the types of pursuit they were engaged in. Older interviewees suggested that 'acceptable' pursuits were more likely to be undertaken by older people, and viewed young people as using the sites for 'less appropriate' activities. In the list above, all but the last three activities were seen as broadly compatible with the quiet reflective atmosphere of a cemetery and were openly admitted to in interview. That is not to say that these activities were without controversy however. A hot topic in many interviews was that of dog walkers:

> I had some physical threats from dog owners who won't pick up poo. One lady's
> dog pooed right where they [the conservation group] were working and J picked
> it up and ran after her saying 'I think this belongs to you'. (Interviewee 2, site
> A, retired male)

On the one hand dog walkers were identified as a group that provided a useful form of policing to ward off the more transgressive and illicit activities of others in the cemetery, but on the other hand stood accused of bad behaviour themselves when they lost control of their dogs or failed to pick up droppings. The more tolerant interviewees accepted the dog walkers' place in the cemetery as either neutral or useful; dog-owners saw it as a largely positive role and those who did not like dogs tended towards wishing them barred from the sites – as is in fact common practice in many burial grounds. Uses of the site that were uniformly welcomed included organised activities such as conservation tasks, historical visits and guided walks. Casting back to the Victorian ideal of educating the masses through the learning landscape of the cemetery we see that this principle is still holding fast. Justification for many activities was their educational worth – even to the detriment of other valued qualities such as peace and quiet or conservation.

Whilst many cemeteries were council property, they frequently had 'Friends of' groups that played a more visible role in their management. The complex layers of ownership and involvement contributed to the sense that these spaces are unregulated, and may have been a factor in the often poor police/warden attendance around the sites. As a result interviewees reported all manner of illegal activities taking place – from grave robbing to shootings.

Littering was also mentioned as a common problem in cemeteries. The nature of things found, however, did also reveal something of the hidden life of the cemetery. Interviewees recounted finding caches of stolen goods: TVs, videos and a safe; the more bizarre type of item, such as traffic cones and even a set of traffic lights, perhaps associated with youthful pranks; and large quantities of discarded bottles, needles and condoms. Graffiti was also identified as a problem by some interviewees, whilst others had a different opinion of what they saw as memorial graffiti:

> I am disappointed that the graffiti is gone. It was people's tributes and reverence for Jim. Their respects and sentiments have been taken away. (Interviewee 1, site D, young female)

The last three pursuits in Table 5.1 – drinking, taking drugs and enjoying romantic pursuits – were more likely to be reported as observed rather than undertaken. The general consensus was that these activities were less desirable in a cemetery, were generally attributable to young people and were likely to take place between dusk and dawn. Interviewees gave accounts of such recreations in a variety of ways:

> Less desirable uses? We have had everything. It is in a zero tolerance, ASBO area. There was a rape, there are problems with needles. (Interviewee 1, site A, retired female)

One interview participant described a couple of incidents from guided walks:

> On a guided walk, J went ahead to mark out the next grave for people to aim for and virtually stood on people making love in the grass. They got bolshie and started throwing stones. Then there is a lady who feeds the rats sausages and chips. On guided walks G has to go ahead and scare the rats off. (Interviewee 3, site A, middle-aged female)

Alongside the stories they recounted, participants also suggested theories for why these things happened in the cemetery in the first place. The most prevalent hypothesis was that the sites lacked a regular police presence and so presented ideal spaces for anti-social behaviours that had been banned in town centres, parks and streets. Backing this up was the evidence that where patrolling of the site increased, either by greater use of the site by dog walkers and other 'respectable' visitors or from law enforcement agencies, the anti-social activities declined:

> Now city patrol attend (although I had to draw them a map to show them where the cemetery was) and the council have events to educate people about proper dog control ... there used to be Lambretta and Mini racing here in the evening so the council then brought in Securicor and the people complained it was like a police state! So the council stopped funding it and then crime and drugs increased. (Interviewee 2, Site A, retired male)

Strategies employed to manage anti-social behaviour included the cutting of long grass and overhanging branches. As the number of secluded spots decreased, so did the levels of drinking and drug taking. One cemetery Friend described how the benches had been moved from the more isolated parts of the cemetery to the busier entrances and crossroads. The move instantly reduced levels of drinking and vandalism and the littering associated with these activities.

Although the theory of convenience and opportunity does seem to encapsulate the driving force behind some users of cemetery sites, it is not the whole story. Some of the cemetery users are attracted to the site because it is a burial place, rather than because it is un-policed. One interview participant proffered her theory that there were certain types of people who found the cemetery setting erotic: 'It is the local fornicatorium you see. And of course you get the Goths that like to screw on graves' (Interviewee 1, site A, retired female).

Whilst sex in a graveyard is not the exclusive preserve of Goths, the interviewee did connect with the idea that the sense of place is a powerful motivator for visitors. Pain et al. (2001) theorise that spaces can reflect, construct and define social activity and identity and in addition that they can be spaces of resistance to society's perceived control. Looking at the internet-based research findings shows that these themes are universal. Although the internet shows a clear case for greater disclosure in the first person, overall the range of activities reported are within the same parameters. What is less evident in these data is the element of conflict that is expressed so frequently in interviews.

Stories of Cemetery Use in Cyberspace

The second strand of my research is an exploration of attitudes to cemeteries and stories of people's activities there as documented in cyberspace. The internet offers a fascinating research field since people both create and obscure their identities, and truths are revealed and hidden in the same way. This is to say that, just as with face-to-face research, people tell the stories they want you to hear and present themselves as they wish to be seen. Dolowitz et al. suggest that a blog exposes 'the intellectual soul of its author or authors' (2008: 101). Whilst this may be manifested as intellectual honesty, it is impossible to tell the difference between that and intellectual creativity in the anonymity of cyberspace; true stories may be hard to find.

For the purposes of my research, however, this hardly matters. My interest is in how cemeteries are described, what they are said to evoke in people and what people say they do there. Whether this is fantasy or reality is not essential to the creation of a sense of place in the same way that fictional accounts of cemeteries create their own valid reality for the site: 'the form and content of imaginative works are related to wider social and geographical relations and processes; they both reflect and affect these wider social relations' (Cloke et al. 2004: 94).

To type the words cemetery, graveyard or burial ground into a search engine is to open the door to thousands and thousands of alternative worlds. In this research I specifically looked at blogs with references to cemeteries or graveyards and explored websites specifically about groups or individuals with an interest in these sites. When newspaper articles or similar features were retrieved by the search engine these were included where they brought useful insights to the field.

An immediate discovery was that websites could be split into two broad categories: those predominately about graveyards and associated themes; and those that happened to include one or two graveyard anecdotes. Both of these provided useful information from their differing perspectives. One would seem to originate from a person with a self-confessed interest in cemeteries, whilst the other is more incidental and reveals stories from a person with what might be described as a 'normal' level of interest rather than a fixation or fascination.

These two groups contrast in how they perceived the cemetery and churchyard landscape. In similarity to interviewees, the range of perceptions continues to fit the stereotypes from creepy to tranquil:

> You really can't find peace and quiet like you can in a graveyard. (Van Skike 2009)

> The mystery and supernatural atmosphere that surrounds graveyards is alluring. People often feel a spiritual or supernatural presence there. (Porter Smith 1997)

> Far from being morbid, a stroll through Père Lachaise is quite wonderful. Its cobbled paths, shady trees and 19th century tombs give an overpowering sense of the romantic and gothic. (Curran 2008)

> Okay, I know this may sound a little bizarre, but I have a slight fascination with graveyards. I always have. I think it has to do with growing up in Richmond, where we don't have any. So every time I go to Nova Scotia I am always

interested to go past all the old graveyards filled with so much history, and so much mystery. (LeBlanc 2008)

I was terrified of cemeteries when I was a kid. In one of our many houses (we moved frequently when I was a kid), our backyard abutted a cemetery, and never once, ever, did I cross the fence line and venture inside. The local kids told tales, of course. Supposedly, one grave was illuminated at night by a lantern, hung by a grieving husband. I find that tale poignant now, but—back then—I was terrified by the thought of some old man keeping constant vigil at his wife's grave. (Lorraine 2009)

Table 5.2   Themes emerging from the data

Categories of sense of place in the cemetery	Description
Supernatural	Associations with the spooky, scary and fearful
Natural	Relating to nature, plants and animals
Human	Pertaining to person-centred aspects such as history and tombstones
Sentimental	Concerning the romantic, melancholy and mysterious aspects of the site
Spiritual	Focussing on the religious and sacred elements

Validation is given to the categories of sense of place that arose as themes in the interviews as they appear again in blog and web data. For simplicity I have put these emotions and interpretations of place into five groups, which encapsulate the themes that arose from the data. These are shown in Table 5.2.

These categories are not mutually exclusive – as the quotations below clearly illustrate – but simply assist in a consideration of the range of places that a cemetery or churchyard represents. As previously mentioned, the sense of place has an intimate relationship with the use of a place. Thus a person who views the cemetery from a human-centred perspective is likely to see it as a practical resource for historical or other studies. Similarly, perceiving the sentimental romance inherent in a cemetery may lead the observer to reflect this in their activity – whether it be photography or poetry. Looking back to blogs for evidence of these uses we see diversity as broad ranging as that found in interview:

Take paper and charcoal or crayons along for gravestone rubbing. Cemeteries are also a very intriguing place for photographs. (Young 2008)

There is no reason to feel slighted just because I had sex in a graveyard. I assure you, the dead people don't care. Oh and by the way, maybe I should have been more specific, but we didn't have sex on a tomb or anything. we [sic] had sex on his car. We were just in the graveyard under the moon. And really, it was quite lovely. The experience is really an amazing one. You should try it some time. (Queen of Spades 2009)

We ate and drank among the tombstones, celebrating life in a place of the dead transformed. (Bouman 2005)

Bouman and many others who describe time spent amongst the graves offer support for the idea that the cemetery is a good space for recreation. Whilst the concept of a 'good place' is subjective, several themes have been identified as typical in popular urban spaces including comfortable seats, suntraps, trees, toilets, viewpoints and shelter from the wind (Whyte 1981). Looking at this list it is clear that burial grounds can often offer some if not all of these features:

As Twitter followers will know, I spent much of today in a graveyard. I was not there for anything to do with zombies or vampires (though the place in question is apparently home to several species of bat), but rather to support a couple of friends of mine who have just put together a new guide book for the place. (Morgan 2009)

One of my absolute favorite drinking spots is Woodland Cemetary in Dayton. StumblingDoug and I go up there in the middle of the night quite often, to drink with the great historical figures of Dayton past: the Wright Bros., Paul Lawrence Dunbar, Erma Bombeck, Matilda Stanley (Gypsy Queen of the US), Adam Schantz (one of the first brewers in Dayton and all around good guy), just to name a few. Wait maybe there are just a few, but hey, it's Dayton fer crikes sake. It's like the highest point in the city and it's pretty old. I have found it to be a definite 'high energy' spot. (Thirstydrunk 2008)

These give us the main types of activities as seen in interview. However a category that got considerably more mention online than in interview – due perhaps to the global nature of the internet – was that of cemetery tourism. Young (2008), Mitra (2009), Payne (2008), Curran (2008), Jones (2008), and Derekbilldaly (2008) all recommend this as an interesting and insightful hobby and suggest both places to visit and what things to take:

When exploring cemeteries come nightfall, each participant should arrive adequately prepared with roughly the following ingredients:

- 1 flashlight or headlamp
- 1 black outfit (to camouflage with shadows and avoid detection)
- 1 stick of charcoal and a spiral notebook (for gravestone rubbings)
- 1 bottle of wine (optional)
- 1 pair of tree-climbable sneakers
- 1 constellation chart
- 1 camera (capable of star-capturing long exposures) (Jones 2008)

Down is not out. Arguably, cemeteries do not figure in the must-do list of backpackers and Gadabouts… With new-found interest in cemetery tourism, headstones are making heads turn. While genealogy is a key driver, others prefer time travel for a *tete-a-tete* with their favourite characters from history – from rocker Jim Morrison and author Oscar Wilde in Paris' Pere Lachaise cemetery to William Shakespeare and Charles Darwin in London's famed Westminster Abbey. Given India's colonial past, a string of states are now trying to cash in on the days of the Raj and more in their very own burial grounds of the yore. (Mitra 2009)

It's no secret that my goal in life is to see a ghost … It is for this reason that I am a champion of ghost tours and graveyards. Full of history, graveyards are one of the most honest and interesting locations a traveler can venture. I understand that some people are squeamish amongst the dead; however, there are few places that encompass such a wealth of local history. (Brown 2009)

This type of cemetery visit, identified in some instances as a pilgrimage (Kirk 2007), is not unique to the blogging and internet community. A large proportion of those interviewed who were members of cemetery Friends groups said that they made a point of visiting other cemeteries and graveyards when on holiday and that they also saw this type of tourism commonly taking place in 'their' cemeteries as well. Tourism tended to be more commonly associated with genealogy and history than the less salubrious pursuits and so we can see it being promoted by both local people (Bryan 2007) and by local and national government (Mitra 2009).

There are more common threads between interviews and blog data than just 'Dark Tourism' (Sharpley and Stone 2009). As previously shown, interviewees and bloggers alike described a wide range of feelings about cemeteries and these were reflected in the uses reported. From the data it is clear that cemeteries and churchyards have the potential for a considerable visitor footfall. As such levels of use increase, the site becomes more attractive to other users: 'What attracts people most, it would appear, is other people' (Whyte 1981: 19). When considering barriers to the use of green spaces this is a positive influence – the presence of other people can make a site feel safer and therefore encourage use and yet it is when people interact and share public spaces that conflict occurs.

Figure 5.1   Visitors to Jim Morrison's grave in Père Lachaise Cemetery, Paris

Source: Photograph owned by the author

Conflict in the Graveyard

With such a wide range of uses being attributed to cemeteries it is not surprising that conflict arises. Where conflict was identified by participants it was overwhelmingly attributed to young people or youths and described in the language of our times as anti-social behaviour. Interviewees ranged in their response to this – from anger, to fear and behavioural changes such as avoiding the cemetery at certain times. They also varied widely in how they thought it should be dealt with; some were keen to see patrols and police presence increase as a simple deterrent, others took a firmer stance and expressed the need for fixed penalties or stronger enforcement. One lone participant took a more charitable view, saying that the churchyard 'should not be a sacred space … it is a space for all' (Interviewee 1, site B, middle-aged male).

Research indicates that young people form a large proportion of the users of cemeteries and churchyards. We see young people visiting cemeteries for their perceived freedom from normal rules. Green space literature augments this explanation with the finding that levels of use are influenced less by the size of a site than by the topography. Bumpy ground, trees and water all increase the attractiveness of a site to users – in particular young people who constitute between 30 and 60 per cent of all green space users. Such features also offer seclusion and unique opportunities for privacy. Combine this with the real or perceived lack of policing and the attraction is clear (Rapoport and Rapoport 1975). As a result of this freedom, visitors may take the chance to drink, have sex or create graffiti with little risk of confrontation. Add in the fact that teenagers in particular are the age group most likely to litter (Nelson 2004) and are the age group most perceived as dangerous (DCSF 2008), and the ire and fear of the other cemetery users becomes clearer.

Littering, hanging around and vandalism do have a negative effect. Evidence suggests that when these activities cause a site to appear run-down and/or neglected, other potential users are deterred, interpreting this as signs of a lack of social control. Indeed this may well be the intention – litter can be a tactical move, a way of claiming a space and exerting power and control over others (Delaney 2005). Many are influenced by the visual effect of littering and related activities and will avoid such spaces out of a fear of crime (Harrison et al. 1995, UGST 2002).

Anti-social Behaviour, Youthful Exuberance or Essential Life Experience?

Whilst the conflict arising between users who indulge in so-called anti-social behaviour and other visitors cannot be denied, and whilst this may have the negative effect of deterring specific groups, I argue that there is still a value to these young people using the site and that their actions should be considered in a measured and rational way. It is well documented that attitudes to young people can be less than charitable and that they are assumed to cause trouble even without evidence (Millie 2009, DCSF 2008). Brown (1998) asserts that the way youth crime is presented and communicated is highly significant to the way in which it is then understood and judged in the community. Participants in my research varied in their tolerance of this kind of youthful exuberance, adding credence to the theory that anti-social behaviour is a contested concept – that one person's anti-social behaviour is another's youthful expression. To push this idea further is to argue that public spaces should be public, un-segregated and therefore open to all. Disorder and difference should be welcomed as a challenge to beliefs and expectations and as a chance to explore the plural norms of behaviour and aesthetics (Valentine 2004).

Furthermore, I would argue that young people can be marginalised by public spaces that offer them little opportunity for self expression, privacy or risk taking. They are corralled in playgrounds and sports centres which are often poorly designed and maintained, and present purely as a means of caging and controlling young people (Aitken 2001). It is also recognised that 'teenagers ... do not have enough say in what is provided' (HM Government 2005: 4, Audit Commission 1996). Worpole (2003b) suggests that since young people have little say in, or control over the environment, they use this fixed environment in creative and non-normative ways to make their own claim on what is essentially an adult landscape. These acts may add a sense of belonging to features that are otherwise part of an architect's legacy – malls and parks that seem the same in every town and offer no sense of rootedness or identity (Cresswell 2004). Adopting specific locations and using them for self-expression and display turns them into what Hendry at al. (1993) describe as a Fourth environment. Perhaps the cemetery, with its unique appearance and local history offers young people the chance to find place amongst our increasingly mobile, rootless culture.

This means questioning those such as Nabhan and Trimble (1994) who are emphatic that recreation in natural spaces is a superior form of leisure. Valentine (2004: 75) points out how 'children often prefer to play in diverse and 'flexible' landscapes', and Hayhurst (2004) argues that if children are denied the chance to experience natural areas they develop poorer social skills and find decision making more challenging. This seems to suggest that if we want to avoid raising anti-social children, we have to let them loose into the natural world and community – into spaces that include cemeteries and churchyards – to develop these skills.

The UK Government's Respect Agenda (Home Office 2006) talks a tough line on anti-social behaviour and celebrates such 'achievements' as issuing 6,000 Anti-social Behaviour Orders (ASBOs) and 170,000 penalty notices for disorder. However, even this control and punishment oriented stance recognises the gains that can come from the exposure of disadvantaged young people to the arts or sport. Perhaps we are coming full circle now as we appear to be returning to the Victorian mores of the cemetery as a didactic landscape – through heritage tours and conservation/environmental projects.

Conclusions

Data from this research indicates how cemeteries and churchyards vary in significance as amenity spaces. Some recreational activities occur in these sites specifically because they are burial grounds. Other users are simply taking advantage of an available green space which has no special meaning to them; the motives and reasons behind each visit are essentially unique to the visitor. In either role, however, it is clear that burial grounds provide a space for recreation that is interpreted diversely and used with great creativity.

The conflicts that arose as a result of these apparently incompatible uses were well reported by interviewees. Most related to what was perceived to be inappropriate and anti-social behaviour, including drinking, having sex and creating general disturbance. The overall perception of this behaviour was that it was encouraged by the relatively private atmosphere of the cemetery or churchyard, where public guardianship and policing levels were generally felt to be minimal. As young people created their own environment in the cemetery, personalising 'their' space with litter and graffiti, they seemed to be fighting back against what is described as the 'end of place' (Tuan 1977) by locating themselves in one of the few remaining places with any sense of uniqueness and identity.

Siding with the young people who use and abuse cemetery spaces is fine from a theoretical perspective since there is clear evidence that they can grow socially and educationally from this use of the natural environment (Harrison et al. 1995, Seymour 2003). Local communities and cemetery visitors might be less enamoured of this opinion if they have been affected adversely by anti-social behaviour. What is significant here, however, is not the experience of the individual, but that of the community. Waiton (2008) argues that the problem is not the anti-social behaviour of troublemakers, but the asocial behaviour of society in general. Rather than avoiding groups of young people that seem intimidating, he suggests that it is better to step out of our own comfort zone and interact with them. Such interactions can have a positive effect in terms of reducing the incidence of anti-social acts but can also break down barriers.

Perhaps the conflict that was identified and described in this research can serve as a point to learn from and as an impetus for change. The more that cemeteries and churchyards are understood as multiple-use landscapes and managed to that end, the better one can hope to improve the sites for all users and circumvent conflict.

References

Aitken, S.C. 2001. *Geographies of Young People: The Morally Contested Spaces of Identity*. London: Routledge.

Amparo Escandon, M. 1999. *Esperanza's Box of Saints*. London: Picador Ltd.

Athkins, D.E. 1993. *The Cemetery*. London: Scholastic Publications Ltd.

Ariès, P. 1994. *Western Attitudes toward Death from the Middle Ages to the Present*. London: Marion Boyars Publishers Ltd.

Audit Commission. 1996. *Misspent Youth…Young People and Crime*. Abingdon: Audit Commission Publications.

Bouman, S.P. 2005. *Cemetery Picnic (Gen. 2:15-17; 3:1-7)* [online]. Available at: http://www.religion-online.org/showarticle.asp?title=3182 [accessed: 30 September 2009].

Brown, K. 2009. *Resurrecting History: Ghosts and Graveyards* [online]. Available at: http://www.ekoventure.com/users/adventure-travelers-united_ states-california-san_francisco-176/blog/articles/1090-resurrecting-history-ghosts-and-graveyards [accessed: 1 October 2009].

Brown, S. 1998. *Understanding Youth and Crime: Listening to Youth?* Buckingham: Open University Press.

Bryan, K. 2007. *Fascinating Graveyards, Stirling, Scotland* [online]. Available at: http://www.europealacarte.co.uk/blog/2007/11/21/fascina-graveyards-stirling-scotland/[accessed: 4 September 2009].

Chevalier, T. 2002. *Falling Angels*. London: Harper Collins.

Ciregna, E.M. 2004. Museum in the garden: Mount Auburn Cemetery and American sculpture, 1840-1860. *Markers*, (XXI), 100–47.

Cloke, P., Cook, I., Crang, P., Goodwin, M., Painter, J. and Philo, C. 2004. *Practising Human Geography* London: Sage Publications Ltd.

Cresswell, T. 2004. *Place: A Short Introduction*. Oxford: Blackwell Publishing Ltd.

Cross, G. 1990. *A Social History of Leisure Since 1600*. Pennsylvania: Venture Publishing Inc.

Curran, A. 2008. *Père Lachaise: The World's Favourite Graveyard* [online]. Available at: http://www.Cluas.com/indie-music/Blogs/French_Letter/tabid/80/EntryId/857/Pere-Lachaise-the-worlds-favourite-graveyard.aspx [accessed: 21 September 2009].

DeBartolo Carmack, S. 2002. *Your Guide to Cemetery Research*. Ohio: Betterway Books.

Department for Children, Schools and Families. 2008. *Youth Taskforce Action Plan* 2008 [online]. Available at: http://www.dcsf.gov.uk/everychildmatters/Youth/youthmatters/youthtaskforce/actionplan/actionplan/ [accessed: 5 October 2009].

Delaney, D. 2005. *Territory: A Short Introduction*. Oxford: Blackwell Publishing.

Derekbilldaly. 2008. *One Graveyard, 8 Travbuddies, A Bottle of Whiskey and a Riot* [online]. Available at: http://www.travbuddy.com/travel-blogs/37486/One-Graveyard-8-TravBuddies-Bottle-1 [accessed: 5 October 2009].

Dickens, C. 1994 [1861]. *Great Expectations*. London: Penguin Books Ltd.

Dolowitz, D., Buckler, S., Sweeney, F. 2008. *Researching Online*. Basingstoke: Palgrave Macmillan.

Dunk, J. and Rugg, J. 1994. *The Management of Old Cemetery Land: Now and the Future*. Crayford: Shaw and Sons.

Francis, D., Kellaher, L. and Neophytou, G. 2005. *The Secret Cemetery*. Oxford: Berg.

Gascone, A.G. 1997. *Grave Secrets*. USA: Troll Communications L.L.C.

Hallam, E., Hockey, J. and Howarth, G. 1999. *Beyond the Body: Death and Social Identity*. London: Routledge.

Harness, C. 1999. *Midnight in the Cemetery, a Spooky Search-and-Find Alphabet Book*. New York: Simon and Schuster.

Harrison, C., Burgess, J., Millward, A. and Dawe, G. 1995. English Nature Research Reports: *Accessible Natural Greenspace in Towns and Cities: A Review of Appropriate Size and Distance Criteria* (Number 153) Peterborough: English Nature.

Hayhurst, R. 2004. Playing safe. *Urbio: Urban Biodiversity and Human Nature*, 6, 14–15.

Hendry, L., Shucksmith, J., Love, J. and Glendinning, G. 1993. *Young People's Leisure and Lifestyles*. London: Routledge.

Her Majesty's Government. 2005. *Youth Matters*. Norwich: The Stationery Office.

Home Office. 2006. *Respect Action Plan* [online]. Available at: http:/www.asb.homeoffice.gov.uk/members/article.aspx?id=10090#section5 [accessed: 6 October 2009].

Jones, S. 2008. *Graveyard Travel: How to Celebrate Life by Visiting the Dead* [online]. Available at: http://www.bravenewtraveler.com/2008/07/22/graveyard-travel-how-to-celebrate-life-by-visiting-the-dead/ [accessed: 2 October 2009].

Kirk, T. 2007. *Paris-Travel-Bargains* [online]. Available at: http://www.studenttraveler.com/mag/01-01/paris.php [accessed: 3 October 2009].

LeBlanc, J. 2008. *Graveyard* [online]. Available at: http://www.jentrance.com/blog/?p=160 [accessed: 3 October 2009].

Linden-Ward, B. 1989. Strange but genteel pleasure grounds: tourist and leisure uses of nineteenth-century rural cemeteries, in *Cemeteries and Gravemarkers: Voices of American Culture*, edited by R.E. Meyer. Michigan: University Microfilms Inc., 293–328.

Lorraine. 2009. *Of Graveyards, Ghosts and the Stories We Tell.* [online]. Available at: http://flyingconfessions.com/blog/?p=231 [accessed: 21 September 2009].

Loudon, J.C. 1981. *On the Laying out, Planting and Managing of Cemeteries and the Improvement of Churchyards.* Redhill: Ivelet Books Ltd.

Lynch, T. 1997. *The Undertaking.* London: Jonathon Cape Ltd.

Millie, A. 2009. *Anti-social Behaviour.* Maidenhead: Open University Press.

Mitra, M. 2009. *India Now Tapping into Cemetery Tourism with Fervour* [online]. Available at: http://economictimes.indiatimes.com/Special-Report/India-now-tapping-into-cemetery-tourism-with-fervour/articleshow/4739273.cms [accessed: 5 October 2009].

Morgan, C. 2009. *The Graveyard Day* [online]. Available at: http://www.cheryl-morgan.com/?p=6260 [accessed: 1 October 2009].

Nabham, G.P. and Trimble, S. 1994. *The Geography of Childhood.* Massachusetts: Beacon Press.

Nelson, S. 2004. *I'm Just a Teenage Dirtbag, Baby!* Wigan: Environmental Campaigns Ltd.

Pain, R., Barker, M., Fuller, D., Gough, J., Macfarlane, R. and Mowl, G. 2001. *Introducing Social Geographies.* London: Arnold.

Payne, P. 2008. Gaijin Bochi: Foreigners' Graveyard [online]. Available at: http://www.peterpayne.net/2008/08/gaijin-bochi-foreigners-graveyard.html [accessed: 5 October 2009].

Porter Smith, A. 1997. *The Appeal of Graveyards* [online]. Available at: http://www.gothicsubculture.com/graveyard.php [accessed: 3 October 2009].

Pratchett, T. 2004. *Johnny and the Dead.* London: Random House Children's Books.

Queen of Spades 2009. in *Why do Some Girls Think its OK to Have Sex in Graveyards?* [online]. Available at: http://answers.yahoo.com/question/index?qid=20090313231940AAYgvc6 [accessed: 25 September 2009].

Rapoport, R. and Rapoport, R. 1975. *Leisure and the Family Life Cycle.* London: Routledge and Kegan Paul Ltd.

Seymour, L. 2003. *English Nature Research Reports: Nature and Psychological Well-being* (Number 533) Peterborough: English Nature.

Sharpley, R. and Stone, P. 2009. *The Darker Side of Travel: the Theory and Practice of Dark Tourism.* Bristol: Channel View Publications.

Sloane, D.C. 1991. *The Last Great Necessity.* Baltimore: John Hopkins University Press.

Stine, R.L. 1999. *Attack of the Graveyard Ghouls.* London: Scholastic Ltd.

Thirstydrunk. 2008. *Graveyard Guzzling* [online]. Available at: http://www. drunkard.com/bbs/viewtopic.php?f=3&t=55929&start=15 [accessed: 24 September 2009].

Tuan, Y.-F. 1977. *Space and Place: The Perspective of Experience.* Minneapolis: Minnesota Press.

Urban Green Spaces Taskforce. 2002. *Green Spaces, Better Places.* London: DTLR.

Valentine, G. 2004. *Public Space and the Culture of Childhood.* Aldershot: Ashgate Publishing Ltd.

Van Skike, A. 2009. *Graveyard Picnics* [online]. Available at: http://blog. buriedmoonstudios.com/2009/02/10/graveyard-picnics.aspx [accessed: 3 October 2009].

Waiton, S. 2008. Asocial not anti-social: the 'respect agenda' and the therapeutic me, in *ASBO Nation: The Criminalisation of Nuisance,* edited by P. Squires. Bristol: The Policy Press.

Welch, D. 1991. *The Management of Urban Parks.* Redhill: Longman UK Ltd.

Whyte, W.H. 1981. *The Social Life of Small Urban Spaces.* Michigan: Edwards Brothers Inc.

Worpole, K. 1997. *The Cemetery in the City.* Gloucester: Comedia.

Worpole, K. 2003a. *Last Landscapes: The Architecture of the Cemetery in the West.* London: Reaktion Books Ltd.

Worpole, K. 2003b. *No Particular Place to Go? Children, Young People and Public Space.* Birmingham: Groundwork UK.

Young, A. 2008. *Explore Local Cemeteries* [online]. Available at: http://www. vagablogging.net/explore-local-cemeteries.html [accessed: 1 October 2009].

Chapter 6

Rest in Peace? Burial on Private Land

Clare Gittings and Tony Walter

Ever since the adoption of Christianity in the early Middle Ages, it has been normal for Britain's dead to be buried in churchyards or other Christian burial grounds (Daniell 1998, Jupp and Gitttings 1999). From the mid-nineteenth century, but with earlier examples in Scotland, cemeteries (i.e. formal burial grounds not attached to a church) have supplanted churchyards as the most common place of burial (Rugg 1997), augmented in the twentieth century by cremation (Jupp 2006). Private burial on your own land, rather than in a churchyard or cemetery, has been and remains rare in Britain. It is, though, legal. The 1850s burial acts that controlled English burial in the name of public health and that still pertain today apply to 'burial grounds', meaning places generally set aside for burial, so the acts do not apply to the occasional grave on private land whose primary use is other than burial (Bradfield 1993).

When burial on private land does occur today, it can be newsworthy – as with the burial of broadcaster Johnny Morris and novelist Barbara Cartland, or when objections from neighbours make the local press. Media coverage can give the impression that burial in the garden or elsewhere on the deceased's land is a recent innovation, but it has been practised, if rarely, for several hundred years. Historical research can throw valuable light on private burial, not least because in certain cases it is possible to trace what has happened to such interments over the intervening centuries. While, of course, what occurred in the past cannot be used to predict with certainty the future of present-day private graves, it does at least raise pertinent issues to consider.

In this chapter, we use eighteenth-century documentary evidence to inform interviews with some who arranged burial on private land in the 1990s and 2000s. We address two related issues; one spatial (where exactly were or are the graves?), the other temporal (did or will posterity leave them undisturbed?). Spatially, there are three possible locations for a private grave: i) the garden immediately near the house, ii) a more remote part of the garden, and iii) elsewhere on one's own estate or on someone else's land. Where resources allow, the second and third more liminal locations were and are preferred, for practical, emotional and symbolic reasons and historically these graves have proved less likely to be disturbed. Eighteenth-century graves were also less likely to be disturbed if the burial had been carried out by more distant relatives. So, the more distant the burial both spatially from the house and socially from next of kin, the more likely the grave will remain secure.

In our research, we examined first the historical material, which raised the possibility of subsequent disturbance, which in turn prompted us to interview some who arrange private burial today to see how concerned they are about the grave's security and more generally what meaning they give to burying on private land. We found similar spatial choices being made, but very different notions of posterity. We follow the same order in this chapter, starting with the historical study.

First, a note about literature. Whereas there is an emerging research literature on woodland burial (e.g. Clayden and Dixon 2007)[1] and on burial of ashes in the garden and other private places (e.g. Prendergast et al. 2006), we know of no academic literature specifically on complete body burials on private land in the UK, apart from Walter and Gittings (2010) where we report how neighbours react; we explore there the concepts of visible and invisible, pubic and private, and boundaries between the two. There are two publications providing practical and legal advice (Bradfield 1993, Speyer and Wienrich 2003), publications which are also read by funeral directors who may have to advise both families on practicalities and local authorities on legalities. This chapter is the first attempt to put private burial into historical perspective, and it is to our historical study that we now turn.

The Historical Study

In a recent historical study of 25 unusual burials in England between 1689 and 1823 (Gittings 2007), 14 cases involving death from natural causes entailed the deceased requesting interment on their own private land. It is these 14 people and their burials, together with two others from 1834, which form the focus of this study. All sources for burials and biographical details are given in the table in Appendix 1.

In this period, although some radical Protestant believers (but not Roman Catholics) had won the right to set up their own burial grounds, in practice many were still laid to rest in Anglican churchyards (Houlbrook 1998: 336–7). Burial elsewhere was often associated with epidemics or punishment, both civil and religious, including excommunication. The 16 people who requested burial on their own property were clearly going against the prevalent practice of their times. So who were they?

All were Protestant, tending toward non-conformity, seeing no need to be buried in consecrated ground. Nine of the 16 lived well beyond the usual lifespan for the times, making them a distinctly elderly group. All but one was male. Laws which caused a woman's property to pass to her husband on marriage restricted

---

1 In the UK, woodland burial grounds offer graves for sale in a setting where trees rather than marked graves and headstones form the primary landscape (Clayden 2003). See also Clayden et al. in this volume.

female scope for such innovation at death; indeed, a married woman was not in a legal position even to make a will without her husband's consent before the Married Women's Property Act of 1882. The 15 men were, or previously had been, employed in a range of different professions. They included a printer, lawyer, soldier, schoolmaster, an Anglican clergyman, an ambassador, two radical political writers, an ironmaster who was a key figure in the industrial revolution, a manufacturer of bricks and tiles, a physician a surgeon and an apothecary. The remaining three were principally landowners although from their wills, it is clear that many of the professional men also owned considerable amounts of land. Seven held titles – two were baronets, one was a knight and four were esquires. Their motivation for choosing burial on their own land is sometimes revealed in, or can be inferred from, their wills. Most frequently it was religious, often leading them to campaign against the notion of consecrated ground. Other motives included a fondness for a particular hilltop view, a desire for isolation, strong feelings about remembrance after death, and classical or biblical precedents.

In the following sections we look at whether all 16 had their wish to be buried on their land met, and if so, whether subsequent generations continued to respect this. To this end, we look carefully at the exact location of the graves, family connections between the dead and the living, the actions of subsequent generations, and predictions about them that did not always prove well founded.

Control

So how much control did they have immediately after their deaths? Were their burial requests respected, despite flying in the face of social or religious convention? In fact, only two of the 16 were not interred according to their wishes. John Horne Tooke, who had forbidden any clergy to visit him, preferring the consolations of Shakespeare in his last days, had his desire for garden burial overruled in favour of churchyard interment. He had already prepared his own grave and black marble inscription in his kitchen garden. Horne Tooke had never married, but his heir was one of his illegitimate daughters who, with other female relatives, decided that he should have a Christian burial in his mother's vault in Ealing churchyard. In overriding his wishes they were supported by one of Horne Tooke's closest friends, believing – probably correctly – that his grave would detract from the value of the house if sold and that his body would at that point be moved elsewhere (Bewley and Bewley 1998).

In the case of William Burnard it is not clear who made the decision not to allow him to lie in the grave he had prepared in his garden in Thame, Oxfordshire. He had made detailed plans for the burial service that he desired, conducted in his schoolhouse by his chosen friends 'according to the dictates of their consciences

... in the most solemn manner'. He left his property to his parents so presumably they had some hand in having him interred in Thame churchyard instead.[2]

Family

Relatives played an important part, not just in determining whether to respect the deceased's wishes for the burial itself but also, subsequently, whether to move the body or let it remain in situ.

All but one of the sample had, unsurprisingly, some reasonably close living relatives whom they mentioned in their wills; Thomas Hollis was unusual among them in having only cousins. He left almost everything, after substantial charitable gifts, to his 'dear friend and fellow traveller Thomas Brand ... from whom a severe plan in Life has kept me much more separate from some years past than otherwise I wished to have been'. Quite a number of others in this study had no direct blood descendants or spouses alive at the time of their deaths; indeed several seem never to have married. Table 6.1 identifies the closest living relatives, as revealed in their wills, for each of the 14 who initially received their chosen burials. It also gives an indication of whether they were subsequently moved, and when, or whether they remain in situ. Obviously the evidence is more clear-cut when a body has been moved and it is far harder to prove that it has not been. The removals listed here resulted in some record of the event, with date given referring to the initial exhumation of the deceased (several were later moved again). Table 6.1 shows that those with spouses and/or direct descendants (children, grandchildren) were more likely subsequently to be moved than were those with less direct relatives such as siblings or nephews.

All four of those whose bodies were moved within about 20 years of their deaths had surviving children. Of these, apart from the highly unorthodox Wilkinson family discussed later, surviving children seem to have been involved in deciding on the change of burial place and to have chosen the comparative safety of the churchyard for their parents' bodies. It was the death of the surviving parent that precipitated this in the cases of the Carteret Webbs, discussed in detail below, and of John Sheffield who was reburied on 16 February 1807 in the churchyard at Downton, Wiltshire, in a joint funeral with his wife (Squarey 1906: 34). However this was not invariably the case; although Alice Liberty outlived her husband by 32 years she was nevertheless interred with him beneath the tomb on their estate when she died in 1809, according to the inscription. In the case of Thomas Backhouse, it was his son's return from abroad to live at the Buckinghamshire estate he had inherited from his father which caused the latter's body to be re-interred in the churchyard at Great Missenden, as recorded in the parish register (Architectural and Archaeological Society 1887–91: 323).

---

2   Maddrell (2009: 43) reports another example, from 1845 on the Isle of Man, of a son burying his father in the churchyard rather than in the hilltop grave he had chosen for himself.

Table 6.1   Long-term outcomes for burials on the deceased's own land: Those with spouse or direct descendants (children/grandchildren), compared to those without

Name and date of death	Survivor's relationship to deceased	Burial still in situ or moved
With spouse or descendants		
Thomas Backhouse, 1800	Son	Moved 1807
John Baskerville, 1777	Wife	Moved 1820
William Liberty, 1777	Wife and daughter	In situ
John Sheffield, 1798	Wife and children	Moved 1807
Sir William Temple, 1699 (only heart buried in garden)	Granddaughter and great-grandchildren	In situ
Sir James Tillie, 1713	Wife	Probably in situ
Susanna Carteret Webb, 1756	Husband and son	Moved 1770
John Wilkinson, 1808	Mistress and their children	Moved 1828
Without spouse or descendants		
Jonathan Dent, 1834	Nephew	In situ
Revd Langton Freeman, 1783	Nephews	Moved between 1880 and 1908
Thomas Hollis, 1774	Cousins	In situ
Sir John Jocelyn, 1741	Brother	In situ
William Martyn, 1762	Sister	In situ
Henry Parsons, 1794	Nephew	In situ

Note: (n=14)
Source: Authors

Location

While different family structures were one important factor influencing the fate of these unusual interments, variations in the choice of burial location – a factor more under the control of the deceased – were also significant. Two different groups emerge. There were those who specified burial within the cultivated, horticultural area close to their houses, often naming the specific part of the garden they had chosen and referring to buildings and other manmade features. Jonathan Dent was most precise in his directions to be interred 'about three feet from the Eastern Wall of my Tenant['s] … Cottage and about midway between the Northern and Southern boundaries of my … garden.' Sir William Temple directed that his heart should be buried 'six feet underground on the South east side of the stone [sun] dial in my little Garden at Moreparke'. The other group elected to be buried elsewhere on their wider estates, sometimes deliberately in a bleak spot. Dr William Martyn chose 'the most barren field … in the most elevated part of it' on his Cornish lands to make his point about the futility of burial in consecrated ground. Thomas Hollis made a similar point in Dorset, being buried, according to his biographer, in 'a grave ten feet deep' in a field 'immediately ploughed over that no trace of his burial-place should remain' (Blackburne 1780: 481).

Of the 16 people in our historical study, nine fall in the garden category (one, as noted above, just a heart burial) while the other seven chose burial elsewhere on their estates. These numbers are again small, so any conclusions can only be tentative. Nevertheless, some very distinct differences emerge when the long-term outcomes of burial in these two possible places of interment are explored, as in Table 6.2.

Table 6.2   Garden versus estate burials: the long-term outcomes of requests for burial on the deceased's own land

Name, date of death and county/place of burial	Long-term outcome: still in situ, burial refused, or body moved, with date
Requested burial in their gardens	
John Baskerville, 1777, Birmingham	Moved 1820
William Burnard, 1834, Oxfordshire	Burial refused 1834
Jonathan Dent, 1834, Lincolnshire	In situ
Revd Langton Freeman, 1783, Northamptonshire	Moved between 1880 and 1908
John Horne Tooke, 1812, Surrey	Burial refused 1812
John Sheffield, 1798, Wiltshire	Moved 1807
Sir William Temple, 1699, heart only buried in garden, Surrey	In situ
Susanna Carteret Webb, 1756, Surrey	Moved 1770
John Wilkinson, 1808, Lancashire	Moved 1828
Requested burial on their estates	
Thomas Backhouse, 1800, Buckinghamshire	Moved 1807
Thomas Hollis, 1774, Dorset	In situ
Sir John Jocelyn, 1741, Essex	In situ
William Liberty, 1777, Hertfordshire	In situ
William Martyn, 1762, Cornwall	In situ
Henry Parsons, 1794, Somerset	In situ
Sir James Tillie, 1713, Cornwall	Probably in situ

Note: (n=16)
Source: Authors

Even allowing for the small number of cases, Table 6.2 clearly suggests that those people who chose burial on their wider estates rather than in gardens close to their houses were less likely to have their bodies subsequently disturbed. Further corroboration of this may be found in four more cases from the original 25 researched but not part of this study (excluded because they did not die of natural causes and/or they did not request burial on their own land), where the deceased was buried on someone else's estate. These were Peter Labilliere and Richard Hull, each interred on Surrey hilltops, John Olliver who was buried close to his windmill on Highdown, West Sussex, and Samuel Johnson whose grave was in his master's woods in Cheshire (Gittings 2007). In all four cases the deceased seems still to be resting there undisturbed, despite not having owned the land.

Posterity and Fate

A range of possible reasons for the better survival of estate burials becomes clearer when we examine in more detail some of the garden burials where the deceased was later exhumed. The aftermath of the death in 1808 of John Wilkinson, the famous ironmaster and a key figure in the industrial revolution, is a sorry tale, highlighting some of the possible pitfalls of an unconventional approach to both burial and inheritance in the early nineteenth century. An account of his multiple burials and reburials shows how easily the desire for a garden funeral could descend into farce through lack of necessary attention to detail.

Wilkinson himself had prepared for his burial by leaving an iron coffin of his own design and manufacture at each of his principal residences and was happy to be buried in the garden at whichever he happened to die; for him what mattered was his metal coffin. On his death at Castlehead in Lancashire, his body was placed in a wooden coffin but at the funeral it was discovered that this would not fit inside the iron one, so he had to undergo temporary interment until a new wooden coffin arrived. He was then disinterred and it was discovered that there was insufficient depth of soil in which permanently to bury the body, until the rock beneath had been blasted. Finally at the third attempt he was buried and a huge iron monument erected over him in the garden, though he did not to stay there undisturbed for long (BBC no date).

John Wilkinson's colourful and eccentric life resulted in him fathering three children in his seventies with his housekeeper, while his childless wife was still alive; he legitimised them only after her death in 1806. One of his nephews had been led to believe he was Wilkinson's heir but now he found himself merely a possible residuary legatee after provision had been made for the housekeeper, with the bulk of the estate destined for the three children. Unwilling to accept this new situation, he contested Wilkinson's will and the case went to the Court of Chancery (Berthoud 1995, Matthew and Harrison 2004). While this may have given John's nephew a minor claim to literary fame as the probable model for 'the man from Shropshire' in Charles Dickens's *Bleak House*, it devastated the Wilkinson inheritance, which was spent on lawyers' fees (Berthoud 1995). By 1828, just when the children were coming of age and should have inherited, the house at Castlehead had to be put up for sale as little else was left of Wilkinson's once great industrial empire (Berthoud 1995, Matthew and Harrison 2004). Fearing that the monument and grave might detract from the asking price, John Wilkinson was disinterred once more and moved, despite his objections to consecrated ground, to Lindale churchyard, with his monument nearby (BBC no date).

Both concern with making money from property and disregard of the religious views of the dead feature in the case of John Baskerville, printer and typographer. He was buried in his garden on the then outskirts of Birmingham in 1775 because of his 'hearty contempt for…the farce of Consecrated Ground', as he wrote in his will. In 1791 his house was burnt in the Birmingham riots, through the body remained in situ. By 1820 demand for building land to accommodate Birmingham's burgeoning population made the presence of a grave an undesirable impediment to builders' profits and Baskerville's body in its lead coffin was removed. Instead of re-interment, the sealed coffin was sold to a plumber and kept in his shop for some years (Matthew and Harrison 2004). In May 1821 the coffin was opened and sprigs of laurel and bay were seen on Baskerville's body (Pardoe 1975: 149). In 1829 coffin and body were placed in the vault of Christ Church Birmingham without any form of ceremony. Even this was not his final resting-place as the church was demolished and in 1898 he was placed beneath the chapel, itself later destroyed, in an Anglican cemetery in Birmingham, remaining to this day in the consecrated ground he so despised (Matthew and Harrison 2004).

The events following Susanna Carteret Webb's interment in 1756 at the age of 45 were worthy of any modern soap opera. She was laid to rest in a cave, possibly following classical or biblical precedents, in the garden of the Surrey property where she had lived with her husband and son. The year after her death, a visitor was shown her coffin, covered in black velvet with silver fittings, near those of two of her infant children, by her grieving widower Philip. He said that he went there daily and was planning to join her there on his death (Larner 1947: 18). However, only a year later, he remarried, this time choosing a much younger woman, Rhoda Cotes, born in 1730, the year in which he and Susanna had married. In his will, made shortly before his death in 1770, he left everything to her 'whatsoever wherever and of what nature kind or property soever', making Rhoda 'sole executrix'; his and Susanna's surviving son, also called Philip, was not mentioned. On his father's death, the younger Philip had the bodies removed from the cave to Godalming Church. He also transferred the lengthy and affectionate monument composed by his father in his mother's memory as a memorial to both of them, reunited in death. This proved to be a wise action. Fourteen months later Rhoda married again and within five years the family's substantial wealth had been so exhausted that the house and grounds had to be sold off (Larner 1947: 17–19).

In some instances where a very long time had elapsed between the initial burial and the exhumation and reburial, it is not now possible to discover exactly when or by whom the body was moved, nor is its final resting place necessarily marked. The Revd Langton Freeman's re-interment is a case in point. Freeman requested an interment emulating ' … as near as may be … our Saviours Burial'. He directed that his shrouded body be laid on a bed in his summerhouse – sealed against intruders – in his Northamptonshire garden. This duly happened in 1783 and his nephew inherited the property. Remarkably, Freeman's body was still there in the early 1880s although the summerhouse was by then in poor repair (Notes and Queries 1880: 106). Between then and 1908 the body was moved to the churchyard and by the 1970s what remained of the building was so dilapidated it had to be demolished (Undertakers' Journal 1908: 211, Haynes 1988: 19).

Mistaken Assumptions

So what did these various testators do to try to ensure their burials would remain undisturbed for eternity, or at least until the Resurrection? Hindsight, revealed in Table 6.2, might suggest that they would have been well advised to concentrate on the location of their burials, choosing remote spots away from houses, in areas unsuitable for urban expansion. However, while most of these testators mentioned choice of location in their wills, it was never in terms of the interment's long-term survival but for a range of other reasons. Instead, they focused on inheritance of the property to secure their earthly resting-place. In doing so, they made at least two assumptions about the future which, unfortunately for many of them, rested on extremely shaky foundations.

The first was a belief that their family line would continue indefinitely, an odd assumption by men who had not themselves fathered children. A number without direct descendants tried to make their familial relationship with their heir closer by forcing them to change their names. Two of those leaving their land to nephews, Henry Parsons and Jonathan Dent, required them to take their uncle's surname. John Wilkinson had already required the same of his illegitimate children, while his litigious nephew also chose to take the surname Wilkinson to help advance his claim to inherit (Berthoud 1995: 4). Others made elaborate plans for how the property was to pass down to subsequent generations, especially in the event of any unexpectedly early deaths in the family. William Burnard possibly had more cause than most to do this in his will, as his initial beneficiaries were his parents. The property was then to pass to his sister and after her death to his nephew 'now aged about 20 years … Subject to this Condition that he shall not sell mortgage or dispose thereof.' In the case of Henry Parsons, his will directed that his nephew and heir, John White, should disinherit his own eldest son Henry in favour of a younger sibling if he should be 'disobedient' and lead 'a bad Course of life'. Sir William Temple was unusual among them all in being able to bequeath property in his will as far ahead as his great-grandchildren.

However, even a passing acquaintance, as surely these gentry families had, with the history of the British monarchy and aristocracy from 1500 onwards would suggest that a stable family line of descent was not so easy to guarantee, as Henry VIII found to his cost when trying to found a lasting Tudor dynasty, and this pattern did not just apply to royalty. Any significant improvements in the infant mortality rate or, indeed, any infertility treatment, were not to occur until the twentieth century (Jupp and Gittings 1999).

Another major assumption made by the testators in this study is that the links between a family, its wealth and the land it owned would remain forever unbroken; the home-loving Sir William Temple even willed that his house be preserved forever unchanged. Again, detailed historical examination suggests that these bonds were not always quite so strong (even in the early modern period) as contemporaries may have liked to believe (Stone 1965: 156–64). Indeed, many of the families appearing in this research had been able to acquire the economic position that they enjoyed at least in part as a result of earlier land mobility. The fortunes of the Carteret Webbs and the Wilkinsons described above show just some of the forces that could part a family from its wealth and land. As a blueprint for the future these assumed links between family, land and wealth were increasingly out of date. This became particularly evident during the agricultural slump of the later nineteenth century and its aftermath when land lost considerable value. So much changed hands between 1880 and 1930 that it has been likened to the two other great land upheavals in English history – the dissolution of the monasteries and the Norman Conquest (Cannadine 1990: 90–103).

It perhaps is therefore not surprising that so few of those buried in all but the most remote locations on their land are still in situ. When Jonathan Dent willed that his heir 'must pay every attention to … keeping such garden in a proper state as a place of Memorial for the dead', he could only state a wish, not definitively shape the future. Substantial economic and historical forces, increasing in strength over the decades, were at work against him and others buried in their gardens. Indeed, all this makes the survival of Dent's grave quite surprising. It can only be imagined, however, what this elderly Quaker would have thought about estate agents advertising on the web, in January 2007, through the Home Sale Network, the presence of his tomb as a selling point for his former house and garden.

If location of the grave and the role of descendants strongly influenced whether these graves continued undisturbed, what about today? Are similar factors still influential? And what factors are important for those arranging burial on private land today?

The Contemporary Study

In 2008, through contacts in the Natural Death Centre (which promotes family organised funerals and natural burial),[3] we conducted unstructured interviews with five people who had arranged private land burials in Britain in the preceding 15 years.[4] It turned out that they all had buried not in their own garden, but on a

---

3   www.naturaldeath.org.uk [accessed: 7 July 2009].

4   Interviewees: *Robin Crichton* buried his wife Trish (and subsequently a cousin) in a wood at the far end of the field adjoining their house; *Heather Johnston* buried her uncle

piece of uncultivated or marginal agricultural land, adjacent to their garden or at a distance, in some cases owned by themselves and in some cases not – comparable to what Table 6.2 terms 'estate' rather than 'garden' burials. These are not so much garden burials as field burials.

Three of the interviewees live in Scotland, which from its Calvinist Reformation in the sixteenth century 'did not recognise one piece of ground as being more holy than another' (Spicer 1997: 177), making possible long before England the development of cemeteries, and thus at least the possibility of family burial grounds away from the kirk. We are unsure whether this different history means that private burial is today interpreted differently in Scotland than in England.

Two of our interviewees had buried their husband, two their wife, and one an uncle and then her mother – in total, three men and three women. In two instances, other burials on the same land were mentioned (of a friend, and a cousin), but not described in any detail. So unlike the historical sample which included a number who had never married, the modern interviews were predominantly with widows and widowers. None mentioned that home burial had been stipulated in a will, and it is too early to know if any will be disinterred and moved. Comparable to the eighteenth century sample, they are middle to upper class with access to land, but with left/green politics. Aged from their 30s to 60s, with spouses having met untimely deaths in youth or middle age, they are more youthful than the historical sample.

Following earlier experience of interviewing people about funerals where they requested we not anonymise the dead whom they wished to memorialise (Walter 1990), our interviewees consented to our proposal to use the deceased's real names. Four of the five (Global Ideas Bank no date, Hale 2005, Johnston 2004, Speyer 2001) had in any case published short articles about the funeral.

An interview was also conducted with a 'green' funeral director who described four home burials; these involved the burial of a son, an uncle, a grandfather, and a lesbian commune member; those of the son and uncle were in the small gardens of ex-council houses. In addition, there are a number of published accounts of garden burial, for example Speyer and Wienrich (2003: 95–103) and Garrett (2001). We draw on these as well as the interviews. As in the eighteenth century, with no data available on the total number or character of garden and estate burials, it is impossible to know in what respects any sample, let alone our very small sample, is or is not representative.

Clear contrasts with the historical sample emerge; the modern cases rarely make assumptions about the continuity of their family line, nor are they really concerned about the grave's long-term security. For most, burial on private land is part of a desire to control the funeral; posterity can take care of itself. The

---

and mother in the field adjoining their cottage; *Richard Hale* buried his wife Angela on the farm of an acquaintance; *Catherine Maxwell Stuart* buried her first husband, John Grey, on a hill on their estate; *Josefine Speyer* buried her husband, Nicholas Albery, in a piece of land they jointly owned with others, and in which there had been one previous burial.

one similarity with the historical sample is that those buried on land outside the immediate garden, especially if it is owned by someone less close than a widow or widower, may well have a better chance of remaining there undisturbed. We will now explore these and related themes, including nature, home, family, personal choice and control.

Planned or Unplanned?

In some instances, husband and wife had discussed their desire to be buried on their own land, but in others there was no such pre-existing thought, let alone plan, but a piece of land belonging to someone else became available after the death. When Richard Hale's wife Angela died suddenly at the age of 56, he considered buying a plot in a woodland burial ground but could not find one that felt right, and after consultation with a green funeral director and a Unitarian minister, it occurred to him that an organic farmer from whom he and Angela had bought produce might be able to help. Arrangements rapidly fell into place.

Heather Johnston's cousin wanted a woodland burial for her father, but the nearest site was too expensive and too far away. Heather and her husband had already agreed that Heather's mother, when she died, would be buried in the field next to their cottage. So they offered to include Heather's uncle, who then became the first burial in their field.

Control

Though we conducted the interviews because we wanted to know about the burial site, interviewees spoke at considerable and to us unanticipated, length about the funeral. It thus became clear that for several interviewees burying on their own, or a friend's, land was part of a passionate desire to control the funeral themselves, rather than have it controlled by strangers (however green their credentials). This is confirmed by a journalist who has interviewed a number of people who have arranged a garden burial (Garrett 2001), and by the funeral director we interviewed who spoke of the two ex-council house garden burials and the burial of the commune member. Choosing the precise grave site, digging the grave, arranging the various parts of the ceremony, having friends and family rather than professionals provide the venue, play the music and speak at the funeral – these were what our interviewees recollected with great fondness:

> I've only been to one green burial, and you really couldn't fault them in any way. It's just that the grave was dug, it was determined for them, you couldn't determine it yourself ... We did it in our own way, in our own place, under our own auspices ... It's a bit like we had a home birth. (Josefine Speyer)

Their passion to control the funeral contrasts with our interviewees' implicit lack of concern about the long-term security of the grave, something they typically failed to mention, unless and until we asked them.

Home

Wanting to be buried at the home where the deceased had lived all his or her life, or had lived with the survivor all their married lives, was a commonly stated motive:

> We lived there all our lives … We both wanted to be buried on our land … At the entrance (to the burial site) I have a poem written on a bronze plaque, which I wrote, saying 'This is where we lived our lives, this is where we gladly died.' (Robin Crichton)

> He was 100, and he was the great granddad of the family, living on the farm, and you know he just didn't want to leave, he'd lived there all his life; wanted to stay close to the family, and the family were wanting that as well. (Funeral director)

In 1989, Johnny Morris buried his wife Eileen 'at the bottom of the four-acre garden they loved, and for the rest of his life visited her grave every evening to tell her about the day's events' (Derby Dead 1999).

Nature

An equally commonly cited, and in some cases related, motive is the desire to be buried in a natural setting. As in the eighteenth century, this might entail burial in the garden, or in a field, wood, or other site detached from the dwelling. Such sites could be part of an estate in an eighteenth century sense. More often, they comprise a field adjacent to the house, or a piece of land some distance away that had been purchased by the couple, or farmland owned by a friend. The natural beauty of the site was mentioned in several cases, and in the one site we ourselves visited – in the Peak District National Park – we can confirm its remarkable beauty. This amplified the appropriateness of this particular site, for Angela Hale was a climber who had loved the hills, more at home there even than in her beautiful suburban garden. All our interviewees were lovers of nature and of the countryside.

If we combine the themes of home and nature, as several who have arranged home burial do, then it is clear that the grave site is very different from the artificial environments maintained by strangers that comprise the typical cemetery or churchyard. We may note, as did Josefine Speyer, a comparison with the natural childbirth movement which also combines the key symbols of home, nature and choice; also with the natural death movement which recasts dying as a natural rather than medical event and promotes choice and dying at home rather than in hospital (Speyer and Wienrich 2003, Walter 1994). Of course, the motif of nature is also central to the commercially-run woodland burial grounds that have proved surprisingly popular in the UK since the early 1990s, but combining the motif of nature with the motifs of home and/or control is what drove our interviewees to bury on their own or a friend's land. That a commercial woodland burial ground could guarantee the long-term security of the grave, whereas they could not, did not seem to concern them.

Family

Apart from Catherine Maxwell Stuart, the twenty-first laird of a stately home owned by her family for many centuries, our modern interviewees demonstrated little sense of belonging to a family that exists over many generations. Their choice of burial on private land was a matter between them and their spouses, a product of the modern conjugal family rather than the eighteenth century propertied family line. When a mother, uncle or cousin was buried on the site, this followed the original intention of spousal burial. Robin Crichton spoke of a longer term, stating confidently that 'the children will certainly all want to go in there', yet added 'but if it dies out, it dies out. But the graves will stay there.' Even Catherine Maxwell Stuart spoke about her first husband's grave on a hill overlooking the house in terms of personal choice, preferring it both to the long-disused family crypt in the local church and to a family burial aisle in a chapel several miles away where her father and grandfather are buried. This is no standard aristocratic mausoleum in which successive generations can expect to be placed (Colvin 1991). Rather, she spoke of the grave as 'a really lovely spot, with a really nice panoramic view, very nice, partly private.'

Our eighteenth-century characters who willed a garden burial, living and dying within the context of an inter-generational family rooted in property, could be characterised in retrospect as having been in the vanguard of the steady rise of individualism in the early modern period (Gittings 1984), but might have been regarded by their contemporaries as eccentric. Our modern arrangers of private land burial, both those we interviewed and those reported in other literature, live in a different world. Primarily they are autonomous, self-acting late modern individuals, operating within a context of a one- or at most two-generation family. Indeed, they are motivated more by spousal love than by individualism, and not at all by any sense of trans-generational family line. They are not eccentrics, they are simply modern individuals who have invested in the intimacy of spousal love (Giddens 1993), taking to the grave values that are central to late modern society, assisted by, in the case of at least one of the ex-council house owners, bloody mindedness, and in the case of our middle to upper class interviewees access to land outside the immediate house and garden. And as individuals who have invested in the (at most) two-generation nuclear family, they are not fooled, as were their eighteenth century forebears, by out-of-date notions of a stable line of family descent.

But they might be fooled by conjugal love. Two hundred years ago, garden graves entrusted to the care of children might fall foul of subsequent lack of finance or lack of care. So today, private land burial motivated by spousal love might prove vulnerable to the potentially lesser loves (or lesser finances) of children or grandchildren. Some of today's private graves, however, are not of the landowner's spouse, but include more distant relatives or even – in the case of Angela Hale – a mere acquaintance of the landowner. If such graves do not require the landowner's spousal love for their creation, they may not require it for their long-term maintenance, and may stand a better chance of long term security. We discuss two examples of this (Heather Johnston, Angela Hale) in the next section.

Land Ownership

So what plans did the cases we examine make to secure the grave site in future generations? Little or none. Some had plans for their children to take over the land, but no plans beyond that. Some had ideas as to how ownership and use might develop, for example, as an unofficial nature reserve or a special place for family celebrations and personal contemplation, but they acknowledged they could not control the future.

There is a general assumption that the presence of a grave in a domestic garden will reduce the value of the house, though this has been disputed by some estate agents (Derby Dead 1999), and we have already noted that Jonathan Dent's historic garden grave was featured in an estate agent's 2007 advertisement for the house. Nevertheless, there is a risk that the presence of a recent garden grave could impede subsequent sale of the house – so what safeguards might be made?

Some with large gardens chose a site away from the house with independent access, so it could be retained by the family if and when the time came, whether sooner or later, to sell the house and the rest of the garden. The Natural Death Centre report one widow, who after the sudden death of her husband, asked her lawyer 'to set aside a part of the large back garden for the grave, with its own access, so that this part would not be sold with the rest of the house and grounds' (Speyer and Wienrich 2003: 96). However, this strategy is not foolproof:

> One couple in their nineties (chose) a spot for burial on the edge of their ten-acre garden where their beloved boxer dogs were buried, with a way to reach the graves from the public footpath, in case the remainder of the property were later to be sold by the family … They felt prepared for their deaths, but their plans were cruelly disrupted when increasing disabilities forced them to move to a residential home, and to sell their home to cover the costs. (Speyer and Wienrich 2003: 101)

The other strategy, similar to eighteenth century estate burial which we have seen was less likely than garden burial to be followed by exhumation, is to bury on a plot entirely separate from the domestic garden, i.e. in a separate field or wood. If this land is agriculturally marginal, and not suitable for housing, it is a reasonable supposition that future farmers will have no problem with one or two unmarked graves. Heather Johnston spoke of the field next to her cottage that now contains her uncle and mother:

> We've actually done a bit more thinking about it and what we're probably going to do is sell it to the farmer, because we know he's quite respectful of the graves; he does graze his sheep on it, and he's also a guy with a lot of integrity and he has got children, he has lads in their twenties who are farming it, so it will continue in his family.

Angela Hale was buried on a tiny flat area on an otherwise steep hillside on an acquaintance's farm in the Peak District. A year later, the farm was sold. The original owner told us 'When it came to selling, the grave didn't put anybody off, as far as I'm aware.' The new owner confirmed this to us, adding 'If it ever became an issue with a future purchaser, I'd go out of my way to tell him that it is absolutely no problem.' Marginal agricultural land already owned and farmed by someone outside the family is probably as good a sign as any that future owners will respect the grave. Assuming that children and grandchildren will sustain ownership can in some cases be as questionable today as it was two hundred years ago.

Conclusion

The advice that those contemplating garden burial may draw from the examples given in the *Natural Death Handbook* (Speyer and Wienrich 2003: 95–103) is twofold. First, those wishing to bury in the garden should pick a spot that can be legally detached and retained in family ownership, should the house be sold. Second, however, the grave will be more secure if the family have access to a separate plot of agricultural land. From our limited survey of those who recently have arranged private burial, we would concur with this, and the evidence from two centuries ago confirms this; those buried on the estate, rather than in the garden, were more likely to stay put.

We would add two more things. First, both historical records and interview evidence suggest that land of little other than aesthetic use may be particularly suitable for private burial. In two of our interviews, what seemed a suitable spot to the landowner (away from the house, of little agricultural or other economic value, but accessible for a mechanical digger and a pedestrian funeral procession) proved also both of great beauty and unlikely to be disturbed by contrary future interests. Second, if the original landowner is not closely related to the deceased, it is likely that subsequent landowners may be willing to continue with the arrangement. This too is supported by historical evidence that land ownership passing to the deceased's children and grandchildren can threaten a grave's security.

Bradfield's guide to do-it-yourself burial (1993: 33) observes that Britain is the only European country that does not re-use graves. He also notes that regular grave visiting in Britain often ceases after ten or fifteen years (Clegg 1989), the time at which in other countries the grave might be re-used. He raises the intriguing possibility that exhumation of a garden grave on selling the house, followed by placing the remains in a more permanent public burial site, far from being problematic as other British writers on private burial assume, has precedents in cultures around the world (Hertz 1960) and could be psychologically appropriate for the family. Grief can entail both a continuing bond with the dead and a letting go, a continuing presence and a manifest absence (Klass et al. 1996), and garden burial of the body followed some years later by public re-burial of the bones could express this well. Given that British law and culture militate against re-burial, however, this particular option is highly unlikely to be taken up, leaving concern for the long-term security of graves as a distinctly British issue. Fortunately for our interviewees – modern British individualists driven by conjugal affection rather than a desire to control posterity – it is not one with which they are overly concerned.

Acknowledgements

We thank those who generously agreed to be interviewed, and the staff of the many Record Offices and Local Studies Libraries who have answered our enquiries and sent photocopies with unfailing patience. Many individuals have generously shared their expertise, including Paul Grantham, Teresa Laden, Malcolm Ramsay, Lesley Rivett and several members of the Natural Death Centre. Any remaining mistakes or omissions are entirely our responsibility.

Appendix 1
Wills and Other Key Sources (1689–1834)

Abbreviations

Cat. = Catalogue
DNB = Oxford Dictionary of National Biography
HALS = Hertfordshire Archives and Local Studies
nd = no date
Pers. Comm. = Personal Communication
RO = Record Office

Name, date, place of residence (and burial place, if different). Places now in London given as [London].	Will, occupation (National Archives PCC, unless otherwise stated)	Other key source(s), mainly biographical (for full details see References)
Thomas Backhouse Esq., c.1720–1800, Haversfield, Gt Missenden, Bucks.	Prob 11/1344 'an old soldier'.	Arch. & Arch. Soc., Bucks., 1863: 147–9 and 1887–91: 322–3.
John Baskerville, 1706–1777, Easy Hill, Birmingham.	Prob 11/1005 Printer and typographer DNB).	DNB; Pardoe, 1975.
William Burnard, c. 1788–1834, Thame, Oxon.	Oxfordshire RO Pec 59/2/31 Schoolmaster.	Oxfordshire RO Transcript (1993) Thame Parish Register.
Jonathan Dent Esq, c.1744–1834, Winterton, Lincs.	Prob 11/1839 Landowner.	Andrew, 1836: 19–22.
Revd Langton Freeman, 1710–1783, Whilton, Northants.	Prob 11/1112 Clerk [in Holy Orders].	Longden, 1938: 5, 127–9; Haynes, 1988: 19.
Thomas Hollis Esq., 1720–1774, Lincoln's Inn [London]. (Corscombe, Dorset).	Prob 11/994 Political propagandist (DNB).	Blackburne, 1780: 481.
John Horne Tooke, 1736–1812, Wimbledon, Surrey (Ealing, Mx.).	Prob 11/1532 Radical and philologist (DNB).	Inscription etc in Bewley, 1998: 268–71; DNB.
Richard Hull Esq., c. 1689–1772, Leith Hill, Surrey.	Prob 11/975 Bencher of Inner Temple and Irish MP (inscription).	Rowe, 1895: passim.

Name, date, place of residence (and burial place, if different). Places now in London given as [London].	Will, occupation (National Archives PCC, unless otherwise stated)	Other key source(s), mainly biographical (for full details see References)
Sir John Jocelyn, Bart., 1689–1741, Hyde Hall, Essex.	Prob 11/713 Barrister (Morris, nd).	Morris, nd: HALS, Gerish Collection; Sawbridgeworth WEA, 1969.
Samuel Johnson, 1690/1–1773, Gawsworth, Cheshire.	No will. Dancing-master and playwright (DNB).	DNB; Inscription in Richards, 1974: 229.
Major Peter Labilliere, d. 1800, Dorking, Surrey (Box Hill, Surrey).	No will.	*Gent's Mag.*, 1800: 693; his papers discussed in Grantham, nd.
William Liberty, c. 1724–1777, Chorleywood, Rickmansworth, Herts. (Flaunden, Herts.).	Prob 11/1031 Tile maker (will). Brickmaker (inscription).	Inscription in Arch. & Arch. Soc., Bucks., 1863: 150; Location in Chilterns Conservation Board, 2002.
William Martyn, c. 1700–1762, Plymouth, Devon (Botus Fleming, Cornwall).	Prob 11/887 Physician.	Inscription in his will.
John Olliver, 1709–1793, Highdown, Goring, Sussex.	W. Sussex RO STCI/44/437 Miller.	Fox-Wilson, 1987; Horsfield, 1835, *II*: 138; Simpson, 2005: 189–200.
Henry Parsons Esq., c. 1710–1794, West Camel, Somerset.	Prob 11/1310.	*Notes and Queries*, 1897: 158.
John Sheffield, c. 1749–1798, Downton, Wilts.	Prob 11/1309 Surgeon and apothecary.	Squarey, 1906: 34–5; *Notes and Queries*, 1897: 158.
Sir William Temple, Bart., 1628–1699, Moor Park, Farnham, Surrey (Westminster Abbey; Moor Park, Surrey).	Prob 11/450 Diplomat and author (DNB).	DNB.
Sir James Tillie, 1645–1713, Pentillie Castle, Cornwall.	Prob 11/537.	Baring-Gould, 1915: 25–33; Maus. & Mons. Tr., nd; Cornwall RO Cat. 'Coryton'.
Susanna Carteret Webb, c. 1711–1756, Busbridge, Godalming, Surrey.	Married woman, not eligible to make a will.	Inscription in Manning and Bray, 1814, *3*: cxliv; letter, in Larner, 1947: 17–19.
John Wilkinson Esq., 1728–1808, Castlehead, Cartmel, Lancs.	Prob 11/1483 Ironmaster.	DNB; Berthoud, 1995: 3–7; BBC, nd.

Note: Quotations reproduce the spelling and capitalisation of the original.

References

Manuscripts

Chichester, West Sussex Record Office: STCI/44/437.

Kew, National Archives: PCC Wills Prob 11, *passim.*

Oxford, Oxfordshire Record Office; Pec 59/2/31; Transcript (1993) Thame Parish Register.

Truro, Cornwall Record Office: Catalogue for Coryton of Pentillie – administrative history; Pillaton Parish Register.

Published Documents

Andrew, W. 1836. *The History of Winterton and the Adjoining Villages, in the Northern Division of the County of Lincoln; With a Notice of their Antiquities.* Hull: A.D. English.

Architectural and Archaeological Society of the County of Buckingham. 1863 and 1887–91. *Records of Buckinghamshire or Notes on the History, Antiquities, and Architecture of the County. II & VI.* Aylesbury: the Society.

Baring-Gould, S. 1915. *Cornish Characters and Strange Events.* London: Bodley Head.

BBC. No Date. *Historic Figures – John Wilkinson (1728–1808)* [Online]. Available at: http://www.bbc.co.uk/history/historic_figures/wilkinson_john. shtml [accessed: 30 August 2006].

Berthoud, M. 1995. John Wilkinson and his family. *The Journal of the Wilkinson Society*, 17, 3–7.

Bewley, C. and Bewley, D. 1998. *Gentleman Radical: A Life of John Horne Tooke 1736–1812.* London: Tauris Academic Studies.

Blackburne, F. 1780. *Memoirs of Thomas Hollis.* London: No publisher.

Bradfield, J.B. 1993. *Green Burial: The D-I-Y Guide to Law and Practice.* London: Natural Death Centre.

Cannadine, D. 1990. *The Decline and Fall of the British Aristocracy.* London: Yale University Press.

Chilterns Conservation Board. 2002. *The Chess Valley Walk* [Online]. Available at: http://www.chilternsaonb.org/downloads/publications/ChessValleyWalk. pdf [accessed: 5 December 2006].

Clayden, A. 2003. Woodland burial. *Landscape Design*, 322, 22–5.

Clayden, A. and Dixon, A. 2007. Woodland burial: memorial arboretum versus natural native woodland? *Mortality*, 12(3), 240–60.

Clegg, F. 1989. Cemeteries for the living. *Landscape Design,* 184(October), 15–17.

Colvin, H. 1991. *Architecture and the After-Life.* New Haven: Yale University Press.

Daniell, C. 1998. *Death and Burial in Medieval England, 1066–1550.* London: Routledge.

Derby Dead. 1999. Dead Celebrities: Johnny Morris. [Online]. Available at: http://www.derbydeadpool.co.uk/deadpool1999/obits/morris.html [accessed: 13 Aug 2008].

Fox-Wilson, F. 1987. *The Story of Goring and Highdown.* Goring: Goring Book Association.

Garrett, A. 2001. Dearly departed – but never far away. *Daily Telegraph,* 8 December [Online]. Available at: http://www.telegraph.co.uk/education/main.jhtml?xml=/education/2001/12/14/tepgrave17.xml [accessed: 13 Aug 2008].

Gentleman's Magazine. 1800. *xxx* (July), 693.

Giddens, A. 1993. *The Transformation of Intimacy.* Cambridge: Polity.

Gittings, C. 1984. *Death, Burial and the Individual in Early Modern England.* London and Sydney: Croom Helm.

Gittings, C. 2007. Eccentric or enlightened? Unusual burial and commemoration in England, 1689–1823. *Mortality,* 12, 321–49.

Global Ideas Bank. No Date. *A Funeral Service and Burial at Home* [Online]. Available at: http://www.globalideasbank.org/LA/LA-22.HTML [accessed: 7 July 2009].

Grantham, P. No Date. *Unconsecrated Burials of Britain.* [Online]. Available at: http://www.grantham.karoo.net/paul/graves/index.html [accessed: 28 December 2006].

Hale, R. 2005. The funeral of Angela Hale. *Natural Death Centre News & Views,* Autumn/Winter, 3.

Haynes, C. 1988. *The History of Whilton.* No place of publication or publisher. Copy in Northampton, Northamptonshire Record Office: ROP 2320.

Hertz, R. 1960 [1907]. *Death and the Right Hand.* London: Cohen and West.

Horsfield, T. 1835. *The History, Antiquities and Topography of the County of Sussex, 2.* Lewes and London: Sussex Press.

Houlbrook, R. 1998. *Death, Religion and the Family in England 1480–1750.* Oxford: Clarendon Press.

Johnston, H. 2004. The natural way burial movement. *Connections: Scotland's Voice of Alternative Health Magazine,* 45.

Jupp, P. 2006. *From Dust to Ashes: Cremation and the British Way of Death.* Basingstoke and New York: Palgrave Macmillan.

Jupp, P. and Gittings, C. (editors). 1999. *Death in England: An Illustrated History.* Manchester: Manchester University Press.

Klass, D., Silverman, P.R. and Nickman, S.L. (editors). 1996. *Continuing Bonds: New Understandings of Grief.* London: Taylor and Francis.

Larner, H. 1947. *Busbridge, Godalming, Surrey a History: Ancient and Modern.* Cambridge: St Tibbs Press.

Longden, H. 1938. *Northamptonshire and Rutland Clergy, 5.* Northampton: Archer and Goodman.

Maddrell, A. 2009. Mapping changing shades of grief and consolation in the historic landscape of St Patrick's Isle, Isle of Man, in *Emotion, Place and Culture*, edited by M. Smith, J. Davidson, L. Cameron and L. Bondi. Farnham: Ashgate, 35–55.

Manning, O. and Bray, W. 1814. *The History and Antiquities of the County of Surrey, 3*. London: J. White.

Matthew, H. and Harrison, B. (editors). 2004. *Oxford Dictionary of National Biography*. Oxford: Oxford University Press.

Morris, E. No Date. *Hyde Hall Sawbridgeworth*. Hertford, Hertfordshire Archives and Local Studies: Gerish Collection, Gerish Box – Sawbridgeworth, part of D/Egr67.

Notes and Queries. 1849–1900.

Pardoe, F. 1975. *John Baskerville of Birmingham: Letter-founder and Printer*. London: Muller.

Prendergast, D., Hockey, J. and Kellaher, L. 2006. Blowing in the wind? identity, materiality, and the destinations of human ashes, *Journal of the Royal Anthropological Institute*, 12(4), 881–98.

Richards, R. 1974. *The Manor of Gawsworth*. Manchester: E.J. Morten.

Rowe, C. 1895. *Leith Hill: A Description of, and List of Places in View from its Summit*. Dorking: Charles Rowe.

Rugg, J. 1997. The emergence of cemetery companies in Britain, in *The Changing Face of Death: Historical Accounts of Death and Dying*, edited by P. Jupp and G. Howarth. Basingstoke: Macmillan, 105–19.

Sawbridgeworth Workers' Educational Association. 1969. *The Story of Sawbridgeworth: The Churches and the People of Sawbridgeworth*. Sawbridgeworth: WEA.

Simpson, J. 2005. The miller's tomb: facts, gossip and legend (1). *Folklore*, 116, 189–200.

Speyer, J. 2001. Nicholas Albery's woodland funeral and memorial celebration, in *Progressive Endings*, edited by N. Albery, S. Wienrich, N. Temple and R. Bowen. London: Natural Death Centre, 95–9.

Speyer, J. and Wienrich, S. 2003. *The Natural Death Handbook*. 4th edition. London: Rider.

Spicer, A. 1997. 'Rest of their bones': fear of death and Reformed burial practice, in Fear in Early Modern Society, edited by P. Roberts and W.G. Naphy. Manchester: Manchester University Press, 167–83.

Squarey, E. 1906. *'The Moot' and its Traditions*. Salisbury: Bennett Bros.

Stone, L. 1965. *The Crisis of the Aristocracy 1558–1641*. Oxford: Clarendon Press.

The Mausolea and Monuments Trust. No Date. *Gazetteer* [Online]. Available at: http://www.mausolea.monuments.org.uk/home.php?admin_page=gaz_text [accessed: 28 December 2006].

The Undertakers' Journal. 1908. Vol. XXIII, No 9 (15 September 1908).

Walter, T. 1990 *Funerals – And How to Improve Them*. London: Hodder.

Walter, T. 1994 *The Revival of Death.* London: Routledge.

Walter, T. and Gittings, C. 2010. 'What will the neighbours say? Reactions to field and garden burial', in *The Matter of Death: Space, Place and Materiality*, edited by J. Hockey, C. Komaromy and K. Woodthorpe. Basingstoke: Palgrave Macmillan.

Chapter 7

From Cabbages to Cadavers: Natural Burial Down on the Farm

Andy Clayden, Trish Green, Jenny Hockey and Mark Powell

This chapter draws on data gathered during a three-year research project which explored the cultural, social and emotional implications of natural burial.[1] In 1993 Ken West, then head of bereavement services at Carlisle Local Authority, opened a new burial section in the town's Victorian cemetery. Whilst this act was in itself unremarkable, as we illustrate below, the design and regulations that were attached to this new burial provision signalled the most significant development in the UK in how the dead are disposed of since the first 'official cremation was carried out' at Woking in 1885 (Grainger 2005). Thus in establishing what has come to be recognised as the first 'natural' burial ground, described by West as 'woodland' burial, a new deathscape was created. Here each grave would no longer be marked by a 'permanent' memorial headstone, instead being located by the planting of a native oak tree. No other markers would be permitted. There would be no headstone or kerb sets and floral tributes would also be discouraged. The identity of the deceased person would be known only to family and friends and the cemetery staff who maintain the burial record and grave register.

Whilst unmarked graves have always been a feature of our cemetery landscapes, provided by the burial authority for the interment of people living and dying in poverty, this new form of disposal was quite different. It represented a positive choice, an acceptance of anonymity in death and required the purchase of a grave. West's woodland interpretation of natural burial also required the family and friends of the deceased person to accept that at some future point, once woodland became established, access to the grave would no longer be possible. Below ground, natural burial has also differentiated itself from traditional funeral practices considered to be harmful to the environment by prohibiting the use of embalming fluids and coffins that are not biodegradable or made from unsustainably managed resources. However, our data from among burial ground managers show inconsistency in the extent to which below-ground environmental aspirations have been either promoted or subsequently enforced.

Since West opened the Carlisle woodland burial ground there has been a dramatic growth in both the provision and diversity of design interpretations of this new disposal option (see Clayden 2003, 2004, Thompson 2002). Why natural

---

1   We are grateful to the ESRC for funding this project between 2007 and 2010.

burial was accepted and developed so rapidly can be attributed to two other key factors. First, the 1994 publication by John Bradfield of *Green Burial: The D-I-Y Guide to Law and Practice,* which promoted natural burial and importantly helped clarify the legal position for new providers, and second, the pivotal role of the Natural Death Centre (NDC) in raising public awareness through its publication *The Natural Death Handbook* which first appeared in 1993 (Wienrich et al. 2003). The NDC has also co-ordinated the Association of Natural Burial Ground (ANBG) providers. There are now approaching 250^2 natural burial grounds in the UK, almost equivalent to the number of crematoria, which stands at 251 (Grainger 2005).

The early expansion of natural burial was dominated by the public sector with Local Authorities re-designating unused areas of their cemeteries for natural burial or, as at Carlisle, using adjacent land designated for future expansion. Local Authorities were also under pressure to respond to the growing public interest in natural burial and to provide choice. Of the total number of sites, currently 56 per cent are provided by local authorities. The publication in 1996 of the *Charter for the Bereaved*, developed by the Institute of Burial and Cremation Administration (IBCA), explicitly encouraged authorities which formally adopted the Charter to provide a 'natural option such as woodland burial' (IBCA 1996: 19) among their grave choices. Whilst the woodland burial concept introduced by West has remained the most dominant form of natural burial, it has rapidly evolved to include new habitat deathscapes such as wildflower meadows, orchards, woodland pasture and burial into mature woodland. However, whilst the focus on the creation or preservation of habitat is an important feature of these deathscapes, there are further dimensions to this shift in the UK's disposal provision. These became apparent during an important early phase of our research when we visited 20 natural burial grounds identified from the ANBG database and chosen from different regions of England and Wales. This activity revealed that alongside a diversity of design interpretations, what was striking about natural burial was the emergence of a new and diverse group of independent burial providers that includes funeral directors, landowners, charitable trusts and farmers.

As this chapter will show, therefore, the distinctiveness of the natural burial ground resides in both its physical and its *social* landscapes and their inherent interaction. Why individuals choose this mode of disposal, either for themselves or someone close to them, and how their immediate and longer term experience of burying naturally is shaped should thus be thought about in terms of both the materialities of a particular environment and the social relationships which underlie its design and development. This approach draws on Ingold's (2000)

---

2   These data have been compiled from a survey of all natural burial grounds as part of an ongoing research project funded by the ESRC. 213 of these sites are now in operation, with a further 37 at planning stage. The data base is building on information supplied by the Association of Natural Burial Grounds, which is coordinated by the Natural Death Centre, a Charitable Trust.

notion of a 'dwelling' perspective which takes account of human beings' embodied engagement with the world around them as a process that unfolds over time. Moreover, and with reference to Ingold's notion of the 'taskscape' within which, he argues, 'temporality inheres in the pattern of dwelling activities' (1993: 153), each burial landscape is replete with worked spaces that have specific processes and meanings attached to their existence. In other words, the meaning or status of a particular landscape is not something human beings impose upon it from outside; rather it emerges over time out of their dwelling within the materialities that comprise particular sites.

Becoming a Burial Ground Provider

The reasons why a diverse group of providers are motivated to open a burial ground are complex, in some cases resulting from a combination of factors which include: a response to a perceived change in public demand as witnessed by media interest in natural burial; the development of a burial style that appears relatively simple to implement and manage and which has a strong resonance with their own knowledge and skills; and finally an apparent lack of clear government regulation which makes the management of private cemeteries far less onerous than those in the public sector. The Local Government Act of 1972 provided a 'comprehensive code for the management, regulation and control of cemeteries' (Smale 2002, see also Ministry of Justice 2009). However this code of practice only applies to burial authorities and does not include private cemeteries.

Here we focus specifically on farmers, whose sites represent one of the largest increases in the private provision of natural burial (37 per cent of all privately managed sites). Data gathered within the farming community give a unique insight into what it means to own and manage a burial ground. Unlike public sector staff who have developed natural burial grounds within the UK's established cemetery culture, the men and women we interviewed from agricultural communities had no experience of designing or managing a burial ground or of working with bereaved people. We therefore explored their motivations for opening a burial ground as well as their experience of welcoming the dead *into* their land, and the family and friends of the dead *onto* their land. We also asked how their unique relationship with place, which may have developed over successive generations and involved working closely on the land, informed how they went about both creating a new and distinctive deathscape and subsequently managing it. Finally we explored the wider impact of this experience on their everyday lives and on the running of their farms. As we argue below, it is the particular complex of motivations, experience and social relationships shared by members of the farming community that underpins and informs the design of a major category of contemporary natural burial ground provision.

The data drawn on here derive from ethnographic research conducted at three farms, in Northumberland, Devon and Powys. Each is jointly managed by a couple and at two sites we interviewed the couple together. Before the interview we visited the burial ground itself, so exploring the physical relationship between the site and the farm and finding out how natural burial was being interpreted in terms of habitat creation, burial and grave marking. This prior understanding of place helped us frame the interview questions and the resulting data are presented below in three sections: motivations for opening a burial ground; design and vision; and management and impact. Together they reveal the different ways in which natural burial is being interpreted and the specific models or conceptions of nature these farmers wished to promote. The data also expose the varying degrees to which they facilitate and engage with the consumers of new deathscapes and how this evolving relationship informed and moulded their understanding of natural burial. We begin with a brief summary of each site.

Seven Penny Meadow Green Burial Field, Northumberland[3]

This site opened in 2002 on the outskirts of Medomsley, a Northumberland village. The land belongs to Alan and Marilyn Willey[4], dairy farmers who have farmed in the area all their working lives. Born on the farm, Alan could trace his family connection to it over four generations. The burial ground itself is several kilometres away from where the couple farm approximately 350 acres and comprises 4.5 acres (1.8 hectares (ha)) of former grazing land adjacent to the main road into Medomsley. At the site entrance a timber gate opens onto a small car park surfaced in stone chippings with space for up to ten cars. Here a timber sign identifies the burial ground and gives the owners' contact details. The almost square field is rough grassland and reasonably level although now ploughed into parallel, undulating rig and furrows.[5] Its boundary is delineated by hedges which contain some mature trees. Over these hedges are views of the surrounding moors and houses in the nearby village.

Only a mown pathway running along the perimeter leading down to the lower eastern edge of the burial ground provides some indication of the field's burial function. Along this edge several garden benches are set back against the hedge and in front of them graves are set in pairs on each of the ploughed rigs. Some are marked by a tree, stone or marble plaque, floral tributes and other personal items but most are difficult to identify. Grass has to be pushed aside to reveal the small grit-stone plaques laid flat on the graves.

---

3   Real site names are used with the consent of the site owners.

4   We have consent to use site owners' real names at each of the three sites we discuss here.

5   Rig and furrow is a medieval system of working the land which was common in upland areas and is closely related to its lowland equivalent of ridge and furrow.

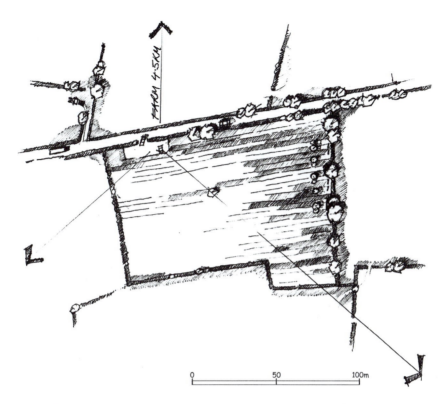

Figure 7.1   Seven Penny Meadows: This image illustrates how rig and furrows are used to align the graves. The emerging woodland is indicated along the eastern edge of the burial field, which is separated from the farmhouse by a distance of 4.5 kilometres (km)

Source: Illustration by Andy Clayden.

Green Lane Burial Field and Nature Reserve, Powys, Wales

This site opened in 2003 near the village of Abermule on land belonging to Ifor and Eira Humphreys, sheep and beef farmers. Ifor has farmed in the area all his working life and moved to this farm in the 1980s. The burial ground is next to the farm, separated only by a narrow lane which provides access to both the farm and the burial ground. The site and its attached nature reserve, which is not intended for burial, comprise 11 acres (4.45 ha) of which 1 acre (0.4 ha) is designated for burial. At the site entrance is a large timber gate, a sign and a tree trunk with a carving of a squirrel climbing a tree. A loose-bound stone driveway leads to a car park with space for 10–12 vehicles and is flanked on one side by a steep bank partially covered with gorse and planted with trees and other native species.

Opposite are views across woodland scrub into the adjacent valley and over to the surrounding mountains. Beside the car park is a picnic table and a plan of the burial ground and nature reserve. Beyond the car park a steep track leads down into the nature reserve, an area of woodland scrub with additional tree planting.

Accessed directly from the car park, the burial field is on a steep grassy bank facing into the valley below and bordered with mature hedgerows and several mature trees. A large Victorian house is visible at the lowest corner of the field, albeit partially screened by new woodland planting that includes different native species. The site's 50+ burials are relatively invisible; there are no memorial plaques, trees or ornaments on any of the graves. Closer inspection reveals small stones set into the turf and painted with a number. These mark the graves. In spring, large clumps of daffodils which correspond with some of the stone markers become visible. Set within the new woodland perimeter planting are several wooden benches and beside some of the trees are small oak stakes fixed with plaques bearing the deceased's name and sometimes a brief memorial message.

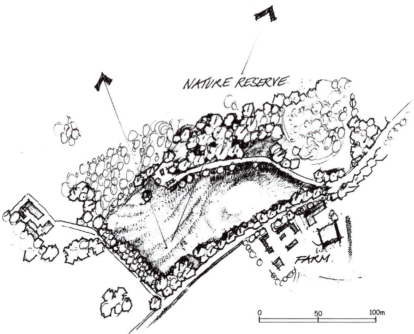

Figure 7.2   Green Lane Burial Field: This image illustrates the very close proximity of the burial field to the farmhouse and indicates the driveway to the car park that connects the burial field with the nature reserve. Also shown is the tree planting that provides partial screening of the adjacent property

Source: Illustration by Andy Clayden.

Crossways Woodland Burials, Devon

This site is located near to the village of Cheriton Bishop in Devon on land belonging to Martin and Julie Chatfield who purchased it 20 years ago when they moved to the area to farm. It is a short walk up a steep lane from their house, which they built. At the entrance is a gate, sign and stone-chipped car park with space for approximately 20 cars, out of which a drive runs centrally through the site up towards a turning point at the far end of the burial ground. Near the turning point several timber benches carry views over the burial ground and out into the surrounding countryside. The burial ground comprises approximately 1.5 acres (0.7 ha) and is sectioned out of a much larger field, separated from it by a new earth bank boundary which Martin has planted with a native hedgerow mix. Existing boundaries are mature hedges containing some established trees that

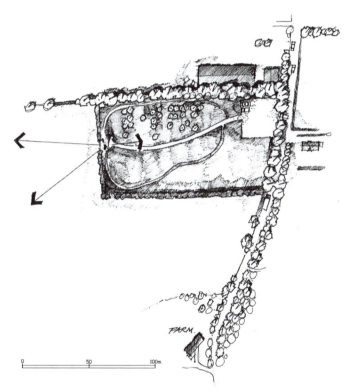

Figure 7.3   Crossways Woodland Burials: This image illustrates the close proximity of the burial field to the farm, the driveway and emerging woodland and the new earth bank and hedging separating the burial ground from the larger field

Source: Illustration by Andy Clayden.

partially screen nearby agricultural buildings. Each grave is marked by a stone plaque laid flat on the grave and a locally sourced native species of tree supplied either by Martin or the family of the deceased.

What Motivates a Farmer to Open a Burial Ground?

Diversifying the farm's income was important for each of the farmers we interviewed when they were deciding to convert land into burial provision. This was not, however, the only consideration. Ifor, at Green Lane, Powys, became aware of natural burial in the mid 1990s after hearing a radio interview with John Bradfield about his publication, cited above (Bradfield 1994). Ifor said he 'thought this is a great idea, this is … for me'. He described himself as 'non religious' and attracted by the absence of religious sentiment; he felt a 'dissatisfaction with the traditional funeral… stuck up in a religious package'. Another influence was his experience of 'several other farmers who've been buried on their own land'. After their first child was born, the couple wanted Eira to remain on the farm rather than returning to managing a shop in the nearby town. Opening a farm shop was considered, but was thought to be too time-consuming and intrusive. Instead, a burial ground allowed Ifor's practical skills of working the land and Eira's experience in management and customer care to be combined. So, although their decision was primarily economic, it also fitted their combined skills and changing family circumstances. Ifor was also motivated by the potential environmental improvement and the opportunity to re-establish a species-rich hay meadow: 'as a farmer we'd all, we all like to be conservationists but equally we need to make money as well'.

Alan Willey at Seven Penny Meadow, Northumberland, shared these environmental aspirations and became aware of natural burial from a BBC programme, *Countryfile*, which reports stories about the changing British countryside. The programme featured an interview with a farmer who had set up a woodland burial ground.[6] Alan saw a chance to diversify his income at a time of falling milk prices, plus an opportunity to return an area of improved pasture[7] back to a 'sort of old fashioned meadow'.

Martin and Julie Chatfield's decision to open Crossways natural burial ground on their Devon farm was stimulated by more personal objectives; a close friend who was dying from cancer wanted to be buried on their land. Martin's experience of his father's cremation was 'so impersonal' that this also prompted them into

---

6   The television programme, *Countryfile* which Alan refers to was also mentioned by several of our other participants, who acknowledged it as their first introduction to natural burial.

7   Improved pasture is land where the soil fertility has been increased possibly through the use of chemical fertilisers in order to increase yields. In this environment many native species will not thrive due to the increased competition of more aggressive grass species.

action. Martin shared Alan and Ifor's environmental aspirations, wanting to establish what he called a 'Devon wood', comprising local native woodland species supplied from his own small tree nursery.

Whilst diversifying farm income was important for all of these individuals, their data reveal more complex and personal motivations inspired by individual circumstances; for example, at Green Lane, Ifor and Eira's desire for a role for Eira that enabled her to stay and work on the farm, and at Crossways the wishes of Martin and Julie's terminally ill friend and the couple's aversion to Martin's father's cremation. In addition, a more intimate relationship with the landscape and a desire to make a positive contribution and commitment to the ecological diversity and sustainability of the landscape were evident, accomplishing goals less easily achieved through intensive farming practices.

What was their Approach to Designing the Burial Ground?

We wanted to discover what models farmers drew on when they were designing these new deathscapes. How did they go about establishing its identity as a burial ground? Early interviews revealed that when asked to describe their approach to 'designing' a burial ground, interviewees did not always recognise their role as a 'designer'. For them design was the preserve of professionally trained people. Yet even when they simply erected a sign outside a field, they put a particular design into practice, one which profoundly altered the qualities of the landscape it then signalled. For example, when we took focus group participants into a site that had formerly been a potato field they trod carefully, fearful of walking on a grave. At a site which had been a conifer plantation, there was no sign, but when a bereaved visitor explained its purpose to would-be dog walkers, they left hurriedly.

Like nineteenth-century urban cemeteries that share the contemporary UK burial landscape with churchyards, natural burial grounds thus reflect particular design aspirations, albeit ones which may emerge through practice – or 'dwelling' (Ingold 2000), rather than being inscribed on paper and then on the land. Here we begin with a discussion of those designs which materialised in early nineteenth-century cemeteries, considering how natural burial grounds both differ from and also share their aspirations.

A tour of any large city will reveal a range of cemetery landscapes that might vary in style, size and ownership, yet share physical characteristics which include: a boundary wall which is often a substantial structure built of stone and or iron railings, large and imposing entrances which declare the site's meaning and function, and finally an internal order frequently set out on a grid and preserves the location and identity of the each grave (Rugg 2000). There may also be substantial buildings such as chapels, entrance lodges, public conveniences and maintenance buildings. Another important characteristic of these landscapes is that they have all been designed by professionally trained architects, landscape architects and surveyors.

Thus in Sheffield, the first cemetery was opened in 1836 by the Sheffield General Cemetery Company and designed by the architect Samuel Worth. Contained by a large stone wall with iron railing inserts, it is accessed via an imposing gatehouse that includes accommodation for the sexton. In addition there is a chapel and registry office designed in a style heavily influenced by Greek and Egyptian architecture fashionable in that era. In 1846 the cemetery was extended to include an Anglican section with grounds set out by the renowned landscape architect, Robert Marnock and a new Anglican Chapel designed by the architect William Flockton. The extension was designed in a picturesque style which included pathways winding their way through planted gardens.

In 1860, Sheffield's first publicly-funded cemetery was opened at Burngreave and also designed by William Flockton and Son[8]. Whilst it shared features with the General Cemetery – a significant boundary wall, entrance gates, lodges and chapels – it differed significantly in style and organisation with a more rigid, formal layout, possibly in response to John Claudius Loudon's influential 1843 book, *The Laying Out, Planting and Management of Cemeteries*, which promoted a more utilitarian layout This desire for greater simplicity and economy is also reflected in the design of the city's first municipal cemetery which opened in 1909 in the rapidly expanding suburb of Crookes. Whilst Crookes cemetery was less grand than its predecessors it still represented a significant investment of resources by the recently formed City Authority and also conforms to the landscape typology identified by Rugg (2000). Set out by the city surveyor, it included a large stone boundary wall, with iron railings, entrance lodges, driveways, a formal burial grid and a chapel, designed by local architect C. & C. Hadfield (Stirling 2009). These design aspirations have figured within historical literature on disposal and memorialisation. Tarlow, for example, lists a range of agendas and imperatives which underpin the sites that have become our contemporary legacy:

> Concerns regarding public health and hygiene, desire to safeguard the corpse against body-snatchers, emulation of foreign or elite fashions, the opportunity to enhance status through erection of ostentatious memorials and the possibility of liberations from the stifling and unsatisfactory confines of burial according to the rite and whim of the established church. (Tarlow 2000: 218)

This insight into nineteenth century resistances to the burial provision of the time resonates with the oppositional stance adopted by contemporary natural burial; for example, by providing an alternative to the crematorium and its current

---

8   The Burial Act of 1853 permitted the local authority to apply for the closure of overcrowded and unsanitary churchyards and also to consider the need to provide a cemetery if it was requested by the local ratepayers. If this need was considered to be appropriate then the Secretary of State was informed and a burial board was established (Brookes Mortal Remains). In 1855 all the Anglican churchyards in the town were closed and in 1858 the first Burial board was established at Brightside and Barlow (Stirling 2009).

associations with potential environmental damage, and the cemetery's apparent anonymity and vulnerability to vandalism and neglect. In addition to the factors listed above, Tarlow also argues that the success of the garden cemetery, which had acquired its characteristic garden style in the early nineteenth century, was 'the creation of an aesthetic and sentimental landscape of remembering which had particular emotional resonances for the bereaved' (2000: 218). These cemeteries were typically located on the outskirts of towns with views of the surrounding countryside and imaginatively planted with native and ornamental trees and shrubs reminiscent of the private parks of the time. Sheffield General Cemetery, for example, retains a fine collection of weeping trees (for example, Willow and Ash), which not only contributed to a picturesque view of the natural landscape that enclosed the cemetery at the time, but also evoked the desired atmosphere of melancholy, thereby stimulating emotions appropriate to mourning.

If the design of the nineteenth century mortuary landscape reflected a particular complex of functional and symbolic agendas, how might their outcomes have influenced the farmers we interviewed as they created a new deathscape? To what extent did these cemeteries provide a model to work *against*? Or are there, in fact, echoes of one model to be found in its successor? We therefore asked our interviewees whether they drew on the layout and design of the traditional cemetery by enclosing the space, defining entrance ways and organising the graves. How had they gone about establishing the identity of the burial ground? As noted, they did not see themselves as designers so we adapted our questions and asked them to describe their 'vision' for the burial ground; in short to explain what they wanted it to be like, now and in the future. We wished to know how they set about this task, who they consulted and what research they had done. We now explore data that describe how they chose where to locate the burial ground, what type they wished to create, and how natural burial was to be expressed physically in terms of boundaries and entrance thresholds, burial and grave marking.

Choosing the Place and Designing the Burial Ground

Each farmer was able to choose one area from within the larger body of their land for a natural burial ground. The fields identified complied with the requirements of the planning authority which would approve the change of use from agricultural land to cemetery. Farmers needed good vehicle access from the highway and adequate parking for a funeral and visitors to the burial ground. Another important factor was the farmers' concern to identify land they considered visually attractive, which would appeal to potential users and, importantly, had views out of the site into the surrounding landscape. Exploiting a visual connection with the wider landscape does echo nineteenth century cemetery design, for example at Sheffield's General Cemetery, yet in the case of a natural burial ground it was seen by the farmers as a key opportunity to reinforce connections with the 'natural' landscape of the farm and its surroundings and the contribution the burial ground could make to the 'natural' environment.

Thus, Ifor and Eira at Green Lane in Powys chose a steeply sloping field with views across the adjacent valley. The land did not 'lend itself to agricultural operations' and was partly wooded with 'huge wildlife potential'. Creating a burial ground would allow income to be generated from land which was difficult to farm, whilst simultaneously improving the farm's environmental profile. Ifor and Eira visited other woodland burial sites prior to opening their own and felt '[we've got] a better site in the first place[and] the resources and time ... to manage it'. Their vision was to create a wildflower meadow with greater conservation value than woodland because of its relative scarcity. It was to be kept as a field so Ifor could take an annual hay crop to feed stock. Eventually, when no longer used for burial, the site might be returned to grazing land. Ifor and Eira successfully applied for a Farm Enterprise Grant which funded passing points on the burial ground's approach lane, a condition of the planning application, and a drive and parking area. None of the farmers interviewed used a design professional to help them with developing their site. Ifor, however, was a member of the local branch of the Wildlife Trust and did consult with them. They conducted an ecological survey of the site and gave advice on planting. This absence of professional design input was not, however, a feature of the other kinds of site we visited. Typically, professionals were consulted at larger sites with greater capital investment and perhaps belonging to a private company, funeral directors or charitable trust. One of the private landowners interviewed who had initially engaged a landscape architect rejected their input when she felt it detracted from the qualities of the existing landscape.

If professional advice was largely absent from farmers' decision-making processes, we need to ask what did inform the way they altered the nature of part of their farm to create a burial landscape. In each case, a 'dwelling' perspective is evident as farmers spoke of their experience of their landscape and knowledge of working the land. The landscapes they created thus differed significantly from the cemetery whilst expressing a continuity of design in that the choices made revealed a desire to work with the existing landscape fabric and to integrate the burial ground with the farm. At the heart of the vision for each of the burial grounds was the habitat farmers sought to create. Establishing this identity is a key factor in promoting these new deathscapes and requires an understanding of the underlying ecological processes. Alan and Ifor wanted to establish a wildflower meadow, and Martin a native woodland. In each case this meant converting land previously used for grazing and therefore improved by increasing the fertility of the soil over many years to increase grass yields. In contrast to improved pasture, wildflower meadows thrive on land low in fertility with reduced competition from more aggressive species.

At Seven Penny Meadow Alan ploughed the field and seeded it with native annuals, including poppies and corn flowers. This was also an opportunity to reinstate the traditional 'rig and furrows', parallel rows of earth mounds and troughs, an upland variation of ridge and furrow that was a common feature of the landscape before the introduction of mechanised farming. These served a very practical role in helping organise the locations of graves which were positioned in pairs on the top of each rig. The naming of the burial ground also enabled identity to be projected and a unique sense of place created. Alan researched historic maps; 'I've gone back to an old plan and that's how I got … the Seven Penny Meadow', the name thus reconnecting the site with the habitat it seeks to recreate and evoking its history. In the first year wild flowers flourished but the land quickly reverted to more vigorous grasses. Alan tried to reduce the soil fertility by scalping or removing the topsoil. Both he and Ifor, at Green Lane also used a native parasitic plant, yellow rattle, which feeds on the more aggressive grasses, reducing their vigour and creating a niche for native species. Both farmers also removed an annual hay crop from the field to prevent the build-up of fertility. These processes indicate that creating a natural burial ground is a deliberate process of steering and managing land 'back to nature', rather than simply abandoning cultivation or maintenance. In trying to establish a woodland at Crossways, Martin followed Ken West's approach at Carlisle, planting a single tree on each grave, sourcing them locally and supplying them from his small tree nursery.

These approaches demonstrate an individual's deeper understanding of their land – the 'taskscape' within which they 'dwelt' (Ingold 1993, 2000) – and their capacity to address site-specific problems and so bring about a particular vision. The challenge of creating native habitats on previously intensively managed land is common to many natural burial grounds but only at farmers' sites have we witnessed the use of flora and scalping to reduce soil fertility. While this does not necessarily demonstrate greater environmental commitment, it does indicate a better temporal understanding of how the land has been managed in the past, and the techniques that work. At a very practical level the farmers also have the skills and machinery that enable them to make these changes. For burial grounds without access to the necessary machinery or the required skills, this approach to managing the land may not be an option.

The establishment of the physical boundary and threshold to each of the three burial grounds also reflects how each farmer worked with the existing landscape fabric and integrated the burial ground with the farm. Thus each site entrance echoes those of adjacent fields with its traditional timber farm gate and a sign. Only at Green Lane Burial Field had Ifor and Eira elaborated the threshold by erecting two large tree trunks sculpted into wildlife motifs that are repeated in an owl sculpture inside the burial ground. Carved into dead wood, this motif could be interpreted as a reference to nature and natural cycles, yet it also serves to frame the nature reserve beyond the burial ground. Only Martin at Crossways needed to construct a new boundary after sectioning off the burial ground from a much larger field. Using topsoil gathered when making the car park, he created an earth bank planted with native hedgerow plants, a form of field boundary distinctive to the vernacular Devon landscape and something Martin felt enthusiastic about. He saw these more permanent boundaries as a way of making the burial ground legible once the woodland had established itself and in the event of the burial ground extending further into the larger field. The bank would also enable visitors to locate a relative or friend's burial place. Ifor and Alan sought to strengthen their field boundaries by planting additional hedgerow trees and Ifor increased the woodland planting along one edge to provide more shelter and screen views from an adjacent property.

In further contrast with the traditional cemetery there were no buildings or facilities at any of the three burial grounds. Although many privately managed natural burial grounds do include a building on site, these are typically modest structures taking their design cue from traditional timber frame agricultural buildings. Martin at Crossways considered erecting a small timber frame structure for shelter during services in bad weather as well as for visitors to the site. Although very different from the large and often overtly religious buildings at traditional cemeteries, these structures do represent a level of investment and a sense of permanence that potentially shifts the way a natural burial ground might be perceived by these new owners, as well as by visitors/users.

Managing the Burial Ground

Thus far we have discussed the design of both cemeteries and natural burial grounds, noting that the concept of 'design' cannot be restricted to professionals, but needs to incorporate the decisions made by these farmers in that their choices also materialised within the landscape. In addition, however, the deathscapes we describe can be seen as forms of 'taskscape' that result from day to day and year on year management in that this ultimately determines whether the original vision is realised. This is also true of cemeteries where, through many decades of 'consumption and use' (Tarlow 2000), burial and memorialising, a landscape will gradually evolve. Of the three Sheffield cemeteries discussed above, only Crookes remains open for burial. By the 1930s, burial space at Sheffield General Cemetery was virtually exhausted and the company struggled to generate sufficient funds to maintain the buildings, graves and landscape. In the 1970s the Local Authority took responsibility and cleared memorials from the Anglican section to create much needed recreation space for the surrounding community. Today a Friends Group manage the cemetery. Despite their commitment and success in securing heritage lottery funding, they struggle to maintain a landscape that defeats efficient, mechanised management techniques. Both its chapels are now boarded shut against vandalism and overgrown, and potentially hazardous areas have been fenced off. Despite Crookes cemetery remaining open, it too has changed. In the 1960s lawn sections were introduced to simplify maintenance and reduce costs (Rugg 2006). However, active resistance to this aesthetic by some bereaved families who have built their own kerb sets has negated any savings the Local Authority might have made. More recently, health and safety regulations which require all headstones to be tested for safety have impacted significantly on the cemetery's character since many headstones are now laid flat on the grave. Part of Ken West's motivation to introduce woodland burial was to solve such problems by removing the need to maintain headstones, simplifying management operations and embracing the idea of a dynamic and naturally evolving landscape which could not be permanently held in check.

Against this background, we ask whether the farmers we interviewed had similar difficulties in realising their vision for a burial ground. Had the processes of consumption and use impacted upon this process? How did they manage a landscape 'back to nature' and what part did bereaved people play in that process? Alan in Northumberland and Ifor in Powys have, to varying degrees, already modified their original idea, not in response to the management of the land, however, but as a result of bereaved people's needs when invited onto their land. Alan, for example, discovered that 'the majority of people are wanting trees now [and] we're finding people that didn't want trees have actually come and stuck a tree on without even telling us'. The small stone plaque he supplies for families to engrave has in some cases been replaced with a much larger marble plaque and 'as soon as one family have put a marble one down another family have took the stone one up and put a marble one down'. As a result, Alan decided he would need 'some ground rules' and put up a notice. He sees the problem emerging when 'families come through an undertaker and we don't get to see them beforehand'. Now he accepts that half the site will eventually be wooded and half wildflower meadow. A striking aspect of this site, as the illustration above indicates, is its distance from the owners' living space; it would seem that Alan and Marilyn's physical absence from the burial ground coupled with bereaved people's physical presence within it has implications for the way this particular deathscape is developing.

At Green Lane in Powys, Ifor and Eira have had less experience of bereaved people wanting to individually memorialise graves. This appears to be partially attributable to the different strategies they adopted: their hands-on approach to the burial ground made easier by its close proximity to the farm; the presentations they have given to different organisations; and their meetings with potential users to explain how the site is managed and what is permitted. The couple visit the site regularly, removing any items apart from cut flowers, and engaging with bereaved people to provide any help needed. For example, Carla[9], whose husband is buried at Green Lane, asked Ifor to help plant bulbs on his grave: 'Ifor came because I couldn't have done it on my own. So he was, I explain to him, I say I wanted to plant, I say the shape of a J for Jim. So we did'. Bereaved families and friends can also purchase a 'permanent' memorial plaque next to one of the existing trees that surround the burial ground, enabling them to name and memorialise the dead person, something popular with over half the users. Maintaining a form of control over what happens on site means that Eira and Ifor have to some extent been successful in realising their vision for the burial ground. Despite these strategies however, they have had to adjust their original concept of a field without grave

---

9 Bereaved research participants' names are anonymised.

markers: 'we started off with no markings at all on the graves, but very soon we decided we needed a location that was visible because we had one or two people come here and we did have one family a bit distressed, they couldn't find the grave'. As already mentioned, Ifor now inserts a small numbered stone on each grave that corresponds with an entry in the burial register.

Martin at Crossways in Devon has not experienced any significant problems with memorial objects on graves, something he attributes to the type of people attracted to his site, 'I think solar lights would be too much ... we're not attracting that sort of person. [We are attracting] people who think differently I think and greens'. This absence of memorialisation may also result from the way he enables people to choose their own tree, with his approval, and place a stone plaque of their choice on the grave and plant bulbs. Unlike Alan's original vision for his wildflower meadow in Northumberland, which he was unable to realise, Crossways provides many opportunities for bereaved people to bring with them their own 'nature' and so maintain a connection with the deceased person by physically marking the grave. What remains unclear is how Martin will manage the woodland, how he will in future decide which trees should be removed to create space for larger and more vigorous specimens – and indeed how bereaved people will react to his decision. Our interviews with bereaved people suggest that for some the tree represents a significant emotional investment; not only is it perceived to be physically connected to their loved one but its continuing growth is symbolic of their life force (Clayden and Dixon 2007). If removed it might prompt a return to more traditional approaches to memorialising, like for example the placing of kerb sets around graves in lawn sections.

Visits to these three sites and interviews with the individuals creating and managing them provide insight into the processes through which particular deathscapes come into being. Importantly, what farmers understood as a vision, rather than a 'design', is simply a starting point. Even where a landscape apparently comprises little more than marker stones hidden within the grass and, for much of the time, no-one working or visiting there, it still needs to be recognised as an emergent landscape. This way of thinking about a deathscape derives partly from the work of Ingold (2000), introduced earlier. As an anthropologist, he argues that the conventional social science belief that people view the world through culturally-specific lenses misleadingly assumes that the world around us is separate from us, something we interact with on our own terms. In contrast, his 'dwelling' perspective requires us to take account of the indeterminate nature of our environment. What that world comes to mean is neither something we impose upon it, nor is it essential or intrinsic to the materialities of landscape and bodies. Instead, a material world made up of vegetation, freshly dug earth, heavy rain, has possibilities or affordances that human beings respond to and develop in particular directions, through embodied engagement (see also Macnaghten and Urry 1998). What data of the kind we present here allow us to ask is how people from different starting points experience, mark or shape a shared environment and how their intersubjective and embodied involvement changes their experience of an unfamiliar type of burial ground.

Impact Beyond the Burial Ground

Thus far we have laid out the processes through which farmers' engagement with their landscapes – as farmers – have resulted in the setting up of particular kinds of burial ground. These have then taken shape through processes of interaction between different individuals and with the materialities of the sites, the bodies buried there and the activities of owners and bereaved people who inhabit these new burial spaces. In the course of the interviews it also became apparent that the processes of setting up these new deathscapes were not contained or restricted to a particular parcel of land. Instead they both draw on and have implications for a broader physical and social landscape. This might be a landscape yet to materialise. For example, Martin at Crossways spoke not just about creating a 'woodland' burial site but also about using their own logs for fuel, 'rejuvenating everything'. Like Alan and his son at Seven Penny Meadow, he plans to introduce a wind turbine, not to make financial savings, but because, in light of his overall concern with the environment, 'it's something you sort of want to do'. However he also envisaged a wind turbine interacting with his work as a burial ground provider in that people with an interest in alternative energy sources might well be drawn to natural burial and would discover his site via the turbine. In this respect, then, a natural burial ground is unlike a cemetery with its singular mortuary purpose, separated off from both everyday life and other life course transitions. Instead, like the churchyard, it has connections with a broader range of belief systems and experiences.

Alongside this breadth of activities within the physical landscape, the social landscapes of these natural burial grounds extend out into a larger community, incorporating providers' previous roles and experiences within quite different settings. Marilyn, for example, at Seven Penny Meadow in Northumberland, had initial qualms about moving into natural burial, seeing herself and Alan as farmers. However, the burial ground they have established helps overcome the loneliness often experienced by women living and working on a farm (see Christensen et al. 1997). Marilyn summed this up when she said: 'no, no I mean I suppose being farmers, you go to market and you meet other farmers. You're not meeting with the public very much. I think it, I think it's interesting'. She added: 'it's nice meeting the families, talking to people, see their views, see what they want, if you can help them in any way, I said to him I'd better be a funeral director now'. Similarly Martin at Crossways in Devon spoke of becoming a funeral director: 'well we're thinking of doing that, yeah that's our next plan. Funeral Directors, yeah, because so many people asking us if we can do it. I think it looks like we just need a fridge ... and my wife's in the medical profession anyway ... so she can handle bodies and we're used to dealing with dead things'. Eira, at Green

Lane Powys, had previously worked in a dress shop in the local town but after her children were born was reluctant to take on work outside the farm. Building on both her previous experience in customer relations and her local networks, she took primary responsibility for the burial ground. Ifor said: 'she mostly deals with the funerals and deals with the people who are coming … she's better at it than me … she's got a better memory than me as well'. A bereaved interviewee exemplified this point when referring to her previous connection with Eira: 'I knew Eira, well not so well, she used to work in a shop, ladies' clothes shop in Newtown before she got married and she had the children and I didn't know Ifor, but he's a lovely chap as well. They're lovely couple. […] Well they're not false, yeah. What you see you get. They're natural couple and that's it. It's, say they are natural couple. You trusted'.

Conclusion

To conclude, we suggest that these data show the farmers we interviewed both drawing from and reinventing their broader physical and social landscapes. If we maintain a 'dwelling' perspective this also enables us to recognise how these interpersonal relationships are not only a response to the affordances of the physical landscape but shape that landscape, both in terms of the burial ground and the farm itself. Thus Ifor at Green Lane Powys, for example, told us that setting up the burial ground had changed his perspective on the farm. He saw people coming to him because he had something unique to offer and this had made him think afresh about his livestock. He now specialises in a rare breed of cattle imported from Australia. Alan at Seven Penny Meadow in Northumberland similarly described the ways in which establishing the burial site had related impacts on the farm as a whole: 'we're doing a fair bit of green waste', he said, 'we take the wheelie bins in from the council'. When asked whether he saw this sitting well with natural burial he said: 'Yeah, that's true. And we've got the long horn cattle up where, on the land, you know, round about the burial ground … English Long Horn. So they're a traditional breed, I believe the second oldest British breed'.

Via these innovations then, farmers' decisions to become burial providers both draw from and impact upon a local and indeed global environment. As regard the future of their sites, here again we find continuities with, as well as diversions from the backgrounds of each couple. Alan, whose family had farmed for four generations, talked about the site's future as a natural woodland, not a meadow, despite its title, Seven Penny Meadow. He said: 'what I would like to see is just left on the farm, just left, just look after itself, it's a natural woodland isn't it?' He went on: 'we've got to do the best we can at our, in our time, and then, you know, the lads, hopefully their family will do the same'. And he noted that one son is good with the livestock and the other helps with the burial ground, deputising as pall bearer on occasion. Whereas Alan's partner, Marilyn said of the site 'its not farming land now', Ifor in Powys said: 'it's a field and it'll always be a field', going on to assert that in future 'it'll be farmed, probably. Whether it'll be managed as a hay meadow is another question because if it's, if you've a grave sinking it gets more difficult to manage'. Martin, at Crossways in Devon, said 'as far as we're concerned, you know, that's it, it's going to be a wood … as far as we're concerned it's going to be there forever … I think it'll just be a wood and there are going to be graves right under it'. Alan and Marilyn at Seven Penny Meadow thus envisaged the burial ground being handed on to future generations within their family; it will return to farm land and they are thinking of putting sheep to graze there. Its ambiguity as a landscape however becomes clear when they talk of using Jacob sheep which are not 'proper' sheep and will not, therefore, signal to visitors that they are trying to farm the land as well as use it as a burial ground.

As these data evidence, therefore, the ways in which the three deathscapes presented here have come into being can be helpfully understood in terms of a 'dwelling' perspective that pays attention to the relationship between physical and social landscape, the personal and family history of the providers, the burial ground itself, and their farm and its wider rural environment. In addition, there is now a new community that has become part of their land. How this nexus of relationships will ultimately unfold is still to be decided as the final section of this chapter has intimated. As shown by the history of the cemetery landscape, exemplified through the case study of Sheffield General Cemetery, the nature, usage and meaning of burial grounds rarely take forms that reflect the original design aspirations for that landscape. Interview data indicate that not only has the evolution of the three deathscapes deviated from farmers' initial approaches to it, but also that their futures remain tied to the particular history and orientation of the individuals involved. It is therefore this mix of dis/continuities, accommodations and innovations that we find marked upon these landscapes of disposal which have so recently emerged within the UK.

References

Bradfield, J. 1994. *Green Burial: The D-I-Y Guide to Law and Practice.* London: The Natural Death Centre.

Christensen, P., Hockey, J. and James, A. 1997. 'You have neither neighbours nor privacy': ambiguities in the experience of emotional well-being of women in farming families. *The Sociological Review*, 45(4), 621–44.

Clayden, A. 2003. Woodland burial. *Landscape Design*, 322, 22–5.

Clayden, A. 2004. Natural burial, British style. *Landscape Architecture*, 94(5), 68–77.

Clayden, A. and Dixon, K. 2007. Woodland burial: memorial arboretum versus natural native woodland? *Mortality,* 12(3) 240–60.

Grainger, H.J. 2005. *Death Redesigned, British Crematoria: History, Architecture and Landscape.* Reading: Spire Books Ltd.

Ingold, T. 1993. The temporality of the landscape. *World Archeology,* 25(2), 152–74.

Ingold, T. 2000. *The Perception of the Environment.* London: Routledge.

Institute of Burial and Cremation Administration (IBCA). 1996. *Charter for the Bereaved.* 1st Edition. Bourne and Grantham: Copytrend/Fytche.

Loudon, J.C. 1843. *On the Laying out, Planting and Managing of Cemeteries.* London: Longman, Brown, Green and Longmans.

Macnaghten, P. and Urry, J. 1998. *Contested Natures.* London: Sage.

Ministry of Justice. 2009. *Natural Burial Grounds: Guidance for Operators.* [online]. Available at: www.justice.gov.uk [accessed: 26 November 2009].

Rugg, J. 2000. Defining the place of burial: what makes a cemetery a cemetery? *Mortality,* 5(3), 259–75.

Rugg, J. 2006. Lawn cemeteries: the emergence of a new landscape of death. *Urban History,* 33(2), 213–33.

Smale, A.S. 2002. *Davies' Law of Burial, Cremation and Exhumation.* 7th Edition. Kent: Shaw and Sons.

Stirling, F. 2009. *Grave Re-use: A Feasibility study.* Unpublished PhD Thesis, University of Sheffield.

Tarlow, S. 2000. Landscapes of memory: the nineteenth-century garden cemetery. *European Journal of Archaeology,* 3(2), 217–39.

Thompson, J.W. 2002. A natural death. *Landscape Architecture.* May: 74–79 and 134–7.

Weinrich, S. and Speyer, J. 2003. The *Natural Death Handbook.* London: Rider.

PART III
Negotiating Space for Memorialisation in Private and Public Space

Chapter 8

The Production of a Memorial Place: Materialising Expressions of Grief

Anna Petersson

In both Western Europe and North America a major change occurred during the eighteenth century concerning our attitudes towards death, leading to new ways of dealing with the deceased and their burial places (Ariès 1994 [1976], McManners 1981, Etlin 1984).[1] During the eighteenth and nineteenth centuries, the successive construction of new cemeteries and new ways of burial physically displayed these shifts in attitudes (Lindahl 1969, Etlin 1984, Lundquist 1992, Wall 2000, Petersson 2004). This materialisation of a new *deathscape* helped authorities and professionals to institutionalise a consciousness among people in terms of what a 'proper place of death' should be like (Petersson 2004: 14–15, 2006: 112–13).[2] The architecture and landscape of the cemetery could also be seen as playing a role in formulating social and individual values (Etlin 1984), creating a design for life (Johnson 2008).

In the last 20 years spontaneous memorialisation, referring to the placing of, for instance, fresh flowers, lighted candles and photos at the sites of motor vehicle accidents, homicides, catastrophes, terrorist attacks and the like, has been acknowledged as also having the power to produce new memorial places and new rituals for veneration and remembrance connected to unexpected and violent death (Azaryahu 1996a, Haney et al. 1997).

This chapter will look at why, how and in what way a spontaneous memorial may develop from an immediate act to a more planned material place of grief and remembrance (Clark and Franzmann 2006, Nieminen Kristofersson 2006, Klaassens et al. 2009). Of special importance for this investigation are the noticeable differences in attitudes to, and growth of, roadside memorials revealed by data gleaned from interviews that I conducted with informants in 2005 on the subject of recent Swedish roadside memorialisation.[3] Analysis will also draw on the

---

1 However, since funeral customs vary between different western societies there is no such thing as a common western or modern way of death, as the sociologist Tony Walter (1996a) notices.

2 For the term *deathscape* see for instance the cultural geographer Kong 1999. The meaning of the phrase a 'proper place of death' is built on the theorist Michel de Certeau's (2002 [1984]) discussion of *proper places*.

3 The interview study was taken as part of the author's ongoing PhD project and will be described briefly later on in this text.

various viewpoints on the production of memorials on public ground found in two applications requesting to erect memorials on sites of individual deaths in the city of Malmö, Sweden, sent to and responded to by the Streets and Parks Department in Malmö.[4] In order to illustrate these perspectives this text will start with presenting some examples from the 2005 interview survey. Thereafter a presentation of the two requests, and the statements issued to them, will be made. I will then move on to discuss the examples in the light of the anthropologist Jean-Pierre Warnier's (2001) idea of 'the three media of symbolization',[5] dealing with the internalisation of a difficult experience, and the philosopher and psychoanalyst Julia Kristeva's (1982) theory of *abjection*, working on the exclusion and rejection of that which is threatening. Throughout this discussion, and this chapter, the material world will be recognised as an important media for expressing, communicating, experiencing, understanding and even debating different experiences of loss and bereavement.

Spontaneous Memorials: Earlier and Recent Studies

Spontaneous memorialisation has become a fairly well documented subject in academic fields such as ethnology, cultural geography, sociology, religious studies and death studies. In studies of spontaneous memorials existential, spiritual and phenomenological ideas have been tied to the sense and meaning of a fatal place (Henzel 1995, Nieminen Kristofersson 2005, Clark and Franzmann 2006, Grider 2006, Petersson 2009a). The impact of political, religious, cultural and social structures and constitutions in the production of a memorial place has also been given attention (Azaryahu 1996a, Ross 1998, Hartig and Dunn 1998, Kong 1999, Doss 2002, Everett 2002, Burk 2003, Petersson 2004, 2006, Clark and Franzmann 2006, Nieminen Kristofersson 2006, Petersson 2009b, Klaassens et al. 2009).

Sometimes the production of spontaneous memorials is considered to spring from bereaved people's discontent with the prevailing burial customs or the formality and strict requirements and regulations of ordinary cemeteries (Hartig and Dunn 1998, Clark and Franzmann 2006), or as an opposition to what I have called a proper place of death (Petersson 2004, 2006: 114).[6] It could also be seen as a way for the surviving relatives to communicate to the outside world about what they consider a 'bad death' (Klaassens et al. 2009: 191–2).[7] At other times the increase in spontaneous

---

4   This study will be described in a few words further on in this text. It was conducted by the author in 2008 within the interdisciplinary research project Designing Places for Memory and Meaning in the Contemporary Urban Landscape, running from 2006 – 2009. LTJ-Fakulteten, Alnarp [Online], available at: http://www.ltj.slu.se/6/O6_for_1.html [accessed: 27 October 2009].

5   Built on the psychoanalyst Serge Tisseron's theory of *symbolisation* (Warnier 2001).

6   See footnote number two in this text.

7   Following Seale and Van der Geest (2004) discussion of the cultural construction of different forms of dying as 'good' or 'bad'.

memorials, like roadside memorials, is understood as a reaction to the impression of death as invisible in modern society (Haney et al. 1997);[8] an international and collective experience of motoring and media culture (Monger 1997, Clark and Cheshire 2003/2004); a current downturn in the belief of institutions, such as the government or established religious associations, in the rise of post modernism and its focus on personal identity (Clark and Cheshire 2003/2004, Clark and Franzmann 2006); the mixing of various cultures and traditions due to immigration and new settlements (Barrera 1997, Everett 2000, 2002); as well as modes of drawing on cultures and religions inherent to a specific geographical location (Henzel 1995, Clark and Cheshire 2003/2004, Petersson 2009a, MacConville and McQuillan 2009).

In a recently published article on Swedish roadside memorialisation, I argue that even though recent roadside memorials in Sweden may be a constellation of both past and present cultures, traditions and ideologies, the phenomenon can still be experienced as something new (Petersson 2009a). Another issue dealt with is how the practice of placing material things associated with the deceased by the site of the grave, the accident site or in the home, may be seen as a way for the surviving relatives to generate the presence of the deceased, charging the memorial place with remembrance and giving it meaning (Petersson 2004, 2009a, 2009b). These findings have been further explored through an examination of the design, location and practice of a cultural event called 'The Altar of the Dead' placed in an urban park in the city of Malmö in November 2006 (Petersson 2009b).

Witnessing the Accident Site: The Start of a Spontaneous Memorial

The centre of attention for the interview study mentioned earlier in this text is the qualitative difference between some strategically chosen interviewees' conceptions and experiences of roadside memorials in Sweden. The study includes semi-structured in-depth interviews with: six close family members from different families bereaved by road traffic accidents; five professional drivers, including four taxi drivers (of which one was a former lorry driver), and one travelling salesperson; three employees at the Swedish Road Administration; and two employees at the Swedish Union of Transportation (of which one was a former lorry driver).[9] The interviewees' conceptions of roadside memorials were further related to their experiences of other spontaneous memorials as well as of cemeteries, home-based memorials and public funeral monuments, with a focus on memorial places and things as material expressions of grief that enable negotiation, encounter and meaning making to take place.

---

8   Described by Ariès (1994 [1976]), Bauman (1992) and Certeau (2002 [1984]), among others.

9   Quotations from these sources are my own translations of the informants' original Swedish.

One of the first themes addressed in my interviews with the surviving family members of traffic accident victims concerned their possible visits to the fatal accident site. The following questions touched upon the potential development of a memorial place at the site of the fatality, regarding memorial activities such as the leaving of flowers or lighting of candles at the site where the accident occurred.

As an example, one informant said that he looked for and left a flower at the accident site in order to work on, understand and finally integrate the death of his son in his life: 'Because just at this stage it [the accident] had not happened, although it had happened. It had not happened to us, although it had happened to us.' This informant further described the recurring visits to the site of fatality as growing into 'a small ritual' that was enacted every night during the first weeks after the fatal car crash. At the site, candles were lighted and open notes were left, to be read by family and friends of the deceased. Half a year after the initial development of the roadside memorial, this interviewee contacted the municipality in the community in which the family is living in order to apply for permission to put up a small plaque. This request was turned down with the words 'what if everyone did that'. The bereaved felt that he did not have the strength to argue, although he did not find the argument convincing, and stated that if he later felt he really wanted a plaque at the roadside memorial he would instead try to contact another municipal employee. After the first period of intense activity, visits to the roadside memorial have become fewer and fewer for this father. This has much to do with the eventual burial of his son and the finishing of his grave. Or, as the informant puts it: 'It is not a pleasant site [the accident site]. So when the grave [was made], and when the burial was [over], it felt good.' Since the headstone was raised, this bereaved father mostly visits the roadside memorial to light a torch on the anniversary of death.

A bereaved mother explained her first visits to the accident site in a similar way: 'I think that it is a way to understand the course of events and so, and to see how it looked, in that way. One can imagine the very incident in some way.' For this informant, the accident site at which three young people where killed, and where she and the other bereaved families laid down wreaths, is not a place she wants to revisit. The interviewee even avoids it and chooses another travel direction when the road section in question has to be passed on the way to the family's country house. The material environment of the accident site is, for this mother, seen as the very cause of the fatal car crash and as she said: 'But then [after the first visits] I did not feel that it was a place that I wanted to honour, in that way.' The bereaved mother further recounted that the family members of the other victims later put up an engraved metal plaque against a pillar carrying the names of those who lost their lives at the site. And, although the informant does not want to memorialise the place herself, she states that she has nothing against this act: 'I thought it was important, also, that all [the names] where there.'

A third informant talked about the accident site as 'the ugliest place in the world.' Even so, it is for this bereaved mother a place that she feels has to be kept clean and even beautiful. 'But still you want it to look nice there, its strange how one is but that, I feel that it, in a way, well it must exist for posterity that it was right here.' This mother further told about the importance of visiting the site of death in order to 'be close' to the place where the life of her daughter ended and to see if there was anything left at the site that might have belonged to the deceased.[10] She further mentioned that she collected leaves from the exact place where her child had laid, which she still keeps in a bowl in a cupboard at home. After the initial visit, the bereaved family members constructed a home-made wooden plaque which, with the consent of the landowner, they put up on the stone wall into which the car had crashed. Since the roadside memorial is situated in a country other than the family's home, their visits to the site are somewhat restricted, although they are still re-visiting five and a half years after the fatal accident. The roadside memorial is also decorated by friends of the deceased who live closer to the site.

The story told by a fourth surviving relative of a traffic accident illustrates a disaster which occurred close to the deceased's home and where the whole community where drawn together at the accident site shortly after the incident. Three and a half years after the death, the roadside memorial which developed at the site is now mostly visited by those other than the closest family members, except on the occasions of the deceased's birthday and the anniversary of death. But, as the informant said: 'I get happy, so to speak, when something is laying there.' The reason for this mother's limited visits to the memorial place, although she likes it when other people come and leave things there, is her feelings of ambivalence towards both the roadside memorial and the accident site as such. This was expressed as a fear of what other people would think if she 'keeps it [the roadside memorial] going' and a general reluctance to making the fatal site into a memorial place. This mother has finally found her own way of memorialising the place by planting flower bulbs before the spring season on a slope in front of the roadside memorial. 'So it is that sort of feeling that some flowers on the slope that is okay with me. A lot of bulbs that come up and so, something that so to speak can come up by itself. That looks fresh also, no dry flowers that lie littering, or so.'

For the fifth interviewee, the initiative of memorialising the accident site was initially taken by those other than the closest family members, and included flowers, lighted candles and teddy bears. Since the traffic accident happened in the same small community in which the deceased were living, many inhabitants knew them and gathered at the fatal site as soon as the word of the disaster got around. Shortly after the growth of the roadside memorial the family put up a wooden cross painted in white with the deceased's names. The site of fatality is still, two years after the deaths, often tended by the parents and grandparents of

---

10   The conception of experiencing some kind of presence of the deceased (see Bennet and Bennet 2000, Davies 2002 [1997], among others), at the fatal accident site or attached to things found at the scene, is further explored in other texts of roadside memorialisation (Hartig and Dunne 1998, Petersson 2004, Clark and Franzmann 2006, Petersson 2009a, 2009b, Klaassens at al. 2009).

the deceased. On the anniversary of the fatal accident, friends as well as unknown community members also decorate the site with flowers and notes; an act that is very much appreciated by the informant since, as she said: 'And then one knows that [they] have them in mind.' As a direct consequence of the fatal accident the community has, together with the Swedish Road Administration, recently made a foot and bicycle path beside the road.

The issue of communicating flaws in the environment where the fatal accident took place is central for the initial development, as well as the continuous upkeep, of another roadside memorial in my study. After having witnessed the accident site and realised that the sun shone in a direction that made it impossible to see properly at certain hours, the final interviewee placed flowers at the site where her husband was fatally injured as a way to 'make people realise that something has happened here and slow down their speed'. She and her children have also made contact with the Swedish Road Administration, which resulted in a decreased speed limit at the road section in question.

When a Death Becomes a Public Event

During the last two decades the Streets and Parks Department in Malmö has received an increased number of applications from private persons wanting to construct memorials on public ground. This development goes hand in hand with the growth of spontaneous memorialisation in Sweden during the same period of time (Gustafsson 1995, Pettersson 2004, Petersson 2009a). There are several examples of spontaneous and public memorials in Sweden, including the floral tributes to the Swedish Prime Minister, Olof Palme, at the site of his assassination in Stockholm in 1986. A bronze plaque was later placed in the pavement at the site and a street was named Olof Palmes Gata [Olof Palme's street].[11] A further example is the murder site of the Swedish Foreign Minister Anna Lindh who was fatally stabbed in Stockholm in 2003. The site where she gave her last public speech is now marked by a glass monument. Other public monuments include that inaugurated in 1997 for the 501 Swedes who died in the ship wreck of MS *Estonia* in 1994; and the immense memorial place created spontaneously after the fire at a meeting hall in Gothenburg which killed 63 young people attending a discotheque in 1998. A decade after the fire, on the 29th of October 2008, a great stone monument was raised in front of the restored building (Petersson and Wingren 2010 forthcoming).

The applications for memorials made to the Streets and Parks Department in Malmö could be divided into three categories: requests for memorials to be erected on the spot of a sudden death; requests for memorials to be erected on the spot of a certain incident or occasion; and requests for memorials to be erected in memory of the performance of a special person in a place associated with this persons life

---

11 For a discussion on the power of commemorative street names, see Azaryahu 1996b.

and/or work. The focus in this text is on two applications concerning memorials on sites of fatalities in the city of Malmö.[12]

The first application related to the knife murder of a nineteen year old man in a public park during the Folk Festival in the city of Malmö in the summer of 1995. At the murder site, a heap of flowers and messages addressed to the deceased soon grew large and the memorialisation lingered on for the whole winter after the incident. The application for a more permanent marker came shortly before the next year's Folk Festival and referred to the probable re-growth of the flower bed, considering the upcoming anniversary of the fatal stabbing, by stating that: 'It is impossible to forget, it is impossible to hide. We must relate [to the memorial] and it is then better to be active and do something with quality.' In addition to this argument the request refers to some well known memorialisations in Sweden, such as the bronze plaque at Palme's murder site, the international art competition that preceded the monument for the Swedes who drowned in the sinking of MS *Estonia*, as well as to the recent increase of roadside memorials on Swedish roads. A project made as a consequence of the murder and called 'A knife in our hearts', which worked against street violence with funds from the Swedish National Institute of Public Health, the former Swedish Civil Ministry and the city of Malmö, was also called attention to.[13] The application finally suggested that a sculptural composition, made specifically for this occasion by the only Swedish artist who took part in the art competition for the Estonia monument, should be raised on the site where the lethal knifing took place.

The other application refers to the fatal traffic accident that killed a well known and respected minister of the church on Malmö beach in 2005. In the request, a sculptured memorial stone with an inscription is mentioned as a possible marker of the accident site. The place of death is further presented as a future meeting point on which a yearly prize will be handed out by the late pastor's memorial fund on his birthday.

In answering the last application, the Streets and Parks Department in Malmö wrote down their earlier customs for dealing with this type of requests in order to better handle future applications.[14] Of importance to the suggestions mentioned is the custom of not focusing on the death of a certain person but on their performance when living. This, since the Streets and Parks Department does not want to take part in commemorating negative occurrences, only positive events. A related practice is to never link the memorial to the place of death. The argument for this

---

12   Quotations from these sources are my own translations of the informants' original Swedish.

13   The project 'A knife in our hearts' produced a film, which was aired on the national Swedish television channel 2, as well as a newspaper and an exhibition.

14   However, these customs only exist as an informal written document, which the author has discussed with representatives from the Streets and Parks Department in Malmö.

is that the memorial place should not remind people of, or associate to, the burial place. A third praxis is to never construct a memorial too soon after a death, since only persons that keep their importance long after their lifetime are entitled to a memorial place. Besides, the deceased must be important to society, either through their deeds or by their status, in order to be honoured by a public memorial.

Subsequently, the first application regarding the erection of a sculptural group for a little known man, at the place where he was killed, was concisely answered in the negative. The fact that the project against violence, which followed from the murder, was quite successful and in a way gave posthumous importance to the departed (or at least to the cause of his death) was apparently not enough. The second request was given both affirmative and negative responses. The negative feedback concerned the place and type of memorial, since neither the place of death nor the sculptural stone were considered appropriate in the light of the criteria of the Streets and Parks Department. Instead a 'simple memorial marker', in the form of a natural stone with an inscription and a plaque, was recommended to be placed somewhere along the beach other than the exact place of death, in memory of the lifetime achievements of the late pastor. Mentioned in the response was that the same type of memorial had earlier been used to honour the works of another important minister of the church in Malmö (who died of natural causes

Figure 8.1   The official monument in remembrance of the pastor Ingemar Simonsson, who was killed in a motor vehicle accident in 2005 on the beach in Malmö, Sweden

Source: Photograph by the author

elsewhere), and that this marker had been placed close to where this pastor often met his people.

Ritualised Activities and Material Expressions of Grief

For the bereaved interviewees in my study, first visits to the accident site seem to be a direct reaction to feelings of unreality and disbelief following the unexpected death of a loved one. By witnessing the place of fatality, the family and friends of the deceased were able to start working on their understanding of the death.[15] Additionally, even though the fatal accident site inevitably is tied to negative feelings of death and disaster, it may still function as an accessible memorial place to grieve at, and care for, before the site of the grave is complete and at especially difficult days such as the deceased's birthday and the anniversary of death. These considerations are also found in other works on spontaneous memorialisation (Horne 1997, Monger 1997, Haney et al. 1997, Hartig and Dunn 1998, Everett 2000, Clark and Cheshire 2003/2004, Clark and Franzmann 2006, Nieminen Kristofersson 2006, Klaassens et al. 2009). In an Australian study, the memorial at the accident site is even seen as more important than the memorial in the cemetery (Clark and Franzmann 2006). Whereas in my study, as well as in a study of roadside memorials in Texas, the roadside memorial is considered less important than the memorial at the site of the grave (Everett 2000, Petersson 2009a, 2009b). My interviews further showed that among the informants bereaved by road traffic accidents, four out of six rated the grave as the main memorial place while the home-based memorial, or simply the home in general, was second; two out of six rated the home as the main memorial place and the grave second; the accident site was considered the least important by all of the bereaved (Petersson 2009a, 2009b).

When estimating the intensity and importance of the various memorials places it is, however, important to note that memorial activity changes over time, as the folklorist Holly Everett (2000: 101–2) notes:

> Immediately following a fatal accident, friends and relatives, as well as witnesses, may focus their activities on the accident site. After a few years, when activity at the roadside appears to decrease or stop altogether, the greater part of memorialisation activity may have moved to the cemetery or the home.

This is also noticeable in my study, where the site of the grave is visited and tended more frequently than the roadside memorial after the initial mourning period has passed, and where candles are lighted at the home-based memorial (which may vary in scale from an arranged place for photographs and personal belongings of

---

15   For a review of the term 'witness', and a fruitful discussion of the act of witnessing, see Maddrell 2009: 684–5.

the deceased to an entire preserved bedroom) at everyday family activities such as eating or watching television as well as at nightfall (see Wojtkowiak and Venbrux, this volume, on home memorials). In my study, the bereaved also told of other significant places which they visit to remember the deceased, such as a favourite restaurant or a special glade for picking mushrooms. Such places, which 'evoke affection' since one has repetitive contact with them and create interpersonal networks at them, could be explained by what the geographer Yi-Fu Tuan (1974: 236–45) calls 'fields of care' (Nieminen Kristofersson 2005, 2006, Petersson 2009a, 2009b).[16]

The recurring activities at these various memorial places may, as the bereaved father above recognised, function as 'a small ritual', in which given patterns and symbolic gestures, like the lighting of candles and leaving of flowers (Gustafsson 1995, Pettersson 1997, 2004), can serve as tools with which to deal with initial intense emotions of grief and difficult thoughts of the accident (Haney et al. 1997, Petersson 2009a).[17] By expressing their feelings through material things the bereaved also make these emotions lasting and visible to others (Durkheim 2001 [1912], Aynsley et al. 1999, Hallam and Hockey 2001, Everett 2002, Petersson 2004, 2009a, Klaassens et al. 2009, MacConville and McQuillan 2009).

Rites can also be considered to affect the 'collective consciousness' of a society by supporting and reaffirming social and moral values and thus preventing them 'from fading from memory' (Durkheim 2001 [1912]: 279–80). Grief rites may further recreate shattered feelings of safety and consistency in a group of people diminished by the death of one of its members. By grieving together, a family can thus counteract the attenuation of the family group and reinstate their feelings of security and togetherness (Durkheim 2001 [1912]). This also concerns groups of people other than the family, such as for instance a community, a network of friends or class mates (Nieminen Kristofersson 2006) or even strangers feeling a connection with those who have died a violent death (Grider 2006).

The sociologist Tony Walter (2006) notes that some disasters even unify an entire nation by threatening not only life, but also moral and social order as well as the feelings of security that people associate with modern society (see also Pettersson 1997, 2004). The production of a spontaneous memorial may thus be seen as an emerging grief rite dealing with feelings of heightened insecurity, eroded cultural values and perceived threats to the continued existence of society (Haney et al. 1997).

In a study of Australian roadside memorials Robert Smith (1999: 105) also recognises that the placing of a memorial at an accident site can be seen as a manifestation of inadequate surroundings: 'Here is also where there emerges

---

16   Whereas the cemetery, in being a formal, sacred and ideal place, may be seen as a 'public symbol' in Tuan's (1974: 236–7) sense of the term.

17   For a wider definition of rite and ritual see Petersson 2004.

the tension with road administrators and politicians over their responsibility for providing an adequate facility, in some cases even their liability.' Smith's point is also valid in my study where the informants from the Swedish Road Administration expressed a concern for the recent increase of roadside memorials on Swedish roads (Petersson 2006)[18] and where some of the bereaved wanted the roadside memorial to give attention to flaws in the road milieu (Petersson 2009a). Similar thoughts are also evident in other studies of roadside memorialisation (Monger 1997, Haney et al. 1997, Everett 2000, 2002, Hartig and Dunn 1998, Clark and Cheshire 2003/2004, Clark and Franzmann 2006, Klaassens et al. 2009, MacConville and McQuillan 2009).

On an individual and psychological level, the creation of a meaningful memorial place, which can be revisited, grieved at and cared for, could also help the bereaved person to transform disruptive feelings of pain and powerlessness into something slightly more graspable and manageable. Such a transformation, involving encounters with the material environment, may be discussed in the light of what the anthropologist Jean-Pierre Warnier (2001: 13–17) calls 'the three media of symbolization'. Warnier's reasoning builds on the psychoanalyst Serge Tisseron's theory of *symbolisation*, in which the subject has to first 'tame' the content and nature of an unexpected and emotionally disturbing experience in order to, at a later stage, be able to bring it to mind (Warnier 2001: 14). To tame an experience, the subject uses three interdependent media, which are all present in various degrees in each process of symbolisation:

1. Through the body and its senses, called by Warnier (2001: 14–15) the 'sensori-motor media', one is able to domesticate a terrible event by physically reproducing and reshaping it. Hence, to imitate or replicate the course of events, without the damaging consequences of the actual occasion, helps a person to weaken the terrifying experience.
2. Warnier (2001: 15) further states that by being a fairly permanent trace of something or someone, which can be looked at, touched or felt over and over again when the transient moment is long gone, the 'media of images' (such as photographs, drawings and other material things) establishes our memory of a certain event and gives us a sense of continuity (see also: Aynsley et al. 1999, Hallam and Hockey 2001).[19]

---

18   However, the southern part of the Swedish Road Administration has recently decided not to remove roadside memorials although they formally are illegal constructions placed on state owned or private property.

19   Richard Schechner (1995: 232), a professor of performance studies, correspondingly speaks of 'low-level ritualising', as in the repetitive broadcasting of images from catastrophe areas on television, as transforming the shock of 'first time' or original violence into something weaker and graspable. See also Walter's (2006) discussion of how the news media reconstructs the moral and spiritual order of a society that death threatens to destroy.

3. Finally, what Warnier (2001: 15–16) calls the 'media of words' helps us recall events, facts, images and ideas at will and to communicate them, as well as critically discuss them with others.

To produce and visit a spontaneous memorial, like a roadside memorial (Everett 2002, MacConville and McQuillan 2009), or a memorial in the cemetery (Francis et al. 2001, 2005) or a cairn (Maddrell 2009), may further be held to aid the bereaved to maintain the memory of the deceased and to create a 'continuing bond' to the departed (Silverman and Nickman 1996: 350, Valentine 2008). To talk about the deceased with persons who knew him or her also helps the bereaved to integrate the memory of the deceased into their lives (Walter 1996b).

Among others, the geographer Avril Maddrell (2009) points out, memorial practices like these reflect a relatively new approach to mourning in modern Western society which reveals a way of living with grief by creating new relations to the deceased instead of trying to seek an end to mourning by letting go of the past (Walter 1996b, Klass et al. 1996, Francis et al. 2001, Everett 2002, Valentine 2008). When the initial mourning period has passed, the relatives and friends of the deceased may no longer need to work on the event of the sudden and unexpected death as often as before, but they will never fully understand and accept it. The spontaneous memorial can therefore continue to fill an important function on particularly traumatic days, such as the deceased's birthday or anniversary of death.

The Abject Spontaneous Memorial

Even though a spontaneous memorial may be a central place of grief and remembrance for the family and friends of the deceased, its very appearance can for others evoke just the opposite of a constructive experience (Ross 1998, Petersson 2004, Petersson 2006, Clark and Franzmann 2006, Petersson 2009b). In their study on the authority of grief, presence and place in roadside memorialisation, Jennifer Clark and Majella Franzmann (2006: 588) give an example of this, when a passer-by criticises a mother tending to the roadside memorial of her son with the words: 'I know this is sad, but it's not fair on me to have to look at these flowers'. As was illustrated earlier in this text, the accident site is, for the bereaved, associated with negative feelings of death and disaster and some of the survivors of road crash victims may even want to avoid the site, even though they initially felt a need to witness it (Klaassens et al. 2009).

In my interviews with informants from the Swedish Union of Transportation, one person told of a fatal car crash in front of her house that changed the view from her kitchen window indefinitely. The roadside memorial, placed on the site of death, is for this informant a constant reminder of the night that the accident occurred: 'I can't let that night with the lads in the car go. I see that car in front of me, and the lads inside with their seatbelts on.' Another interviewee, a former lorry driver, explains his aversion to roadside memorials in a similar way: 'As a professional driver and colleague to this [man]. I don't know him, but it [he] is within [the workforce], it was a lorry driver that had caused it [the fatal accident]. [...] he was reminded every time he drove by [...] eh, daily and so on. So he must have had it really tough, to work on this [his] trauma in that way.' The involuntary reminding of the accident, inflicted by the sight of the roadside memorial, seemed for both of the interviewed informants to be the very cause of their failure to work on the trauma. They even expressed, in an accusing way, the reflection that the family of the deceased had not 'moved on'. For these interviewees, the common suggestion that 'life goes on' was seen as rejected by the bereaved through their continuous maintenance of the roadside memorial (Klaassens et al. 2009).[20]

The reactions to spontaneous memorials mentioned above may perhaps be seen as tied to ways of using repression to handle one's fears of death and of life's temporariness and fragility, which could be further explained by what the philosopher and psychoanalyst Julia Kristeva (1982) calls *abjection*.[21] For Kristeva (1982), abjection is something that is incomprehensibly and confusingly horrible from which one does not cease to separate from. It is 'a land of oblivion' that is constantly remembered (Kristeva 1982: 8). The repression of this 'forgotten time' can suddenly break through in a moment of revelation, terribly clear in all its ambiguity, when something contradictory to the expected suddenly appears or when the expected is turned upside down (Kristeva 1982).

The notion of abjection is suggested by Kristeva (1982) as coextensive with social and symbolic orders (like law and religion), on an individual as well as a collective level. Abjection should further be seen as a universal phenomenon, constituted in the wake of the human ordering of life, although what is considered abjection varies from one 'symbolic system' to another (Kristeva 1982: 68). Kristeva (1982) additionally proposes that abjection accompanies all religious structures and even if it always refers to the exclusion of a substance it determines different forms in different religions; it appears as a rite of defilement and pollution in paganism and as exclusion, taboo or transgression in monotheistic religions. The various means of purifying the abject thus runs through the history of religion (Kristeva 1982).

As an example, most cultures and societies have considered the dead body as a threat to the orders and structures of society and have therefore used various rituals of purification and transition in order to control it. As Kristeva (1982: 109) points

---

20   A similar debate of what is considered expected and normal expressions of grief is found in a study of the second press discussion of salvaging the victims of the MS *Estonia* (Reimers 2000) as well as in a study of crisis groups and spontaneous support in connection with the catastrophic fire at a meeting hall in Gothenburg (Nieminen Kristofersson 2006).

21   To *abject* means to cast off or away, to cast out, exclude or reject. An *abject* is an outcast, exile, a degraded or downtrodden person. *Abjection* is the state or condition of being cast down or the act of casting off or away. *Oxford English Dictionary* [Online]. Available at: http://dictionary.oed.com [accessed: 17 August 2009].

out: 'Burial is a means of purification'. Rites of passage, such as the funeral rite, allow for the dead body to go through a purifying *preliminal* phase, an ambiguous *liminal* phase, and a reintegrating *postliminal* phase in which the corpse is let back into society when confined to the burial place (Gennep 1960 [1909], Turner 1970 [1967]).

Kristeva (1982) also suggests that without a proper context the dead body does not signify death as cultivated; it merely shows us the improper or unclean side of life that we constantly try to reject in order to live. However, it is not a lack of sanitation or health that causes the abjection, states Kristeva (1982), it is the disturbance of identity, system and order.[22] When 'seen without God and outside of science', in other words without the specific setting of religion or science, the dead body is merely 'death infecting life' (Kristeva 1982: 4).

A similar communicative difficulty may appear when we come into contact with death outside the official context of a cemetery, a hospital, or a morgue (Petersson 2004, Walter 2007, Petersson 2009a). Such as when the everyday setting of a spontaneous memorial somehow collides with the message of death and grief that is conveyed by the fresh flowers and lighted candles placed by a site of fatality. In addition, while you can consciously avoid a visit to the cemetery, or at least prepare yourself for an expected encounter, the unpredicted sight of a roadside memorial may suddenly bring about repressed feelings of pain and anger connected to previous experiences of death (Petersson 2004, 2006, 2009a, 2009b). And so 'The borders of the cemetery, originally enclosing the churchyard to separate the consecrated earth from the unconsecrated continues, in current secular and large-scale cemeteries, to keep death in order inside well-trimmed hedges, straight grids of paths and proper grave lots' (Petersson 2004: 57).

Material expressions of grief may however not only be seen as threatening some people's borders between the space of life and death, they can also bee seen as threatening one's ideals of society, culture and tradition. In her landmark work from 1937 the Swedish ethnologist Louise Hagberg (1937: 488–9) discusses the modernisation of Swedish cemeteries and tombstones at the end of the nineteenth century, with words such as 'disgraceful' and 'banal' compared to the 'beautiful old memorials' [my translations]. A current example of this was seen when a distinguished landscape architect, at a recent church antiquary conference in Sweden, suggested a distribution of folders to grave owners containing recommendations for certain kinds of vegetation considered as part of our cultural heritage and considered by her as more appropriate in old cemeteries than the existing planting placed on the graves by the deceased's relatives (Petersson 2004). This attitude is also visible in the response from the Streets and Parks Department in Malmö advocating a 'simple memorial marker' instead of the sculptured stone preferred by the bereaved of a well known deceased pastor in Malmö.

---

22 This brings to mind the way the shift from a religiously ordered to a scientific world view changed our relationship to death and the afterlife and consequently the way we care for and where we place our dead (McManners 1981, Etlin 1984, Petersson 2004).

For the informants in my study, the practice of decorating the grave with photographs and items such as toys was considered to be a recent custom influenced by other cultures and traditions, due to migration and long distance travel, as was the practice of memorialising the site of a fatality (Petersson 2009a). In a discussion of grave decorations in a magazine for cemetery issues, a cemetery worker sums the issue up: 'It seems like pure settlers' joy to some. Everyone has to be an individualist and the imagination has no limits. They plant trees and put out statues. It seems as if they want to use the grave as a cult place or an altar.' [my translation] (Kyrkogården 2002: 6, Petersson 2004: 46). Hence, in some cases we could perhaps see the condemnation of spontaneous memorials, in part, as a manifestation of resistance to the unfamiliar.

Conclusion

By their very placing, as well as their symbolic expression, spontaneous memorials bring about reflections on the sacred and the profane and trigger questions of public and private interests. The look and location of spontaneous memorials sometimes even challenge social, cultural and ideological notions of what a proper place of death should be like. Spontaneous memorials may furthermore evoke unwanted thoughts of the ever-present powers of death and turn everyday life upside down by pointing to its temporariness and fragility. The unexpected and violent deaths that spontaneous memorials commemorate could also be seen as revealing cracks in the moral and social order that people associate with modern society and thus threaten people's feelings of security and belonging to that society. The custom of the Streets and Parks Department in Malmö of not letting public memorials be associated with the burial place, nor commemorate harmful events such as a violent death or a fatal accident, could be seen as a measure of precaution taken to prevent various types of negative experiences, or feelings of abjection, described above.[23]

For the bereaved, the accident site is a significant place of negotiation and interaction: for some just the first time after the death, when witnessing the site where the fatal accident took place in order to bring the incomprehensible event in to their minds; for others, the production and maintenance of a memorial place at the accident site, as well as the recurring visits to it, may function as a grief rite as well as creating a potential field of care. The materialised expressions of grief engaged in this practice can also help the bereaved to maintain the memory of the deceased and create a continuing bond to the departed.

Bearing all this in mind, we may conclude by recognising spontaneous memorials as significant places for expressing, communicating, experiencing, understanding and sometimes even debating different ideas of death and grief.

---

23   For a similar way of 'purifying' the public sphere, although tied to the elimination of religious symbols instead of symbols of death, see Howe 2009.

As the architectural historian Richard Etlin earlier stated (1984: ix), the material environment of the eighteenth century cemetery did not simply mirror a new understanding of death; it also played a role in forming nascent ideas and emotions. I will end this chapter by proposing that this is also the case with the material expression of spontaneous memorials.

References

Ariès, P. 1994 [1976]. *Western Attitudes Toward Death: From the Middle Ages to the Present*, translated by Patricia M. Ranum. London and New York: Marion Boyars.

Aynsley, J., Breward, C. and Kwint, M. 1999. *Material Memories. Design and Evocation*. Oxford and New York: Berg.

Azaryahu, M. 1996a. The spontaneous formation of memorial space. The case of Kikar Rabin, Tel Aviv. *Area*, 28(4), 501–13.

Azaryahu, M. 1996b. The power of commemorative street names. *Environment and Planning D: Society and Space*, 14, 311–30.

Barrera, A. 1997. Mexican-American roadside crosses in Starr County, in *Hecho en Tejas—Texas-Mexican Folk Arts and Crafts*, edited by Joe S. Graham. Denton, Texas: University of North Texas Press, 278–92.

Bauman, Z. 1992. *Mortality, Immortality and Other Life Strategies*. Cambridge: Polity Press.

Bennet, G. and Bennet, K.M. 2000. The presence of the dead: an empirical study. *Mortality* 2(5), 139–57.

Burk, A.L. 2003. Private griefs, public places. *Political Geography*, 22, 217–33.

De Certeau, M. 2002 [1984]. *The Practice of Everyday Life*. Berkeley: University of California Press.

Clark, J. and Cheshire, A. 2003/2004. RIP by the roadside: a comparative study of roadside memorials in New South Wales, Australia, and Texas, United States. *Omega – Journal of Death and Dying*, 48(3), 229–48

Clark, J. and Franzmann, M. 2006. Authority from grief, presence and place in the making of roadside memorials. *Death Studies*, 30, 579–99.

Davies, D. 2002 [1997]. *Death, Ritual and Belief: The Rhetoric of Funerary Rites*, second edition. London and New York: Continuum.

Designing Places for Memory and Meaning in the Contemporary Urban Landscape. *LTJ-Fakulteten, Alnarp* [Online]. Available at: www.ltj.slu.se/6/O6_for_1.html [accessed: 27 October 2009].

Doss, E. 2002. Death, art and memory in the public sphere: the visual and material culture of grief in contemporary America. *Mortality*, 7(1), 63–82.

Durkheim, É. 2001 [1912]. *The Elementary Forms of Religious Life*, translated by Carol Cosman. Oxford and New York: Oxford University Press.

Etlin, R.A. 1984. *The Architecture of Death. The Transformation of the Cemetery in Eighteenth-Century Paris*. Cambridge, Massachusetts and London: The MIT Press.

Everett, H. 2000. Roadside crosses and memorial complexes in Texas. *Folklore*, 111, 91–103.

Everett, H. 2002. *Roadside Crosses in Contemporary Memorial Culture*. Denton, Texas: University of North Texas Press.

Francis, D., Kellaher, L. and Neophytou, G. 2001. The cemetery: the evidence of continuing bonds, in *Grief, Mourning and Death Ritual*, edited by J. Hockey, J. Katz and N. Small. Buckingham and Philadelphia: Open University Press, 226–36.

Francis, D., Kellaher, L. and Neophytou, G. 2005. *The Secret Cemetery*. Oxford: Berg.

Grider, S. 2006. Spontaneous shrines and public memorialization, in *Death and Religion in a Changing World*, edited by K. Garces–Foley. Armonk, New York: M.E. Sharpe, 246–64.

Gustafsson, G. 1995. Svenska folket, Estonia och religionen, in *Två undersökningar om Estonia och religionen*, Religionssociologiska studier 1. Lund: Lunds Universitet, 7–46.

Hagberg, L. 1937. *När döden gästar. Svenska folkseder och svensk folktro i samband med död och begravning*. Stockholm: Wahlström & Widstrand.

Hallam, E. and Hockey, J. 2001. *Death, Memory and Materiel Culture*. Oxford and New York: Berg.

Haney, C.A., Leimer, C. and Lowery, J. 1997. Spontaneous memorialisation: violent death and emerging mourning ritual. *Omega – Journal of Death and Dying*, 35, 159–71.

Hartig, K.V. and Dunn, K.M. 1998. Roadside memorials: interpreting new deathscapes in Newcastle, New South Wales. *Australian Geographical Studies*, 36(1), 5–20.

Henzel, C. 1995. Cruces in the roadside landscape of Northeastern New Mexico. *Journal of Cultural Geography*, 11(2), 93–106.

Horne, D. de L. 1997. Treatment of pain, fear and loss following a road accident: a case study, in *The Aftermath of Road Accidents: Psychological, Social and Legal Consequences of an Everyday Trauma*, edited by M. Mitchell. London and New York: Routledge, 194–5.

Howe, N. 2009. Secular iconoclasm: purifying, privatizing, and profaning public faith. *Social and Cultural Geography*, 10(6), 639–56.

Johnson, P. 2008. The modern cemetery: a design for life. *Social and Cultural Geography*, 9(7), 777–90.

Klaassens, M., Groote P. and Huigen, P.P.P. 2009. Roadside memorials from a geographical perspective. *Mortality*, 14(2), 187–201.

Klass, D., Silverman, P.R. and Nickman, S.L. 1996. Preface, in *Continuing Bonds: New Understandings of Grief*, edited by D. Klass, P.R. Silverman and S.L. Nickman. London and Philadelphia: Taylor & Frances, xvii–xxi.

Kong, L. 1999. Cemeteries and columbaria, memorials and mausoleums: narrative and interpretation in the study of deathscapes in geography. *Australian Geographical Studies*, 37(1), 1–10.

Kristeva, J. 1982. *Powers of Horror: An Essay on Abjection*, translated by Leon S. Roudiez. New York: Columbia University Press.

Kyrkogården, 7. 2002. Garnsvikens begravningsplats, 4–6.

Lindahl, G. 1969. *Grav och Rum: svenskt gravskick från medeltiden till 1800-talets slut*. Stockholm: Almqvist & Wiksell.

Lundquist, K. 1992. Från beteshage till trädgård – kyrkogårdens historia, in *Kyrkogårdens Gröna Kulturarv*, Stad och land 103. Alnarp: Movium.

MacConville, U. and McQuillan, R. 2009. Remembering the Dead: Roadside Memorials in Ireland, in *Dying, Assisted Death and Mourning*, edited by A. Kasher. Amsterdam and New York: Rodopi, 135–55.

Maddrell, A. 2009. A place for grief and belief: the Witness Cairn, Isle of Whithorn, Galloway, Scotland. *Social and Cultural Geography*, 10(6), 675–93.

McManners, J. 1981. *Death and the Enlightenment. Changing Attitudes to Death among Christians and Unbelievers in Eighteenth-century France*. Oxford and New York: Oxford University Press.

Monger, G. 1997. Modern wayside shrines. *Folklore*, 108, 113–14.

Nieminen Kristofersson, T. 2005. *Places of Dying, Rituals of Mourning*. Paper for the Inaugural Nordic Geographers Meeting: Power over Time–Space, Lund, Sweden, 10–14 May 2005 [Online]. Available at: http://www.keg.lu.se/ngm/html/papers/paper_nieminen.pdf [accessed: 27 July 2009].

Nieminen Kristofersson, T. 2006. Från olycksplats till minnesplats, in *Katastrof! Olyckans geografi och antropologi*, Ymer 126, edited by T. Lundén. Stockholm: Svenska sällskapet för antropologi och geografi, 269–80.

Oxford English Dictionary [Online]. Available at: www.dictionary.oed.com [accessed: 17 August 2009].

Petersson, A. 2004. *The Presence of the Absent. Memorials and Places of Ritual*, PhLic diss. Lund: Department of Architecture, University of Lund.

Petersson, A. 2006. A proper place of death?, in *Architects in the Twenty-First Century – Agents of Change?*, Nordic Association for Architectural Research Annual Symposium 2006, edited by K. Rivad. Copenhagen: The Royal Danish Academy of Fine Arts, School of Architecture, 110–17.

Petersson, A. 2009a. Swedish *Offerkast* and recent roadside memorials. *Folklore*, 120, 75–91.

Petersson, A. 2009b. The Altar of the Dead: a temporal space for memory and meaning in the contemporary urban landscape, in *Nature, Space and the Sacred: Interdisciplinary Perspectives*, edited by S. Bergmann, P. Scott, M. Jansdotter, and H. Bedford Strohm. Farnham: Ashgate, 131–44.

Petersson, A. and Wingren, C. (2010, forthcoming). Designing a memorial place: Continuing care, passage landscapes and memories of the future. *Mortality*.

Pettersson, P. 1997. *Implicit Religious Relations Turned Explicit. The Church of Sweden as Service Provider in the Context of the Estonia Disaster*. Paper for the Conference of the International Society for the Sociology of Religion (ISSR), Toulouse, France, 7–11 July 1997.

Pettersson, P. 2004. *Ten Years after the MS* Estonia *Disaster. The Development of the Church of Sweden as Public Provider of Rituals – A Complementary Welfare Function.* Paper for the Seventeenth Nordic Conference in Sociology of Religion, Reykjavik, Iceland, 19–22 August 2004.

Reimers, E. 2000. *Vad lindrar de sörjandes hjärtan? Aktörer, ämnen, motiv och värderingar i det andra pressamtalet om bärgning av offren från Estonia*, Medie- och kommunikationsvetenskap 4. Umeå: Institutionen för Kultur och Medier, Umeå Universitet.

Ross, C. 1998. Roadside memorials: public policy vs. private expression. *American City and County* [Online]. Available at: www.americancityandcounty.com/mag/government_roadside_memorials_public. [accessed: 31 January 2006].

Schechner, R. 1995. *The Future of Ritual. Writings on Culture and Performance.* London and New York: Routledge.

Seale, C. and Van der Geest, S. 2004. Good and bad death: introduction. *Social Science and Medicine*, 58, 883–5.

Silverman, P.R. and Nickman, S.L. 1996. Concluding thoughts, in *Continuing Bonds: New Understandings of Grief*, edited by D. Klass, P.R. Silverman, and S.L. Nickman. London and Philadelphia: Taylor & Francis, 349–55.

Smith, R.J. 1999. Roadside memorials: some Australian examples. *Folklore*, 110, 103–5.

Tuan, Y.-F. 1974. Space and place: humanistic perspective. *Progress in Geography: International Reviews of Current Research*, 6. London: Edward Arnold and New York: St Martin's Press.

Turner, V. 1970 [1967]. *The Forest of Symbols: Aspects of Ndembu Ritual.* Ithaca: Cornell University Press.

Valentine, C. 2008. *Bereavement Narratives. Continuing Bonds in the twenty-first Century.* London and New York: Taylor & Francis.

Van Gennep, A. 1960 [1909] (translated by Monika B. Vizedom and Gabrielle L. Caffee). *The Rites of Passage.* Chicago: The University of Chicago Press.

Wall, B. 2000. *Gravskick i förändring, tradition och visioner.* Stockholm: Svenska kyrkans Församlingsförbund.

Walter, T. 1996a. Funeral flowers: a response to Drury. *Folklore*, 107, 106–7.

Walter, T. 1996b. A new model of grief: bereavement and biography. *Mortality*, 1(1), 7–25.

Walter, T. 2006. Disaster, modernity, and the media, in *Death and Religion in a Changing World*, edited by K. Garces-Foley. Armonk, NY: M.E. Sharpe, 265–82.

Walter, T. 2007. *Space, Place and Death.* Paper presented at the 8th conference on: The Social Context of Death, Dying and Disposal, Bath, UK, 13–15 September, 2007.

Warnier, J.-P. 2001. A praxeological approach to subjectivation in a material world. *Journal of Material Culture*, 6(1), 5–24.

Chapter 9

Bringing the Dead Back Home: Urban Public Spaces as Sites for New Patterns of Mourning and Memorialisation

Leonie Kellaher and Ken Worpole

Introduction

This chapter seeks to describe, understand and explain the rise of new forms of memorialisation emerging in the UK urban context, focusing particularly on the increasing temporal and spatial separation of the forms of bodily disposal and the rituals associated with the commemoration of the deceased. It is divided into four sections: i) re-positioning commemoration – the wider context of urban memorialisation; ii) traditional and modern forms of disposal and memorialisation; iii) explaining shifts towards 'cenotaphisation'; and iv) bringing the dead back home.

Re-positioning Commemoration. The Wider Context of Urban Memorialisation

Memorial benches, along with dedicated trees and their inscribed plaques, have long been appearing across the UK to commemorate someone of public importance in a prominent place though sometimes this also occurs in privileged sites such as National Trust gardens, private golf courses and other places defined by membership or association. However, what now engages and sometimes confronts people as they go about their everyday journeys across local, urban space are tributes to the more recently deceased for whom no special distinction is claimed other than an association with the locality. Unlike the more conspicuous phenomenon of roadside memorial shrines marking accidental fatalities or violent criminal deaths, little attention has been paid to these lower-key forms of contemporary memorialisation. And yet, benches, trees and plaques, with their more varied distributions and situations than appears to be the case for roadside shrines, suggest a new wave of patterning and inscription of public space across towns and cities that is more far-reaching and may be rooted in wider communitarian and societal impulses. This is the principal argument of this chapter.

We can distinguish between these two forms of contemporary memorialisation on several counts. Firstly, that the positioning of roadside shrines and of benches and trees speaks of two distinct liminalities. The kerbs and verges on which spontaneous shrines are invariably established are civil engineering devices to separate traffic and people, and are marked and re-defined by the shrine. Attention is drawn to them as the shielding barriers they are intended to be, but at the same time they are revealed as the failed thresholds they have become in these particular tragic instances. The transitional state bridging the life-changing event and some future state of re-incorporation in society, which Van Gennep identified as liminality, is so compressed, if not expunged by the shrine, that its near absence may block the moves towards re-incorporation for the survivors, or indeed the victim, that Van Gennep's Rites de Passage lays out (Van Gennep [1909] 1960). The shrine may extend spatially and, with objects accumulating over time, become a fixture pointing an accusatory finger at the contingent world through the subsequent detritus of decaying floral tributes and rain-sodden toys (Maddrell 2009b).

In contrast, the spaces in which dedicated benches and trees are situated can be read as liminal in a more graduated sense, providing an interval in which the deceased and the survivor may start to take their places, albeit with changed identities, back in their community. These memorials are frequently positioned on footpaths, rather than arterial highways, and sometimes on open grass in places with a vista. Rest, tranquility and a certain measured continuity appear to be intended. For both benches and trees, their installation is seldom spontaneous, requiring permissions and sometimes guiding oversight from regulators such as the local authority, as well as financial and emotional support from family and community. The artifacts themselves are ubiquitous, having everyday qualities that would be equally appropriate in an ordered domestic setting. The – generally wooden – benches are crafted from the material provided by their alternative, the growing tree; one carries a message of stability and stasis, as well as human investment in its shaping and emplacement; the other signifies a trust that nature will grow and regenerate so that the memory can be sustained and develop.

These contrasting liminalities carry quite different meanings for memorialisation. The roadside shrine represents an end point and appears to have more in common with the emotionally charged milieu sometimes associated with pilgrimage (Danforth 1989, Walter 1999) whereas the bench or tree in the urban setting is to be passed, by-passed, visited and tended with more or less devotion and frequency. It is true that benches and trees are places at which floral tributes are placed on anniversaries, as well as photographs, cards and sometimes open letters, but these are generally restrained, temporary commemorations. It seems fair to say then, that the bench-tree-plaque phenomenon has more in common with an established grave plot in a cemetery (Francis et al. 2005, Worpole 2003) than does the roadside shrine which, as has already been observed, represents a sudden void or rupture created in the social world, not a translated presence (Hallam and Hockey 2001, see also Petersson, this volume).

The processes that generally lead to bench, tree and plaque emplacements may belie preceding personal and emotional turmoil, but the appearance of the memorial signals something of a carefully considered investment in a public demonstration that the deceased is remembered and that this is to be measured out over time in a way that involves others, known and unknown. Such involvement may simply be one of catching the eye and perhaps the imagination. It may be a restrained call for sympathy and solidarity, but mostly the low-key nature of the memorial and the modest gestures of remembrance that are enacted from time to time call for a sharing of the knowledge that a certain person was connected with, and is remembered as, part of the particular place. Many memorials implicitly convey a desire to offer comfort and beauty to the setting and thus to make a contribution from the deceased and their survivors to the public realm.

Having suggested that benches, trees and plaques are the stuff of long-standing cemetery culture, these 'gifts' to the public well-being also suggest something different; something that is not sequestered and apart in the way that cemeteries are. One starting point for explaining what these urban memorials signify is, nonetheless, the correspondence with cemetery practice already mentioned. We are drawn to ask whether these new proliferations and clusterings are associated with – and to some degree related to – the changing situation for many cemeteries, as land space and burial, along with other forms of deposition for the remains of deceased populations, has become problematic. But explanation cannot rest simply upon this possibility. Previous research (Francis et al. 2005, Worpole 2003) and preliminary observations from current field-work[1] indicate another feature associated with benches and tree memorials: it is that the corporeal remains are not to be placed at the same spot as the memorial; indeed many authorities refuse explicit requests. They may admit that ashes are sometimes scattered 'illicitly' near the memorial – and research with mourners seeking to place ashes confirms this – but the practice is not officially sanctioned as appropriate for park-like public spaces (Hockey et al. 2007). Nonetheless, such dissociations between the corporeal remains and the memorial may be taking hold as a likely trend for future memorialisation practice, and for which we are proposing and exploring the concept of 'cenotaphisation'.

It is within these frames that this chapter explores the space that memorial benches, trees and plaques occupy in past, present and future formulations as to how the dead can be imaginatively located and recalled by those who survive. We explore how places of abode for the dead, as represented materially by their physical remnants and symbolically in the more or less coded, more or less proximate, tributes arranged by bereaved survivors, may permit recollections that enhance remembrance for contemporary, urban, ostensibly secular societies.

---

1   Local fieldwork undertaken in the London Boroughs of Hackney and Islington, 2008–9.

Edging Towards a Fitting Place

While our research on memorial benches and trees is local, there is sufficient evidence in the public domain for many towns and cities, to suggest that the trend has important implications in the much larger question as to how people dispose of and commemorate the dead in contemporary urban Britain. This is to say, how many people in the UK now struggle to find the right fit for the deceased individual and the social person to occupy in an afterlife no longer shaped by commonly held religious beliefs (Davies 1997). There seem to be two closely linked elements in this effort. The first is a fitting disposal decision that will permit a second element to work, and to work well into the future that the survivors will experience; in other words memory and memorialisation. This is particularly relevant to what we see as the growing spatial and psychic separation between the rituals and symbols of memorialisation and the physical disposal of the material remains of the dead.

The latter process now ranges from burial in a cemetery or churchyard, dispersal of cremated remains by interment or scattering at the originating crematorium or elsewhere, or burial in a woodland cemetery in a grave which does not allow for any permanent form of marking or memorial. Our earlier research concerning historical and contemporary practices in burial sites and cemeteries, where marking and identifying the place of deposition was an urgent imperative, has alerted us to the likely strength of a trend towards separation of mortal remains and memorial marker that contradicts earlier, near universal impulses to mark the presence of mortal remains as precisely as possible. This trend may still be obscured within a quotidian ordinariness in which the dead are brought closer to home in a new way that may transform public urban space by shifting its meaning from the municipal to the vernacular to the numinous (Maddrell 2009a).

What we are endeavouring to understand is whether, and if so how, new practices may be related to preceding custom and beliefs. Traditionally, the corpse had to be contained in order to limit its polluting influences – physically and spiritually. The urge to protect the remains must always have been present at the personal level but increasingly it enters public aspects of disposal. We now ask how these containments may be effected and experienced as the dead are increasingly released from the sequestered and bounded spaces reserved for them up until the end of the twentieth century. Are the human remains still 'contained' and protected within the ground where they are interred or where the ashes finally come to rest? What does the process of inscription and, finally, the ongoing process of memorialisation now 'contain' and protect? At one time containment and protection occurred in exactly the same place, but today they are increasingly separated, spatially and over time. Where do the dead abide today?

Traditional and Modern Forms of Disposal and Memorialisation

In this section we look back at the changing historical forms of disposal and memorialisation within the broad public discourse of traditional and modern rituals surrounding death, disposal and commemoration. The changes represented by the historical sequence encompassing the village churchyard, the Victorian cemetery, the modern crematorium and recently established woodland burial sites reveal the progressive bifurcation; part temporal, part spatial, of funerary rituals, enactments and processes of memorialisation that were once highly integrated. In the village churchyard, the norm for burial was 'in perpetuity', often marked by inscription at the site, so that the consecrated grave plot became the site of ongoing personal and community memorialisation. The same was true of the Victorian cemetery, though this entailed a stricter and legislated insistence on the grave space being inviolate for all time (as it was realised that the same churchyard space may have been used over the centuries for many burials, and this was now regarded as unacceptable). With the rise of cremation begins the process of a separation between the bodily remains and the place of inscription, notably with books of remembrance some distance from the ashes scattered in the crematorium grounds. Later, as the twentieth century moved towards a close and people began to take ashes away for private ceremonies, spatial and temporal dissociation accelerated. Finally, with woodland burial, though the grave space is a focus for memory, little or no inscription is allowed and, as with cremation, forms of inscription and memorialisation begin to occur in ceremonies and rituals enacted closer to the urban home such as in the commissioning of memorial benches or the planting of trees.

The Village Churchyard

Traditionally it might be assumed that a churchyard, cemetery or burial ground is a clearly demarcated and bordered space consecrated or in some other way made sacrosanct within, and governed by, formal and informal rules, customs and practices which do not obtain outside the walls or entry gates. Burial in a village churchyard in Britain has always occurred within a theological framework assuming that the body stays where it is 'laid to rest' in perpetuity, awaiting a time when the spirit ascends to a higher sphere, to final resurrection or achieves some other kind of transformation. Thus the churchyard becomes a kind of settled, almost essentialist community, and a moral anchor of the community around it. The burial ground was both spatially integral and central to the townscape in which it was a foundational element (Worpole 2003).

In pre-reformation times and then in post-reformation Britain up to the late eighteenth century, perpetuity was synonymous with a theologically declaimed eternity into which the soul of the departed entered at death to endure purgatorial suffering and await final judgement. Burial and its rituals were to reflect this interlude, with ecclesiastic authorities determining and controlling the beliefs and practices that underpinned the disposal of corporeal remains. Integral to these practices was a belief in resurrection – the central tenet of Christianity. Ritual to commit to earth the body, which had held the soul and in life had enacted the good and evil that would be judged, signified preparation for the eternity to follow resurrection. Thus the principle that corporeal remains should be undisturbed operated to reassure virtuous and law abiding members of society that they were well placed to be favourably judged and reconstituted, with their physical bodies intact once more, for an eternity of celestial bliss. For the others, damnation could be prematurely visited upon them by the ecclesiastic imposition of an unconsecrated grave, leading to insecure burial that could not ensure full corporeal and spiritual resurrection. Grave robbery and the use of the stolen body for medical anatomy also posed a serious threat to the resurrectionary ideal that mortal remains should rest complete and undisturbed until the day of reckoning (Richardson 1988).

Nevertheless, the pre-reformation and then the post-reformation established church has constantly practised the re-use of grave space in what has often been a finite and very small area of ground commonly known as 'God's acre'. Pre-reformation, pressure on burial space in a predominantly rural society may not have necessitated much re-use. However, according to the landscape historian Oliver Rackham, 'the average English country churchyard contains at least ten thousand bodies.'(Rackham 1986: 344) This pragmatic approach would inevitably have entailed some disturbance of the remains, but the theological and spiritual meanings of perpetuity and eternity persisted whilst the churchyard's resource – earth – continued to be sustainable and available to successive generations as a permanent and continuing site of memory and communal identity. This stands in contrast to the Victorian cemetery which, while adopting the same commitment to burial in perpetuity, disallowed any re-use that disturbed the remains. This made it unsustainable over time and increasingly marginal to, and separate from, the living community, as its resources were depleted or over-used so exhausting every last available space.

The Victorian Cemetery

The idea that the deposition of remains, especially burial of the body, should be an act that persists 'in perpetuity', never to be disturbed, is now a deeply rooted belief for the general UK population. Customs elsewhere that require transfer of skeletal remains to ossuaries, or otherwise to vacate rented grave space after a set period, are regarded as macabre curiosities. The phrase 'in perpetuity' is frequently quoted by members of the public in relation to the rights they believe they hold for their grave plot, despite the fact that grave deeds will no longer generally confer such rights. The assurance that the corpse would be both contained and protected for all time may have relatively recent origins, but it has enjoyed a firm entrenchment in public perceptions since the nineteenth century.

Accounts of how medieval and post-reformation burial grounds, notably in London and Paris, were overwhelmed by a volume of corpses for which they were unfitted are familiar (see for example Etlin 1984). These scandals before the early nineteenth century, as population densities accelerated and issues of public health came to the fore, led to the establishment of cemeteries at the peripheries of the new metropolitan and urban centres. The cemetery, run by emergent public authorities, by the joint stock companies or by religious bodies concerned for their members' religious and socio-ethnic sensibilities, became the last resting places for an expanding majority, whether in individual or communal graves. Whilst administrative municipality, commerce and religion remained in some kind of balance when it came to dealing with dead, the ecclesiastic authorities were ultimately no longer deemed able to manage the scale of need amongst poor, destitute, old and sick people and the churches gradually ceased to have controlling sway over custom and practice.

Burial was displaced from churchyard to cemetery following the nineteenth century Burial Acts. The 1857 Act legalised interment in municipal cemeteries so that the individuality of the deceased was secured and could now be purchased in a grave-plot 'in perpetuity' (Rugg 2000). Churchyards that had accommodated the dead for centuries were side-lined as income from burials declined. These shifts, codified and secured legislatively, brought opportunities not only for the entrepreneurs who had invested in the joint stock companies, but also for the new middle classes to display their family and dynastic wealth in monumental style. Thus the spiritual under-girdings of burial in perpetuity were re-interpreted to become fixed in the corporeal. The established church surrendered control to the new burial authorities and to the joint stock companies and by the middle of the twentieth century local authority municipalisation dominated cemetery activity (see Jupp 2006 for an account of these shifts, particularly up to the mid-nineteenth century). Although there were many aesthetic changes in disposal patterns and styles within the cemeteries, infilling of paths and the use of reserve land for laying out lawn grave areas, disposal continued to be predominantly burial up until the 1960s when cremation finally overtook burial as the preferred mode of disposal in the UK (Worpole 2003).

The Crematorium

The emergence of cremation at the end of the nineteenth century and its popularisation until 2000 is comprehensively covered historically, theologically and legislatively in Peter Jupp's book *From Dust to Ashes: Cremation and the British Way of Death* (2006). Hilary Grainger, in *Death Redesigned* (2005), provides an equally comprehensive account from the perspective of architectural history (see also Grainger, this volume). She points out that more than half the crematoria in Britain were built between 1950 and 1970 and this indicates the post war character of these places, the practices they permitted and the dispositions they shaped for the three decades of the twentieth century that followed this burst of municipal provision. Most crematoria were built in existing cemeteries and were thus at the periphery of conurbations, though they were not evenly distributed. Many served, and in some cases continue to serve, very large catchment areas. This meant that as cremation took hold and it was customary to leave the ash remains at the originating crematorium in the newly established Gardens of Remembrance, many bereaved families returned to homes at considerable distances from the ashes of people who had been very close in life.

Such spatial separation appears to have been accepted as par for the cremation course as the advantages of cremation were enumerated in Britain in the 1950s and 1960s. It was clean, it left 'Land for the Living', and for much needed new housing, and it encouraged a simplified mourning code judged to be appropriate to a recovering British society. Ash remains are, quite clearly, a distinct transformation of the corpse and these too held new possibilities; they were the corpse reduced to manageable proportions, were portable and ultimately divisible. In further contrast to the decomposing and polluting body, they were dry and purified. Their disposal did not require the same avoidance of pollution and though contained, temporarily or permanently, they could be released to air or water, rather than committed to earth.

Up until the 1970s, most family members left the ashes at the originating crematorium for the staff to scatter in the grounds, or to be placed in a leased niche, an ash grave in earth, or in an urn or casket; it was usual for the ashes to remain *in situ*. From the 1970s onwards however, increasing numbers of families have been opting to claim the ashes to take them away for a private ceremony some time after the cremation (Kellaher et al. 2005). It may be that the uneven distribution of crematoria already mentioned, required many families to go through their bereavement without the consolation of being close to the remains, or even knowing precisely where they were located. This distancing in turn could have generated or consolidated memorialisation that symbolically located the dead closer to home. Alternatively, along with other emergent characteristics claimed for post-modern society, some bereaved survivors took matters into their own hands and in the seventies started to reclaim the ash remains and remove them from the crematoria. It is true that many families were – and some remain – unsure as to how to effect a final and fitting disposal. What is manifestly clear, however, is that increasing numbers of people were no longer content to leave disposal within the control of the authorities. Most recently, in the twenty-first century, a number of innovative disposal and memorialisation practices have emerged. Whilst memorialisation features on proliferating numbers of web-sites, the more strikingly physical but still new disposal approach of woodland burial needs to be included as the most recent development in what may be an evolutionary schema.

Woodland Burial

The recent and rapid expansion of woodland burial sites in the UK (Clayden et al. 2010) – most having been authorised and established over the last decade – suggests yet another development that is germane to this thesis. It is that in its 'pure' form the burial site is generally not marked by an identifying plaque; where such markings are allowed, they are made of bio-degradable materials intended to naturally disintegrate leaving no material trace. We ask whether this is another aspect of the trend towards the separation of remains and identifying marker or memorial, even though this kind of grave is not empty. The traditional hallmark of the cemetery, as a place of inscription and communal memory, is nonetheless made redundant in this form of interment.

A Series of Moments and a Sequence of Places

In the disposal contexts available to people in Britain over the last two centuries and longer, we have started to trace how the separation of mortal remains from the most privileged sites of commemoration, has evolved into a contemporary form of what we wish to term 'cenotaphisation'. This has hitherto been a minority practice sanctioned for circumstances such as death in battle, or for those assumed dead whose bodies are never found, but now arguably is acquiring a degree of mainstream acceptability in memorialising practices, and creating the urban 'deathscapes' which we are now witnessing. By cenotaphisation we particularly focus on the spatial and temporal separation of memorialising practices from the deposition of remains.

Thus there is now for many families no one 'moment' of farewell or one place to which memory is attached, but a series of moments and a sequence of places. The culmination is a memorial place, more or less public, which is no longer materially connected with the mortal remains. We might also point to even more recently generated forms of memorialisation, not least on internet sites, some of which show the memorial benches and trees on which we focus here and which represent a culminating point of cenotaphisation; an empty grave. The progressive transformations of practice and belief that lead towards a trend that may be consolidated in twenty-first century Britain are now discussed in more detail.

Explaining Shifts Towards Cenotaphisation

In the previous section we examined the historical shifts in both the place and the form of disposal and commemoration, evolving to the present situation whereby a separation of disposal and memorialisation takes place. In this section we seek to explain the major social and cultural factors which appear to have produced these changes.

While there is no single explanation to the evolutionary process we describe, in which memorialisation becomes increasingly distinct from the final resting place of bodily remains, there are a number of contributory factors to this trend. The first of these is the perceived decline of the Victorian cemetery.

Cemetery Decline as an Influence

After inner-city churchyards were closed for health reasons in the nineteenth century, the large urban cemeteries came to dominate and, while new churches continue to be built in urban centres, they do not have immediately adjacent burial space on site. Even many rural churchyards have been declared 'closed' to further interments as a result of visible over-crowding. Thus the Victorian cemetery and its twentieth century municipal successors emerged as the principal disposal place for the majority of the population. While there is a broad assumption that historic cemeteries in UK towns and cities show evidence of neglect and decline, hard evidence detailing the extent and long-term social costs of such apparent deterioration remains difficult to find (Dunk and Rugg 1994). In early 2001 the House of Commons Environment, Transport and Regional Affairs Committee met and took evidence on the matter of cemeteries. On 21 March 2001, the Committee published its 'Report and Proceedings of the Committee, together with the Minutes of Evidence'.

While statistical evidence remained lacking, opinion was forthright on the matter from many quarters. The Institute of Burial and Cremation Administration (IBCA) stated in its evidence that 'the condition of our municipal cemeteries is generally very poor particularly if compared to the high standard achieved in cemeteries provided and managed by the War Graves Commission.' (Environment, Transport and Regional Affairs Committee 2001: Vol. II 12–13) It further noted that:

> In particular, the heritage value of cemeteries is disappearing daily. This occurs primarily through neglected Victorian chapels, drainage systems and perimeter walls and gates. Such neglect often occurs even though the structures are listed. Memorials are seen as mowing impediments and destroyed, and the integrity of Victorian and Edwardian design and planting schemes ignored. (Environment, Transport and Regional Affairs Committee 2001: Vol. II 12–13)

The Garden History Society provided evidence which stated that:

> The poor condition of cemeteries rules many of them out as areas for public amenity: it is difficult to determine to what extent their abandonment as places for quiet enjoyment is due also to a cultural shift in attitudes towards mortality, but we suggest that their condition has played a large part in discouraging this sort of use, which is not only legitimate but positive. Indeed, the condition of cemeteries represents a deplorable wasted resource in terms of urban greenspace. (Environment, Transport and Regional Affairs Committee 2001: Vol. II 136)

Furthermore, as The Association of Gardens Trusts noted:

One other problem facing the older cemeteries is over use. As the space is finite and burials are continuous, many cemeteries have lost their original dignity and aesthetic appeal as a result of over intensive use. This is particularly a problem with the sinuous woodland cemeteries where a balance between burial grounds and walks was originally intended. (Environment, Transport and Regional Affairs Committee 2001: Vol. II 172)

As a result, the Committee concluded that 'while many local authorities continue to provide a decent level of maintenance in their cemeteries and a few are excellent, the condition of many cemeteries is unacceptable.' They concluded with the bald statement that, 'Unsafe, littered, vandalised, unkempt, these cemeteries shame all society in their lack of respect for the dead and the bereaved.' (Environment, Transport and Regional Affairs Committee 2001: Vol. I xxiii)

While much concern has been expressed at the difficulties facing modern cemetery managers, few legislative or indeed financial resources have been made available to address the problem, which continues to influence and determine public choice as to the preferred means of disposal, usually detrimentally in the case of cemetery burial. Whilst is true that some religions still prohibit cremation – such as the Orthodox Jewish and Muslim faiths – woodland burial has brought earth interment back into public favour for other groups in contemporary Britain. Even so, the cemetery remains a contested or problematic space for many today (see Deering, this volume). Cemeteries are also problematic in that they are often made up of the earlier Victorian sections, where burials took place six or seven generations ago, and the interwar and post war sections where burials are of people who remain within the personal memory of adult descendents, some of whom continue to visit and tend plots on a regular, sometimes very frequent basis (Francis et al. 2005). Cemetery management must take account of these two burial aesthetics and sets of personal and historic/heritage connections. The interests of the living do not always sit comfortably alongside those who have passed beyond personal knowledge or remembrance into a heritage realm to be conserved for an anonymous posterity.

In order to manage these complex settings and the different sets of interests, along with local and central legislation and ecclesiastic law, burial authorities have found themselves increasingly entangled in asserting prohibitions and establishing mechanisms for controlling the public that comes to mourn, commemorate, reflect or visit for historic or ecological enjoyment. As a consequence, grave owners may experience alienation and discouragement and seek other mourning and memorialising options for future decision making. It should be noted that today, only 28 per cent of UK disposals take the form of earth burial (The Cremation Society of Great Britain 2009), clear evidence that, for whatever reason, cemeteries are no longer the principal focus for funerary rituals.

Furthermore, most municipal cemeteries have not enjoyed the same renewal of interest and status as have public parks over the past two decades, though in earlier times cemeteries and parks were managed to the same high standards by the same maintenance and gardening regimes. Many Victorian city parks and cemeteries were designed by the same architects or landscape architects. The rise in esteem and quality of public parks, partly as a result of political pressure and partly as a result of large-scale lottery funding, has perhaps made them more attractive to people as places where commemoration might be more appropriate and uplifting. Doubtless this was certainly not anticipated by either park or cemetery managers, but funding opportunities and programmes of urban renewal which have privileged parks, have shifted affection away from cemeteries and back into the naturalistic urban domain as sites of attachment and meaning.

Equality in Death as an Influence

Over the centuries and more recently, the decades in which we have situated this evolutionary schema, equality in death has been both a truism and a justifying prompt for disposal activity. At the same time, distinctions had always operated in the traditional churchyard, with prime positions and impressive monuments overshadowing more modest memorials and common or pauper graves positioned inconspicuously, but equality in death had been the theological rule and publicly accepted sentiment associated with death (Worpole 2003). With the establishment of cemeteries, equality as a lifetime aspiration extended to the treatment of the corpse in death. Thus, with class differentials shaping society in the nineteenth century, equality took on political and public health connotations and became attached to the beliefs and sentiments associated with death. But as with churchyards, cemeteries were also the place for the extreme ostentation that we now think of as the 'Victorian Way of Death' (Curl 2000).

The equality ideal continued to resonate into the twentieth century and is still recited in cemeteries, not least because it contains an undeniable truth. Burial and mourning ideals evolved from elaborate mourning to the twenty-first century practices that now attach to cemetery-based disposal, largely because the idea of all being equal in death was driven home in the early twentieth century by the terrible casualties of the First World War. By 1918 the necessity for remains to be disposed of uniformly touched all social classes and most families. Consequently, dispossession of the body of a close family member was no longer restricted to the disgraced and the pauperised, as no remains of war dead were repatriated. Families had to mourn in the absence of material remains, let alone any knowledge as to the exact circumstances of death.

Strange (2005) discusses the way in which only a muted pattern of mourning was possible for paupers who had lost control and knowledge as to the fate of the body of close family members. Dispossession of the body and ignorance as to its burial place is identified as a crucial key or cue to mourning activity. When all families were placed in the same position, without bodies to bury and thus fix in place and memory in the contemporary custom, the silence and restraint forced upon the pauper became the norm that persisted almost to the end of the twentieth century. The First World-War was, moreover, the event and period when 'equality in death' came to be overlaid by secular meaning, made explicit in the design for war graves. This design, with all it came to connote, was adopted in the interwar years with some enthusiasm, as cemeteries opened up reserve land and established the acres of uniform lawn grave sections that are now viewed as a hallmark of municipality. The simple gravestone design, along with a grid layout that exposed and expressed the enormity of loss and the equal distribution of death, signalled another kind of judgement that had to be made in the devastated present, thereby eclipsing any last judgement and eternity of religious tradition.

The first beneficiaries of the ensuing simplification of burial included working people who did not own their homes and for whom the grave plot, with its small grave-garden, represented a first property acquisition. This group still speak with satisfaction at such achievements, recalling that an identifiable grave only came within financial reach of nearly all families after 1945 (Francis et al. 2005). However, it is these characteristics of grid uniformity, accentuated by restrained tree planting, which are now frequently castigated as a bleak, sterile wasteland of regimented, grey and toppling monuments, dangerously susceptible to vandalism and other disorders. As such, cemeteries are contrasted, unfavourably, with churchyards which are perceived as lying at the heart of rural communities with small monuments set haphazardly amongst mature trees and shrubs. These are the moderate, balanced settings of the English picturesque where death may be approached, seen as co-existing with and protecting endangered wildlife. Even more tellingly, it is the anonymity of the municipal cemetery that is increasingly rejected in favour of cremation, so that ashes may be placed somewhere 'safer', 'warmer' and, emotionally if not geographically 'closer to home' (Kellaher 2008).

Ashes – Miniaturisation, Portability and Mobility – as an Influence

Cremation was not simply a new form of disposal but brought with it significant shifts in the nature of ritual practice and memory. When cremation was legitimised at the turn of the nineteenth century, only the well connected, often with secular and humanist leanings, embraced it as their preferred disposal method. But as cremation overtook burial it appears that the bereaved gradually came to appreciate and capitalise on certain advantages it offered. Firstly, it was cheaper than burial and did not involve purchase and maintenance of a grave plot. It was judged 'cleaner' than burial. By this people referred as much to the purifying processes that transformed the corpse in such a way that ideas of persistent pollution could be put aside. It was, initially, a one-stop practice, the remains being disposed of, often without family witness, by crematorium staff. It seems to have aligned itself to the post war, stripped-down 'contemporary' ethos of the fifties and sixties. Only in the seventies did family members act in ways that suggest the miniaturisation and portability of the body as ash remains was being regarded as a helpful adjunct to mourning. Small ash graves began to be offered in Gardens of Remembrance with inscribed plaques, sometimes marking the place where ashes were buried or lodged, sometimes elsewhere in the grounds. The favourable space implications of miniaturisation were echoed in 'Land for the Living' themes (Grainger 2005, Jupp 2006) and the clearance of monuments from cemeteries accelerated as a simpler disposal approach was urged upon the public.

Over the last two decades of the twentieth century and into the twenty-first, ash remains have been treated to more and more elaboration. Tributes are piled on to, around and above the small ash graves and other sites in crematoria grounds. This can sometimes lead to disputes between cemetery managers and bereaved relatives who lease these sites (for periods of ten to 25 years). The accelerating trend for family members to opt to retain the ash remains rather than leaving them with the crematorium can sometimes be an explicit consequence of dispute about the type of monument or tributes permissible, or a tacit resistance to authority. It is also a consequence of families taking advantage of the purified, portable, divisible and miniaturised character of ash remains.

In many instances it seems that family members seek ways of placing their dead, and managing their losses, that they judge 'fitting'. In working out what and where the final disposal might be, it is clear that many bereaved people think long and hard. They take into account the past life of the deceased, their relationships with those who have died and those who survive and the places in which these associations have been enacted (Kellaher et al. 2010). A case can be made that families, no longer bound by or even knowledgeable about religious beliefs of an afterlife, enter imaginative states in which memory can be cultivated and endlessly developed, because they are in total control of the material and symbolic placement of the remains. Cremation has permitted this shift towards repossession of the dead and their retention within family, local place and private memory. At the same time, the impulse to extend remembrance to the world outside the family persists. Survivors want the name of the deceased to be seen and noticed; they choose the most eye-catching place they can afford in the Garden of Remembrance, and they may put the name and image on the internet. They want the deceased to be acknowledged in immediate circles as well as more widely.

Bringing the Dead Back Home

The dead and the bereaved are now, for many, released from the theological, ecclesiastic and municipal boundaries within which they have been retained and their mourning managed for centuries. There is undoubtedly uncertainty as to how far, or how close to home and the everyday worlds of living and working the dead should be (see Gittings and Walter, this volume). The proliferation of informal memorials, however, suggests that the bereaved need the dead to be close at hand at least some of the time, either within the dwelling, its garden or in some significant public place.

Home to the Domestic Sphere

Repossession of the dead may entail a strong sense of reclamation; from the authorities, ecclesiastic, municipal and perhaps also medical, that shape complex modern life. Over 60 per cent of sets of ashes (Pharos 2007) are now taken away from crematoria by family members for private, individual disposal, sooner or later, with more or less ceremony and ritual. When this happens, and in some regions this average can translate to four-fifths of sets of ashes, the remains may be kept at home, inside the house whilst a decision is thought through. Few sets of ashes are brought home with the intention that they should stay inside the house indefinitely. Rather, they rest awhile, perhaps until a partner dies and the two sets can be mingled and placed somewhere meaningful, which can include return to crematoria, cemetery, burial ground or Gardens of Remembrance, or to an existing family grave. Equally, and more or less permanently, they may be placed in the garden. But in a highly mobile society, families are aware that moving house and home may be inevitable and take decisions accordingly.

Personal mementos, if not small shrines of pictures, photographs and flowers, have long been maintained at home in remembrance of the deceased (Hallam and Hockey 2001, see also Wojtkowiak and Venbrux, this volume). Whether or not the mortal remains are buried in the cemetery, placed in the garden or dispersed in a more distant private location, the home and the family must always have been a central site for memory. But in the past there also appears to have been a need for the deceased to be acknowledged beyond the immediate family, for them to be incorporated within a public realm of some kind. Whilst private grief and personal memory is one aspect of mourning it seems hitherto to have been matched or balanced by a public facet that secures the dead in a place beyond personal and present time within which private grief is most sharply experienced as close and personally affecting. Furthermore it can be argued that growth of local memorialisation in public places is a way of anchoring memory to place, and inscribing the familiar streetscape with the names of those who once lived there, which may also be interpreted as a form of resistance to the rapidity of change and standardisation of the public realm (Minton 2009).

Home to the Communal Sphere

One key feature of these new forms of memorialisation has to do with proximity; to home, to community, and to familiar geographies and special places. A tree may be planted in the street in which the deceased formerly lived, not usually accompanied by a plaque, but known to be special by those who commissioned and planted it. Memorial benches in parks are usually clustered around particular landscape features, such as lakes, river walks, or looking out across a view of the city. Some of these memorials may be as close as 50 metres to the former home, if there is an ornamental park or footpath close by. In many cases the local park may have accrued a special status as a stable place, a piece of nature, in a fast-changing urban milieu. It is therefore perceived as an ideal as a place for the marking and retention of memory. Given also that parks have, as already mentioned, gained a renewed status in public policy in recent years, this makes them even more attractive as sites of local meaning and attachment.

This emphasis on locality has both functional and symbolic aspects. A tree or bench can be placed so close to home that it may be visited or passed every day, which is certainly not the case with a grave or columbarium niche located in a cemetery or crematorium many miles away. It can be guarded or tended as necessary. In more symbolic terms it can be seen as a gift to the neighbourhood, adding amenity and pleasure for others, and thus enhancing the value of community whilst commemorating a devastating absence or acute loss. These local emplacements can be accompanied by rituals too, such as annual gatherings to mark the anniversary of a death, with perhaps the laying of a wreath, or the attachment of open letters or poems to the bench or tree on other occasions deemed 'special' by the bereaved. Yet there is no doubt that an over-concentration of memorials in certain kinds of public places is now beginning to be resisted, both by local authority managers, as well as some members of the public; further proof that such forms of memorialisation at a local level are gaining popularity. A number of local authorities are beginning to develop regulatory policies on the conditions to be met with regard to the planting of memorial trees or commissioning of commemorative benches in public spaces. For example, a senior officer of South Lanarkshire Council noted how:

> Members of the public in South Lanarkshire have told the Council, 'With the greatest respect, memorabilia surrounding memorial benches and trees make a picnic in the park depressing for all… the council will continue to run the commemorative tree and bench scheme but proposes to restrict the installation of future memorials of trees, benches and plaques to cemeteries, the crematorium and woodland burials grounds.' (Clelland 2007, 10–11)

Conclusion

The principal argument put forward in this chapter is that contemporary funeral culture in the UK suggests a new wave of patterning, in which the disposal of people's mortal remains are increasingly spatially and temporally dissociated from the one or several points at which the deceased are memorialised. Proliferations of commemorative trees, benches and plaques across the everyday urban landscapes of towns and cities, along with other observations concerning the status of cemeteries and the emergence of new practices around ash placements and burial, have prompted our coining the term 'cenotaphisation'. This signifies the empty grave, the absence of corporeal or, more likely, cremated ash remains, at or near the marker, to name or identify the deceased and inscribe memory in public space which may itself be re-inscribed. Whilst cenotaphisation may not have become mainstream, that such a separation of mortal remains and marker is a preferred option for even a minority of bereaved people is such a remarkable reversal of long-standing practices that it attracts attention and deserves comment.

In exploring whether cenotaphisation indicates an accelerating trend, we have argued that there is evidence for its roots in wider communitarian and societal impulses that entail new constellations of belief about an afterlife and how disposal can be aligned with the needs of the bereaved by accommodating the dead fittingly. This means finding a 'fit' so that in their new state of being dead, the deceased can be envisaged as in life; doing and being, in familiar or longed-for places that the bereaved can also envisage and may share. As one underlying influence for cenotaphisation, careful selection of the place or places for disposal of remains becomes an option because ashes are miniaturised and portable. Importantly, they can be made as mobile as the person was in life; their disposition can reflect the good life and possibly, the good afterlife.

It is arguable that attempts to reach a good 'fit' start at the funeral where the emphasis is moving away from mourning the deceased at the site and time of the final disposal, to celebrating the life of that deceased person on another occasion, often unconnected with any aspect of the disposal of the bodily remains. For example, increasing numbers of people are choosing to organise and conduct their own cremation services, rather than relying on standardised religious or humanist rituals arranged by funeral directors. Such personalised services frequently foreground tributes to the life and achievements of the deceased, rather than employing or even adapting, traditional religious rites, texts and rituals.

These practices, as we suggested at the beginning of the chapter, seek consciously or otherwise, to re-integrate the dead back into the living community, and the various devices of informal memorialisation now employed, help to consolidate and ground this process. One obvious aspect to secularisation is that long-standing beliefs in an after-life, and indeed a final place, to which the dead may fully come to rest, may no longer be widely held. Instead mourners have to devise personal rituals and ceremonies by which they reassure themselves that the dead are at home somewhere, no longer metaphorically wandering abroad unanchored and unclaimed. They are temporally 'contained', in memory, whilst being spatially free to be and do as fits their new state of being dead. Burial and commemorative practices from the pre-reformation village churchyard to the woodland burial site and the commemorative bench, via the Victorian cemetery and the twentieth century crematorium, have been charted to suggest an incremental pattern towards cenotaphisation. In that the key elements of all funerary practice: disposal of the bodily remains, marking of the site of the bodily remains, identifying inscription and subsequent memorialisation, have been combined and re-combined to meet changing social, religious and cultural changes suggests, we argue, an evolutionary and transformative process. However, where once all elements were neatly enfolded within each other, they are now distinct and often separated spatially and over time. As noted, this appears to be a marked departure from practices over centuries if not millennia. Yet even where the final site of the bodily remains and the final site of memorialisation are widely separated, what is clear is that the bereaved are now much more active participants in a process that brings the dead person to the spaces of home and familiar locale. It may be the case that the material cultural props are placed at a distance; it seems equally the case that this permits new conjunctions of meaning that 'contain' the dead within the boundaries of familial and communitarian protection to the same degree that they were once physically contained and protected within earth and community based institutions such as the church and municipality. In a world of geographical mobility and social change, we suggest that the impulse, indeed imperative, to contain and protect the dead continues to be deeply embedded in everyday bereavement practices.

Cenotaphisation, as a series of moments and a sequence of places through which disposal and memorialisation can now be enacted, is rooted in fundamental impulses to contain and protect both the dead and the living from harm or evil in this world and in an after world. At the same time, cenotaphisation seems to permit a mobility that contradicts 'containment' of the dead since they are free to be 'dead' in the way only they can know. But the bereaved no longer bow their heads and let go of the dead. They now forge continuing bonds and, we argue, through inscription across the public realm, can at times inhabit or share in part an afterlife with the deceased through places that are familiar to the close circle of family and friends, as well as to the communities that once shaped a shared life, and now offer to shape the absence so that the bereaved can find a place for themselves and the deceased that 'fits'.

References

Clayden, A. Hockey, J. and Powell, M. 2010. Natural burial: the de-materialising of death?, in *The Matter of Death, Space, Place and Materiality*, edited by J. Hockey, C. Komaromy and K. Woodthorpe. Basingstoke: Palgrave.

Clelland, S. 2007. Memorials of the moment. *Green Places*, 39, 10–11.

The Cremation Society of Great Britain. 2009. *International Cremation Statistics 2007* [online]. Available at: www.scgw.demon.co.uk/CremSoc5/Stats/Interntl/2007/StatsIF.html [accessed: 21 January 2010].

Curl, J.S. 2000. *The Victorian Celebration of Death*. Gloucestershire: Sutton Publishing.

Danforth, L.M. 1989. *Firewalking and Religious Healing: The Anastenaria of Greece and the American Firewalking Movement*. Princeton: Princeton University Press.

Davies, D.J. 1997. *Death, Ritual and Belief: The Rhetoric of Funerary Rites*. London: Cassell.

Dunk, J. and Rugg, J. 1994. *The Management of Old Cemetery Land*. Kent: Shaw & Sons.

Environment, Transport and Regional Affairs Committee. 2001. *Cemeteries*. Volumes I and II, House of Commons, London: HMSO.

Etlin, R.A. 1984. *The Architecture of Death: The Transformation of the Cemetery in Eighteenth-century Paris*. London: MIT Press.

Francis, D., Kellaher, L. and Neophytou, G. 2005. *The Secret Cemetery*. Oxford: Berg.

Grainger, H.J. 2005. *Death Redesigned, British Crematoria: History, Architecture and Landscape*. Reading: Spire Books.

Hallam, E. and Hockey, J. 2001. *Death, Memory and Material Culture*. Oxford: Berg.

Hockey, J., Kellaher, L. and Prendergast D. 2007. Of grief and well-being: competing conceptions of restorative ritualisation. *Anthropology and Medicine*, 14(1), 1–14.

Jupp, P. 2006. *From Dust to Ashes: Cremation and the British Way of Death*. Basingstoke: Palgrave Macmillan.

Kellaher, L. 2008. The past, the present and the future of burial, in *Death our Future: Christian Theology and Funeral Practice*, edited by P. Jupp. London: Epworth Press.

Kellaher, L., Hockey, J. and Prendergast, D. 2005. In the shadow of the traditional grave. *Mortality*, 10(4), 237–50.

Kellaher, L., Hockey, J. and Prendergast, D. 2010. Wandering lines and cul-de-sacs: trajectories of ashes in the United Kingdom, in *The Matter of Death, Space, Place and Materiality*, edited by J. Hockey, C. Komaromy and K. Woodthorpe. Basingstoke: Palgrave.

Maddrell, A. 2009a. A place for grief and belief: the Witness Cairn at the Isle of Whitehorn, Galloway, Scotland. *Social and Regional Geography*, 10(6), 675–93.

Maddrell, A. 2009b. Mapping changing shades of grief and consolation in the historic landscapes of St. Patrick's Isle, Isle of man, in *Emotion, Place and Culture*, edited by M. Smith, J. Davidson, L. Cameron and L. Bondi. Aldershot: Ashgate.

Minton, A. 2009. *Ground Control: Fear and Happiness in the Twenty-First-Century*. London: Penguin Books.

Pharos International. 2007. *The Official Journal of the Cremations Society of Great Britain*, 73(2).

Rackham, O. 1986. *History of the Countryside*. London: Dent.

Richardson, R. 1988. *Death, Dissection and the Destitute*. London: Penguin Books.

Rugg, J. 2000. Defining the place of burial: what makes a cemetery a cemetery? *Mortality*, 5(3), 259–76.

Strange, J.-M. 2005. *Death, Grief and Poverty in Britain, 1870–1914*. Cambridge: Cambridge Social and Cultural Histories, Cambridge University Press.

Van Gennep, A. 1909. (translated by Monika B. Vizedom and Gabrielle L. Caffee). *The Rites of Passage*.London: Routledge and Keegan Paul.

Walter, T. 1994. *The Revival of Death*. London and New York: Routledge.

Walter, T. 1999. *The Mourning for Diana*. Oxford: Berg.

Worpole, K. 2003. *Last Landscapes: The Architecture of the Cemetery in the West*. London: Reaktion Books.

Chapter 10

Memorialisation of US College and University Tragedies: Spaces of Mourning and Remembrance

Kenneth Foote and Sylvia Grider

Introduction

This chapter focuses on the spaces of mourning and remembrance arising from events of violence and tragedy faced by United States (US) colleges and universities.[1] Recent tragedies at Texas A&M University, Virginia Tech and Northern Illinois University and on other campuses have gained much attention, especially the spontaneous formation of spaces for mourning and, later, the creation of permanent memorials (Figure 10.1). This chapter sets these events in chronological and spatial perspective by raising two key questions. First, what are the historical precedents for these memorial spaces – that is, how do the responses to contemporary university tragedies compare to earlier tragedies in the US? Second, what are the dynamics of the formation of memorial spaces, both spontaneous and permanent? Of particular interest are how spontaneous shrines are transformed into permanent memorial spaces and whether the proximity factor of on-campus and off-campus deaths results in different memorial practices.

Such tragedies are, of course, a worldwide phenomenon and have claimed students of all ages. Events such as the Columbine High School shooting in Colorado, the massacre at Dunblane Primary School in Scotland, and the Beslan school siege disaster in North Ossetia received worldwide coverage, but there have been hundreds of other tragedies caused by fire, natural disasters, shootings, building collapse, and other misfortunes. Rather than analyse all of these at once, we focus here on one subset – tragedies affecting US institutions of higher education from 1903 to the present. Our aim is not to dismiss the other tragedies but, rather,

---

1   US colleges and universities are two-year and four-year, post-secondary institutions of higher education offering associate, bachelors, masters and doctoral degrees. The term college is usually applied to institutions offering associate and bachelors degrees. The term university is usually applied to institutions offering graduate degrees. The terms are however not always used consistently, for instance the US Air Force Academy is a college and the Colorado School of Mines is a university.

Figure 10.1   Examples of permanent memorials resulting from campus
tragedies: a) site of Oklahoma State airplane crash (2001) near
Strasburg, CO (*Source*: Photograph by the author); b) tree of life
sculpture honouring faculty killed at the University of Arizona
College of Nursing (2002) (*Source*: Photograph reproduced with
the permission of Janice J. Monk); c) dedication (1999) of Tower
Garden honouring victims of 1966 University of Texas (Austin,
TX) mass murder (*Source*: Photograph by the author); d) benches
honouring students shot at the University of Central Arkansas
(Conway, AR) (2008) (*Source*: Photograph reproduced with the
permission of Michael Yoder).

to use these college and university tragedies as a window on the dynamics of one type of contemporary space for mourning and bereavement.

Theoretical Context

A comparative historical and spatial perspective on college and university tragedies can contribute to three interrelated areas of research into deathscapes: 1) the rise of vernacular memorials and spontaneous shrines as responses to tragedy; 2) the connection between campus memorialisation and broader trends in public commemoration and monument building; and 3) and changes in rituals surrounding death and bereavement in modern Western societies. We argue that careful study of antecedents and precedents of these phenomena can help set our understanding of the dynamics of contemporary deathscapes on firmer theoretical and empirical foundations.

Vernacular memorials (commonly referred to as spontaneous shrines) are today one of the most distinctive features of tragedies, on college campuses and elsewhere. Over the past twenty years or so they have also become the focus of considerable research (Grider 2001, 2006, 2007, Margry and Sánchez-Carretero 2007, Santino 2004). Case studies dominated the earliest research, among the first being the detailed examination of the aftermath of the 1989 Hillsborough disaster which claimed 94 soccer fans (Walter 1991). Few researchers have synthesised the entire spectrum of vernacular memorialisation, which includes not only spontaneous shrines but also roadside memorials marking the sites of fatal car wrecks, 'ghost bikes', urban memorial murals, online 'cybershrines', and such specialised memorials as the AIDS quilt. Roadside memorials, perhaps because they are so common and widespread, have attracted the most attention, much of which is limited to online websites featuring photographs of these small memorials.

Only in the past few years have longer scholarly analyses begun to appear, including Everett's (2002) *Roadside Crosses in Contemporary Memorial Culture* and a collection of essays entitled *Spontaneous Shrines and the Public Memorialization of Death* edited by Santino (2006), who is credited with coining the term, 'spontaneous shrine'. Santino, a folklorist, has championed the performative aspect of vernacular memorialisation. Another collection of essays is in press: *Grassroots Memorials: The Politics of Memorializing Traumatic Death* (Margry and Sánchez-Carretero in press). This volume will be perhaps the first to take a theoretical position regarding vernacular memorialisation, namely, the role of politics and protest in the vernacular memorialisation process. Furthermore, this volume brings to the forefront the question of terminology and takes issue with the most commonly used term, 'spontaneous shrine', since research has shown that many of these memorial forms are carefully planned and cared for and not necessarily spontaneous, and because the religious connotations of the term 'shrine' does not always capture the secular meanings and symbolism of the memorials. Some researchers have taken an unusually critical position toward this phenomenon (Doss 2002, 2008, Sturken 2008). In her recent monograph, *The Emotional Life of Contemporary Public Memorials*, Doss labels the rise of vernacular memorials a 'mania' or 'the contemporary obsession with issues of memory and history and an urgent, excessive desire to express, or claim, those issues in visibly public contexts. Contemporary acts, rituals, or performances of memorialisation are often exorbitant, frenzied and extreme – or manic' (2008:7). However, Doss neither contextualises her claim historically nor considers more than a handful of high-profile memorials to support her position. All this literature points to the value of using a broader historical sample of tragedies as a means of better contextualising current trends.

College and university tragedies can also offer insight into issues of public memory and commemoration, areas of research which have seen a dramatic rise in interest over the past decade (Azaryahu 1996, 1999, Dwyer and Alderman 2008a, Foote and Azaryahu 2007). Much of this recent research has focused on the ways in which discourses of the 'past' are constructed socially and expressed materially in landscape, public memorials and heritage sites. Within the literature, the most attention has focused on the political dimensions of public memory; how the past is re-presented to express hegemonic relations of power and authority. But other research has stressed that memorials can arise from many motives and express the interests of many different constituencies, not just the most powerful. Public memory is nonetheless an 'invented tradition', a selectively embellished, even mythologised version of events which serves social or political ends (Hobsbawm and Ranger 1983, Cosgrove and Daniels 1988, Norkunas 1993).

These traditions create what Schwartz (1982) has termed a 'register of sacred history'; a set of shared historical experiences and attitudes which define and bond a community. Public memory is then part of the symbolic foundation of collective life and often lies at the heart of a community's sense of identity. The question of 'who we are' becomes an issue of what we share and do together as a community and, more often than not, this sharing involves locating history and its representations in space and landscape.

This marking of 'sacred history' is often a key feature of college campuses. Their buildings and landscaping often play an important role in fostering a sense of community. This is precisely why campus tragedies can present such conflicts over memory and commemoration. On-campus violence and tragedy violate the norms of community life and can leave stark evidence of this violation. While the impulse to efface this evidence is often strong, it is typically countered by an urge to honour the victims and reassert community values. These conflicting desires, both to forget and to remember, mean that fitting such tragedies into a campus's sense of tradition can be very difficult. Sometimes a tragedy can be cast in a positive light, but the urge to ignore or efface it may be strong.

Our final reason for exploring these tragedies is the light they shed on changing ritual practice in contemporary Western society. Over the past generation there seems to have been a rise in the West of increasingly multi-religious ritual spaces such as sites of pilgrimage, memorials, war cemeteries, *lieux de mémoire* as well as other private, semi-public, and public places of localised rituality. The rise of these new sites has had, in part, the effect of sometimes marginalising more traditional ritual spaces, but it has also affected the form and function of these ritual spaces, including both their material forms and their performative elements (Bell 1992, 1997, Foote 2010, Grimes 2000, Post et al. 2003, Walter 1994, 1999). As Lukken (2005) has noted, contemporary society has seen the rise of secular 'rituals in abundance' even though traditional religious ceremony is declining in many nations. Although traditional religious rituals are not in decline in the US to the extent they are in many areas of Europe, are we still seeing in campus tragedies the emergence of new ritual forms and spaces?

Although these growing bodies of research on contemporary memorialisation are now extensive, relatively few studies are comparative and most include only a few specific cases. Furthermore, relatively few studies consider broad historical trends in commemoration, generally focusing instead on the most recent and well documented instances, such as the death of Princess Diana, the school shootings at Columbine High School, and '9/11'.

In this chapter, we have chosen to undertake a comparative study of a distinct category of American memorialisation practices over the past century to determine how they have changed over time. The memorialisation of disasters and catastrophes involving American colleges and universities both on and off campus provide a solid context for addressing these issues because these institutions, as a whole, exhibit a very strong sense of community and identity that spreads beyond the campus itself, to former students as well as supporters of athletic teams and other school activities. As one writer noted after the loss of five Bluffton University students in 2007: 'Bluffton and its baseball team have become part of an indescribable community that continues to provide support from every direction' (Duling 2007: 8–9).

Significantly, American college campuses are home to many memorials, ranging from athletic stadiums and student activity centres (often termed 'unions') honouring veterans of the world wars (and commonly named Memorial Stadium or Memorial Union), to other honourifically named buildings, as well as specifically created memorial structures erected on various campuses. As one example, the Harry Elkins Widener Memorial Library at Harvard University, one of the preeminent libraries in the US, honours an alumnus who died on the RMS *Titanic* and was funded by his mother.

Many colleges also have ceremonies to honour the loss of past or present members of their community who are recognised, for instance, at commencement or other annual ceremonies. Texas A&M University has had, for example, the tradition of 'Silver Taps' since 1898, a memorial service which occurs on the first Tuesday of the month if students have died in the previous month (Texas A&M 2009b). Other deaths are honoured during the annual 'Muster' ceremonies held now worldwide as reunions for members of the A&M community on 21 April (Texas A&M 2009a). Our interest is in examining how these memorial traditions relate to the growing number of spontaneous shrines marking traumatic deaths of students, such as those created after the fatal collapse of the bonfire at Texas A&M University (1999) and the mass murders on the campuses of Virginia Tech (2007) and Northern Illinois University (2008). Are these recent memorials related to existing memorial cultures or do they represent something new?

Although we focus here on college and university tragedies, we will set them in the context of national and international trends in the memorialisation of tragedy and violence. For example, the commemoration of the victims of mass murder is a difficult issue worldwide, not solely on college campuses. Some of the first known spontaneous memorials in the US to honour victims of mass murders are very recent, such as those in San Ysidro, California (1984) and Killeen, Texas (1991); previously such events would not have been marked. At the international level considerable debate continues over the appropriate memorialisation of the victims of the Holocaust and other genocides (Ashworth and Hartmann 2005, Lennon and Foley 2000, Williams 2007). Furthermore, events on college campuses influence and are influenced by the way that similar tragedies are treated when they occur in primary and secondary schools, among other close-knit communities and among the general public. This means that, although we are comparing college and university responses in this chapter, we must necessarily discuss them in the broader context of national and international events.

Case Studies and Method

Our analysis is based on 67 campus tragedies dating back to 1903 (Table 10.1). Originally we planned to study school tragedies both inside and outside the US across all grade levels, but the cases were too numerous and too varied to analyse in one paper. Over the past century, there have been hundreds of school tragedies caused by blizzards, fires, aeroplane crashes, traffic accidents, mass murders, earthquakes, and many other disasters. We chose to focus here on college and university tragedies in the US, for two reasons. First, as noted above, colleges and universities often cultivate a strong sense of community and shared purpose among students, staff and faculty. This means that, when disaster strikes, a sense of community loss often results. Second, focusing on the US allows us to consider how patterns of commemoration and memorialisation change through time in a single country. This is not to say that all events occurring in different historical periods and widely different regions of the US can be readily compared, but by focusing on colleges and universities in the US, we hope to gain a better sense of how one type of deathscape has changed through time.

This analysis focuses on the largest and highest profile incidents (incidents attracting national or international news coverage) that claimed lives on and off campus. These are events that are commonly memorialised, but we have also included for comparison many less well-known tragedies. We sought to include as many cases as possible that: 1) occurred on or adjacent to a college campus;

Table 10.1   US college and university tragedies and memorialisations

* indicates that total includes the killer.

Date	University	Event	Deaths
1903, 31 October	Purdue University (West Lafayette, IN) *Memorial Gymnasium (now Felix Haas Hall). Dedication of a tunnel in the football stadium on centennial. http://en.wikipedia.org/wiki/Purdue_Wreck*	Off campus train wreck (Indianapolis, IN)	17
1927, 22 January	Baylor University (Waco, TX) *Marker at railroad crossing. Monument on campus. Remembrance at annual homecoming. Copeland (2006). http://www.wacohistoryproject.org/Moments/Immortal_10.htm.*	Off campus bus hit by train	10
1935, 12 December	Columbia Presbyterian Hospital School of Dental and Oral Surgery (New York, NY) *No known memorials.*	On campus shooting	3
1936, 4 June	Lehigh University (Bethlehem, PA) *No known memorials.*	On campus shooting	2
1949, 13 November	Ohio State University (Columbus, OH) *No known memorials.*	On campus shooting	1
1954, 15 May	University of North Carolina (Chapel Hill, NC) *No known memorials.*	On campus shooting	1
1955, 11 January	Swarthmore College (Swarthmore, PA) *No known memorials, but recent movie: Alston (2006).*	On campus shooting	1
1960, 29 October	California Polytechnic State University (San Luis Obispo, CA) *Memorial plaques at two campus locations. Memorial Plaza. http://en.wikipedia.org/wiki/Cal_Poly_football_team_plane_crash*	Off campus plane crash (Toledo, OH)	22
1965, 18-19 July	University of Texas (Austin, TX) *No known memorials.*	Off campus murder (Austin, TX)	2
1966, 9 July	University of Colorado (Boulder, CO) *Memorial plaque on rock (2006). Urban legend of haunted death site (Macky Auditorium). http://www.bouldercountyparanormal.com/macky.htm*	On campus murder	1
1966, 14 July	South Chicago Community Hospital (now Advocate Trinity Hospital), Nursing School (Chicago, IL) *No known memorials.*	Off campus murder (Chicago, IL) by Richard Speck	8
1966, 1 August	University of Texas (Austin, TX) *Tower Garden Memorial dedicated in 1999, completed in 2007. Lavergne (1997). http://www.utexas.edu/events/tower/*	On campus mass murder	15*
1966, 12 November	Rose-Mar College of Beauty (Mesa, AZ) *Not an institution of higher education. No data collected.*	On site killing	5

Date	University	Event	Deaths
1968, 8 February	South Carolina State University (Orangeburg, SC)	On campus shooting by local and state police	3
	Memorial space on campus, gymnasium named for the slain, precipitating off-campus site marked (2000), annual remembrance ceremony. Bass and Nelson (1999), Cram and Richardson (2009), Williams (2008) http://www.orangeburgmassacre1968.com/ http://en.wikipedia.org/wiki/Orangeburg_massacre		
1969, 17 January	University of California (Los Angeles, CA)	On campus shooting of non-students	2
	No known memorials.		
1969, 28 November	Pennsylvania State University (University Park, PA)	On campus stabbing	1
	Urban legend as haunted death site (Pattee Library). http://en.wikipedia.org/wiki/Betsy_aardsma		
1970, 4 May	Kent State University (Kent, OH)	On campus shooting by National Guard	4
	Memorial at death site (1971, 1975), annual candlelight walk and vigil, Center for Peaceful Change, stained glass windows in library, May 4 Scholarships, memorial sculpture on campus (1990), various sculptures and markers on other US campuses. Bills (1982), Graham (2006), Weiss (2008). http://en.wikipedia.org/wiki/Kent_State_Shootings		
1970, 14-15 May	Jackson State University (Jackson, MS)	On campus shooting by state police	2
	Renamed campus plaza, monument. http://en.wikipedia.org/wiki/Jackson_state_killings, http://www.may41970.com/Jackson%20State/jackson_state_may_1970.htm, Spofford (1988)		
1970, 24 August	University of Wisconsin (Madison, WI)	On campus bombing	1
	Memorial plaque (2007) at bomb site (Sterling Hall). http://en.wikipedia.org/wiki/Sterling_hall_bombing		
1970, 2 October	Wichita State University (Wichita, KS)	Off campus plane crash (Silver Plume, CO)	31
	Memorial monument on campus, small memorial at site, http://webs.wichita.edu/?u=MEMORIAL&p=/history/. Musical tribute "Waltzing in Heaven" http://en.wikipedia.org/wiki/Wichita_State_University_football_team_plane_crash		
1970, 14 November	Marshall University (Huntington, WV)	Off campus plane crash (near Huntington airport)	75
	Memorial fountain on campus, bronze statue on campus, cenotaph at Spring Hill Cemetery in Huntington, Memorial Highway (0.75 miles of State Route 75 and 11.91 miles of US Route 52), renamed city street, plaque at crash site, plaque at East Carolina University (opponent in last football game), memorial bell tower planned, commercial movie Nichol (2006). http://en.wikipedia.org/wiki/Southern_Airways_Flight_932.		
1976, 12 July	California State University (Fullerton, CA)	On campus shooting	7
	Grove of trees near site of murders in library.		
1977, 13 December	University of Evansville (Evansville, IN)	Off campus plane crash (near Evansville airport)	29
	Memorial Plaza with fountain. http://collegebasketball.rivals.com/content.asp?CID=750603		
1978, 15 January	Florida State University (Tallahassee, FL)	On campus bludgeoning	2
	The last murders by serial killer Ted Bundy, who also killed women at other universities. Photos of victims are kept in sorority house, memorial cross at one victim's church, macabre "anti-memorials" in Florida the day Bundy was executed in 1989. http://en.wikipedia.org/wiki/Ted_bundy, http://www.sptimes.com/News/112899/Floridian/Margaret_Bowman_1957.shtml		
1978, 18 August	Stanford University (Stanford, CA)	On campus shooting	1

Date	University	Event	Deaths
	No known memorial.		
1979, 5 October	University of South Carolina (Columbia, SC)	Off campus shooting (Columbia)	2
	No known memorial.		
1979, 14 December	University of Washington (Seattle, WA)	On campus shooting	1
	No known memorial.		
1983, 17 December	Cornell University (Ithaca, NY)	On campus shooting	2
	Memorial trees and plaque.		
1984, 13 October	California State University (Fullerton, CA)	On campus shooting	1
	Marker (1986) and scholarship honoring victim, Edward Cooperman. http://physics.fullerton.edu/student_awards.html. The marker honors others who were murdered or committed suicide in the same period.		
1988, 21 December	Syracuse University (Syracuse, NY)	Off campus terrorist plane bombing (Lockerbie, Scotland)	270
	Thirty-five Syracuse students and eleven students from four other US universities were among the victims. Memorial on campus, in Lockerbie and in Arlington National Cemetery; annual "Remembrance Week"; Remembrance Scholarships for Syracuse students, Lockerbie/Syracuse Trust Scholarships for UK students, "Dark Elegy" statuary assemblage. Various memorials in Scotland.		
1989, 25 July	University of Washington (Seattle, WA)	Off campus shooting	2
	No known memorials.		
1991, 1 November	University of Iowa (Iowa City, IA)	On campus shooting	6*
	Memorial plaque, rock, and tree at Physics Department. Memorial walkway. Memorial at Women's Resource and Action Center.		
1992, 14 December	Bard College at Simon's Rock (Great Barrington, MA)	On campus shooting	2
	Spontaneous memorials, memorial Peace Grove, annual remembrance ceremony.		
1995, 26 January	University of North Carolina (Chapel Hill, NC)	Off campus shooting (Chapel Hill)	2
	One victim honored in same memorial as 1996 fire. No other known memorials.		
1996, 12 May	University of North Carolina (Chapel Hill, NC)	Off campus fire (Chapel Hill)	5
	Memorial at entry to arboretum donated by senior class of 1997.		
1996, 15 August	San Diego State University (San Diego, CA)	On campus shooting	3
	Memorial with tables near engineering building.		
1996, 17 September	Pennsylvania State University (University Park, PA)	On campus shooting	1
	Spontaneous memorials and Peace Garden near site of shooting. http://www.campusmaps.psu.edu/buildings/peace.shtml		
1996, 16 October	Purdue University (West Lafayette, IN)	On campus shooting	2*
	Spontaneous memorials and memorial fund to prevent drug abuse.		
1997, 21–22 Dec.	University of Colorado (Boulder, CO)	Off campus murder (Boulder)	1
	Spontaneous memorial, memorial tree and plaque, scholarship and campus safety fund to honor victim, Susannah Chase.		

Date	University	Event	Deaths
1998, 7 October	University of Wyoming (Laramie, WY)	Off campus murder (Laramie)	1
	Spontaneous memorial, campus bench, web memorial, foundation. Loffreda (2001) http://www.uwyo.edu/News/shepard/default.htm, http://www.matthewshepard.org/site/PageServer		
1999, 18 November	Texas A&M University (College Station, TX)	On campus collapse of bonfire construction	12
	Spontaneous memorials and permanent memorial on site of bonfire collapse (2004). Smith (2007), Grider (2000) http://www.tamu.edu/bonfirememorial/		
2000, 13 January	Kenyon College (Gambier, OH)	Off campus van crash (Coshocton, OH)	1
	Memorial fund honoring student victim, Molly Hatcher.		
2000, 19 January	Seton Hall University (S. Orange, NJ)	On campus arson at Boland Hall	3
	Spontaneous memorials and garden.		
2000, 10 February	Prairie View A&M University (Prairie View, TX)	Off campus van crash (Karnack, TX)	4
	No known memorials.		
2000, 28 June	University of Washington (Seattle, WA)	On campus shooting	2*
	Memorial lecture and endowment fund		
2001, 27 January	Oklahoma State University (Stillwater, OK)	Off campus airplane crash (Strasburg, CO)	10
	Spontaneous memorials, memorial at crash site (2001) and on campus memorial. http://osu.okstate.edu/Colorado/		
2001, 27 January	Dartmouth College (Hanover, NH)	Off campus murder (Etna, NH)	2
	Zantop Memorial Garden and Zantop Memorial Lecture, http://www.dartmouth.edu/~earthsci/zantop.html.		
2001, 18 May	Pacific Lutheran University (Tacoma, WA)	On campus shooting.	2*
	Memorial on campus, web memorial, http://www.plu.edu/about/tours/ar7.html, http://www.plu.edu/print/holloway/		
2002, 16 January	Appalachian School of Law (Grundy, VA)	On campus shooting	3
	http://www.asl.edu/memorial/resolution_204.php, http://www.asl.edu/memorial/index.php. No other known memorials.		
2002, 28 November	University of Arizona (Tucson, AZ)	On campus shooting	4*
	Spontaneous memorials. "Tree of life" artwork at Nursing College. Victims also honored on campus Women's Plaza of Honor.		
2003, 9 May	Case Western Reserve University (Cleveland, OH)	On campus shooting	1
	Scholarship, service award, and leadership award created in name of victim, Norman Wallace.		
2006, 26 April	Taylor University (Upland, IN)	Van struck by truck (near Marion, OH)	5
	Spontaneous memorials, vernacular memorial at accident site (I-69 near Marion). Prayer Garden on campus.		
2006, 3 September	Shepherd University (Shepherdstown, WV)	On campus shooting	3*
	Spontaneous memorials, memorial bench near site of shooting.		
2007, 2 March	Bluffton University (Bluffton, OH)	Bus crash (Atlanta, GA)	7
	Spontaneous memorials, on campus memorial at baseball field, memorial in Atlanta.		

Date	University	Event	Deaths
2007, 16 April	Virginia Tech (Blacksburg, VA) *Spontaneous memorials. Permanent memorial, http://www.vt.edu/remember/.*	On campus mass murder	33*
2007, 2 April	University of Washington (Seattle, WA) *Endowed scholarship.*	On campus shooting	2
2007, 4 August	Delaware State University (Dover, DE) *No known memorials.*	Off campus shooting (Newark, NJ)	3
2007, 21 September	Delaware State University (Dover, DE) *No known memorials.*	On campus shooting	1
2007, 30 September	University of Memphis (Memphis, TN) *Spontaneous memorials. No permanent memorial currently planned.*	On campus shooting	1
2007, 28 October	University of South Carolina (Columbia, SC) and Clemson University (Clemson, SC) *Spontaneous memorials at both campuses. Sorority and web memorials at USC. No permanent memorials to date. http://www.sc. edu/forevertothee/news_1028b.html, http://www.sc.edu/forevertothee/index.html*	Off campus fire (Ocean Isle Beach, NC)	7
2008, 18 January	Duke University (Durham, NC) *Fellowship created honoring victim, http://abhijit.mahato.pratt.duke.edu/*	Off campus murder	1
2008, 8 February	Louisiana Technical College (Baton Rouge, LA) *Spontaneous memorials and ceremonies, memorial trust fund, memorial garden.*	On campus shooting	3*
2008, 14 February	Northern Illinois University *Spontaneous and web memorials. http://www.niu.edu/memorial/index.html Permanent memorial planned.*	On campus shooting	6*
2008, 5 March	University of North Carolina (Chapel Hill, NC) *Spontaneous memorial, permanent memorial planned, scholarship to honor victim and student body president, Eve Carson. http://www.unc.edu/eve/index.html*	Off campus murder (Chapel Hill)	1
2008, 26 October	University of Central Arkansas (Conway, AR) *Spontaneous memorials, permanent memorial (benches). http://www.uca.edu/remember/*	On campus shooting	2
2009, 21 January	Virginia Tech (Blacksburg, VA) *Spontaneous memorial.*	On campus knifing	1
2009, 7 February	University of Houston (Houston, TX) *No known memorials.*	On campus shooting	1
2009, 10 April	Henry Ford Community College (Dearborn, MI) *No known memorials.*	On campus shooting	2*

2) occurred in an off-campus site closely associated with a university (such as a sorority or fraternity house); 3) involved a group of faculty or students off campus on a university activity (a travelling sports team or students studying abroad); 4) claimed at least one life beside the perpetrator.

Beginning with cases we studied previously (Foote 2003, Grider 2005), we gathered more examples from newspapers, the web, online databases such as Lexis/Nexis and those developed by newspapers, and other print sources. We did not attempt to inventory all individual on- or off-campus deaths of college students, faculty and staff over the same period; there are too many. With approximately 4,391 colleges and universities (Carnegie Foundation 2009) there are hundreds of thousands of students, faculty, and staff involved in higher education in the US. Each year some of these die of natural causes, accident, and homicide. Few of these result in memorialisation other than at the gravesite, but some do. For this reason, we have included some individual deaths in this analysis. But, as we note below to qualify our findings, memorialisation of individual deaths tends to be the exception rather than the rule.

Information about the outcomes of the tragedies was gathered primarily from secondary sources because many of these events are very thoroughly documented in newspapers, articles, books, and the web. In cases where information was scarce, we queried news and information offices, archives, libraries, and administrative officers. We sought information about all types of remembrance including vernacular memorials; permanent memorials and markers; scholarships and endowments; annual or periodic commemorations; living memorials and other ways the victims were remembered, mourned, or memorialised.

Findings

Memorialisation of college and university tragedies has taken many forms, and the repertoire of responses seems to have increased through time. Permanent memorials – art, statuary, gardens, reflecting pools and ponds, buildings, benches, plaques – are perhaps the most common and have a long history among US colleges within the larger tradition of honouring heroes, victims and martyrs with place names, street names, and building names. Since the late 1980s or 1990s, spontaneous memorialisation has become very common and now occurs after almost all tragedies. These memorials vary in size and form, but usually include flowers, candles, notes, signs, posters, religious objects, and other mementos left at a tragedy site. Spontaneous memorials, however, did occur earlier, for instance at Kent State. Even though the campus was closed immediately after the shootings, in the following months and years people left candles and other mementos to mark where the four students were slain. There were also candlelight vigils and other memorial services, not only at Kent State but on other campuses throughout the country. But the lack of documentation often makes it difficult to know for certain whether spontaneous shrines occurred in the past as often as they do today.

In Table 10.1, we use the phrase 'no known memorials' to qualify the results of our research. Particularly for earlier events, it is difficult to say with certainty that no memorialisation occurred. This is especially true of spontaneous memorials because they are so rarely documented. Sometimes they appear in news photographs or are noted by reporters, but in other cases no record exists of what happened in the immediate aftermath of the tragedies. Also, not all campuses keep inventories of markers and memorials, so the existence of some may have been forgotten.

Increasingly, 'living memorials' have become more common, including scholarships, fellowships, endowments, and professorships, the point of which is to develop ways of engaging students and faculty in keeping the memory of victims alive. A noteworthy example of this practice are the *Remembrance Scholarships* established at Syracuse University and the Lockerbie/Syracuse Scholarship Trust award scholarships in honour of the victims of the 1988 bombing of an airliner over the Scottish town of Lockerbie. Other events have been memorialised in books, poetry, art work, movies, posters and television. Memory, in this sense, includes all the ways in which an event or its victims are presented and represented through time in whatever form or media. That is, the invention of historical traditions involves more than the creation of permanent, spontaneous and living memorials, but is instead the way an event is retold and relived through time. On some campuses, the collection, documentation, and archiving of shrine artefacts becomes yet another venue for shaping the memory of the event. Syracuse, Texas A&M, Virginia Tech, and Oklahoma State are some of the universities which have created special artefact collections, in addition to their document collections devoted to their tragedies. Increasingly these memorial forms have moved into cyberspace as memorial websites and blogs and into wikis in ways that make private grieving accessible to a worldwide audience.

All of the accidental tragedies in our sample, as opposed to many of the murders, have been commemorated with the exception of the Prairie View Texas A&M van crash of 2000 (Figure 10.2). These include train, air and bus crashes, fires, and other similar events dating back to the Purdue University train crash of 1903. With the exception of the Taylor University tragedy and Lockerbie bombing (which claimed students from several universities), all of the transportation accidents involved sports teams. These teams, along with fraternities and sororities, are often among the most tightly knit sub-communities on campus.

On the other hand, murders on or off campus have only recently been acknowledged publicly. The South Carolina State University (1968), Kent State University (1970) and Jackson State University (1970) shootings are the first such events we can identify which were memorialised, both in the immediate aftermath (within a year) and later. These were key events in the civil rights movement and in the protests against the Vietnam War and, as a consequence, highly political in connotation. The first memorialisation of non-political murders on a US campus was for the victims of the mass shooting at California State University at Fullerton in 1976.

Figure 10.2   The Immortal Ten memorial at Baylor University (Waco, TX) honouring the members of the campus basketball team killed in 1927. An example of the many memorials which honour accidental campus tragedies
Source: Photograph reproduced with the permission of Baylor Photography.

There seems to be no patterning in terms of institution size, whether a college is public or private, or by other general institutional characteristics. There are not enough cases in our study to explore these relationships in detail. Five of the sampled institutions have religious affiliations (Seton Hall, Bluffton, Taylor, Pacific Lutheran and Baylor) which could potentially influence commemorative practices, and all have memorialised their campus tragedies but, again, the number is too small for meaningful comparison with secular institutions. Four of the campuses are historically African American colleges (South Carolina State, Jackson State, Prairie View A&M and Delaware State). The memorialisations at South Carolina State and Jackson State seem to be important precedents for the marking of murders on campuses, especially since they involved civil rights activism. The Prairie View A&M and Delaware State tragedies have not yet led to memorialisation.

Off-campus tragedies often lead to two memorials, one at the site of death and the other on campus. This is particularly true of the accidents involving travelling sports teams. Marking the actual death site is, of course, a defining characteristic of spontaneous shrines in general and roadside memorials in particular. The marking of sites of plane crashes may therefore be related to the vernacular practice of memorialising car wreck fatalities at the site. For example, the memorial marking the 1970 Wichita State plane crash that killed 31 features a small white cross which can be seen from the highway and, at first glance, does resemble a roadside memorial. However, many students die off campus and are not memorialised on campus. It tends to be the tragedies claiming more lives (which includes many of the sports team accidents) that result in memorialisation. Lockerbie, the largest of the tragedies considered here, is exceptional because it resulted in additional off-site memorials, such as the one dedicated in 1995 in Arlington National Cemetery but, of course, it claimed hundreds of lives.

The tradition of honouring major losses among a university's community has a long history, predating some of the memorialisation which occurred on campuses after the First and Second World Wars. At the same time, two changes have occurred, though gradually. First, as noted above, is that instances of homicide and mass murder did not begin to be marked until the late 1960s starting with South Carolina State University (1968) and continuing with Kent State and Jackson State (1970) (Figure 10.3). Through the 1970s and 1980s such events were occasionally honoured until, by the 1990s, they were marked regularly.

Second, through time, the trend has been toward increasingly public displays of grief as evidenced in the rise of vernacular memorialisation. But, again, this happened gradually. Outpourings of grief were often, though never exclusively, focused on the funeral, gravesite, and home. Though these continue to be important, death sites and public spaces now seem equally important venues for expressing and sharing grief. Although the basic form and content of spontaneous shrines memorialising campus tragedies have remained fairly stable over time, there has been a rise in their frequency and the speed with which they are created. These trends seem to reflect changing patterns of vernacular memorialisation in US society at large.

The tendency is for contemporary memorialisation, whether spontaneous or permanent, to occur at the place of death. Other sites include places significant to the victims or sites of meaning to the university. For example, the memorial to Elauri Jaquette, a student killed on the University of Colorado at Boulder campus in 1966, was not placed at the murder site but along a stream where she enjoyed studying. Following the murder of three members of the nursing faculty at the University of Arizona in 2002, memorials were created both at the College of Nursing and as part of the campus Women's Plaza of Honour dedicated in 2005. On many campuses, like Arizona, meaningful sites for new memorials have evolved over many years through a process of accretion near previous memorials, in proximity to buildings marking a campus's history, or sites associated with major campus rituals and celebrations of community (Foote 2003, Dwyer 2004). These preferences for situating memorials do not seem to have changed much across the cases considered here.

The form of these spaces of mourning has also changed through time. Permanent memorials have generally moved away from figurative representations of victims to more abstract symbols of loss and vernacular memorials, though

Figure 10.3   Campus memorials from the civil rights and Vietnam War
period: a) Jackson State University (Jackson, MS) (*Source*:
Photograph by the author); b) death site of one of the Kent
State students, now permanently marked (*Source*: Photograph
by the author); c) Kent State University (Kent, OH) (*Source*:
Photograph by the author); d) South Carolina State University
(Orangeburg, SC) (*Source*: Photograph reproduced with the
permission of Owen Dwyer). These memorials set precedents
for honouring victims of on-campus violence

often sharing many characteristics, have grown in size and complexity in recent years. These changes seem to follow wider contemporary trends in public art and vernacular memorialisation. With respect to permanent memorials, traditional symbols are still common; for example the use of trees as symbols of life; plaques on rocks as symbols of permanence and eternal remembrance; gardens, pools and benches as places for reflection or prayer. The tendency is to include only the names of the deceased and perhaps birth and death dates, with no mention of the event which caused the deaths. Figurative art continues to be used on some occasions, for example in Baylor's Immortal Ten memorial or the bas-relief portraits of the Bonfire Memorial at Texas A&M University. Perhaps the trend is not to develop entirely new symbols for death, dying and bereavement, but rather to blend traditional symbols into complex new arrangements. The ring of the Bonfire Memorial encircles the footprint of the stack of logs that collapsed, but also represents a very common symbol of community, love, and unbroken faith. The gateways for the individual victims, which face each of their hometowns, recall imagery of the passage from life to death as well as the form of ancient funeral stele. Gateways at other sites also act as grottos or as the side-chapels of churches, allowing space for privacy and reflection (Figure 10.4).

Discussion

Rather than being a recent phenomenon, spaces of mourning and remembrance on American college campuses seem to be related to precedents dating back a century or more. Universities have a long history of celebrating the accomplishments of their faculty and students as well as marking their losses. The precedents are most clearly seen with respect to accidents, which with only one or two exceptions have all been memorialised. The site of the tragedy (on-campus or off-campus) does not seem to have a major impact on mourning and remembrance. Rather, how an event is seen to impact the university community is more important, which involves many factors as discussed below.

Substantial change has occurred, however, in the range of events which result in spaces of mourning, the increasingly public nature of mourning, and the speed with which memorialisation takes place. The most striking of these changes involves honouring victims of individual homicides and mass murder. Until relatively recently, such events went unmarked. Now the victims are honoured in the same ways as those of other tragedies.

But this change did not occur all at once. Although tracing the roots of this trend will require further research, our findings seem to indicate a transition beginning in the late 1960s spurred by the civil rights and anti-war events at South Carolina State University (1968), Kent State University (1970) and Jackson State University (1970). These were highly politicised killings and led to annual commemorative rituals and a number of markers and memorials through time. These seem to be the

**Figure 10.4   Memorial at Texas A&M University (College Station, TX) on the
site of the bonfire collapse of 1999, an example of how symbolic
forms are blended together in some contemporary memorials.
These visitors to Jamie Hand's gateway are from her hometown
and members of the baseball team on which she played in high
school**

Source: Photograph by the author

first instances we can identify in which violent death on a college campus resulted in new spaces of mourning.

But Kent State and Jackson State came at the end of a very violent decade. During the 1960s the John F. Kennedy, Martin Luther King, Jr., and Malcolm X death sites became sites of pilgrimage and mourning. The Vietnam War and the civil rights movement resulted in highly politicised places frequently marked by violence and protest. Many of the sites of violence of the civil rights movement have now been marked, too (Dwyer and Alderman 2008b).

At the same time, the change cannot be traced to a single watershed event. For instance, the anti-war bombing at the University of Wisconsin in 1970 – the same year as Kent State and Jackson State – resulted in a very modest marker only many years later. The victims of the 1966 mass murder at the University of Texas at Austin were not honoured publicly until 2000.

The impact of South Carolina State, Kent State and Jackson State are important, but we see only a gradual increase in spaces of mourning designed to honour victims of violence. The 1970s and 1980s were decades that were periods of transition, with some violent events commemorated (California State Fullerton and the Lockerbie bombing, for instance), but most were not. However, by the 1990s the victims of almost all events of violence are commemorated. So a transition spanning the period from the late 1960s to the 1990s resulted in victims of murder and mass murder being honoured like victims of other tragedies.

The increasingly public character of grieving can be seen in the rise and prevalence of spontaneous memorials. Although we cannot state our conclusion with complete certainty since some earlier memorials may not have been documented to the extent they are today, the trend toward marking events of tragedy and violence with vernacular memorials apparently began during the 1980s. Previously, such symbols of loss and grief tended to be confined to the more private domain of funerals and gravesites. From the 1980s onward, such symbols moved increasingly into the public realm to honour, for example, the victims of killings at the University of Iowa (1991) and Bard College at Simon's Rock (1992). Other precedents outside of academia include the site of the McDonald's mass murder in San Ysidro, California (1984); the Luby's Cafeteria shootings in Killeen, Texas (1991); the Branch Davidian deaths outside Waco, Texas (1993); and the Oklahoma City bombing (1995). Many scholars have proposed that the vernacular memorialisation seen at the Vietnam Veterans Memorial, completed in 1982, was a key factor in the widespread public acceptance and practice of vernacular memorialisation, but our data seems to indicate a somewhat earlier origin.

Outside the US, major school tragedies included the École Polytechnique Massacre in Montreal in 1989, the victims of which were widely memorialised in Canada, and the Dunblane Primary School massacre (1996) in Scotland. This period also saw, among others, major episodes of public grieving after the Heysel Stadium disaster in Brussels (1985), the Hillsborough Stadium disaster in Sheffield (1989), and the death of Princess Diana (1997). Further research is needed to determine the extent to which events such as these in Canada, Europe and elsewhere have influenced or been influenced by the patterns of memorialisation explored in this chapter. We suspect that the precedents for permanent memorialisation in other countries may date back further historically than is commonly assumed, just as has been found among the cases examined here. Patterns in other countries may also be traced to unique local and national contexts, just as the civil rights and anti-Vietnam War movements seem important in shaping how violence is now commemorated on US campuses. What is unmistakable, however, is the similarity in the US and in other western societies of vernacular memorialisation from the mid-1980s onward.

This change in vernacular memorialisation is mirrored in the speed with which tragic events are commemorated, especially the creation of spontaneous shrines within hours or days of the event. Whereas the victims of accidental events have almost always been honoured relatively quickly, this speed is now common after almost all tragic events. This trend mirrors the point made by Post et al. (2003) that, through time, a repertoire of rituals and memorial activities has evolved, such as candlelight vigils, silent processions and vernacular memorials, which are now commonly used after tragedies of all sorts.

The more general acknowledgement or recognition (both within and outside academia) of violence and tragedy means that the victims and consequences of some tragic events – once ignored – are now being recognised retrospectively. A point raised by Foote (2003) is the importance of key anniversaries in the transformation of sites of tragedy, particularly those anniversaries 30–50 years after an event, such as the Second World War or the Holocaust, when the survivors with first-hand memories seek to leave lasting testimony. Thus, we see major memorials being created only recently for the victims of the 1966 University of Texas mass murder and the 1927 Baylor University bus crash.

It is difficult, however, to generalise from these patterns. Certainly large, highly politicised events help shape general trends in how commemorative rituals and spaces are created, but even individual homicides and deaths can have an effect. Highly visible deaths such as the murder of UNC Chapel Hill's student body president Eve Carson (2008) and of popular senior Susannah Chase (1997) at the University of Colorado have resulted in memorials, but most others have not. Those events which are seen to involve 'outsiders', such as the murder of his Iranian girlfriend at the University of Washington in 1989 by a deserter from the Iranian army who had entered the US illegally, or are interpreted largely as 'accidents' which could have occurred anywhere (but just happened to occur on a campus) are less likely to result in memorial spaces. On the other hand, the internationally publicised collapse of the bonfire on the campus of Texas A&M University in 1999, which killed twelve students, was memorialised immediately by several large spontaneous shrines, and a permanent memorial was dedicated five years later.

Perhaps the key idea is determining what events come to be viewed as ruptures of the campus community, and subsequently result in memorials and other forms of remembrance. Sometimes this is the death of the student body president, a crime that has political connotations, or an accident that strikes at the heart of a venerated campus tradition.

There is, inevitably, a nether world of commemoration and remembrance; events which fall between the categories discussed here. In the decades before it was acceptable to acknowledge victims of violent death, some events were remembered in campus legends and other types of oral tradition, for example, the sensational murders of female students at Penn State in 1969 (stabbed in the library stacks) and Colorado University in 1966 (bludgeoned in an auditorium practice room). Both of these murders have become subjects of urban legends and 'haunted' tours; small markers decades after the murders honour the victims.

In such cases, these acts of remembrance may be transgressive. They insist on recognising events that expose social and cultural tensions which remain unacknowledged, for example, the sometimes extreme male violence against female victims in the homicide cases we surveyed. As more memorials to female victims of male violence are created (though these memorials tend to be silent as to cause of death), it may make this violence more visible and, perhaps, spur action.

Conclusion

Contrary to Doss's (2008) assertion of a recent 'mania' for memorials in the US, we see in campus tragedies patterns that have remained remarkably consistent over more than a century with changes occurring gradually from the late 1960s onward. The most important of these have been to: honour a greater range of events (such as homicide and mass murder); acknowledge grief more publicly through vernacular memorials; and initiate memorialisation sooner than in the past. However, we suggest these trends with caution. The nature of our sample, the largest and most high-profile campus tragedies, means that our cases are skewed toward those more likely to lead to memorialisation and under represent the many other deaths that go unmarked. We suspect that overall public memorialisation remains a relatively rare outcome.

This conclusion cautions against generalising too far beyond the cases considered here to other school tragedies outside higher education, to campus tragedies outside the US, or to tragedies in general. Colleges and universities in the US often cultivate a strong sense of community that, when faced by tragedy or adversity, leads naturally to reaffirmation of the community's sense of common purpose. This same sense of common purpose is not always present when tragedy strikes in other places or groups. Even US public primary and secondary schools are different from college and university campuses because enrolment is generally compulsory. Efforts to create a sense of 'school spirit' can succeed, but colleges and universities have a head start because students volunteer to matriculate, often because they are attracted to a particular college's sense of community.

But this observation also helps explain why, in some cases, non college-related tragedies result in little public memorialisation; commercial aeroplane and other transportation accidents, for instance. In these situations, the lack of a sense of community among survivors and victims' families means responses remain private.

Our findings raise as many questions as they answer. They imply looking more carefully at the 1960s and 1970s as a key period in the development of contemporary patterns of memorialisation in the US. They also suggest the need to look very carefully at the ways different expressions of memory are interwoven. Whereas some scholars focus on vernacular memorialisation, others on permanent monuments, and still others on the representation of tragedy in film, television, the news media, and in popular culture, we have undertaken a broad approach which considers how these forms interact both to express and to shape contemporary attitudes toward death, dying and bereavement. Perhaps the most important next step for research in this area, as mentioned above, is to explore how the patterns seen in the US influence, reflect, compare and contrast with those in other countries and cultures.

References

Alston, M. Director. 2006. *The Killer Within*. Discovery Films.

Ashworth, G.J. and Hartmann, R. 2005. *Horror and Human Tragedy Revisited: The Management of Sites of Atrocities for Tourism*. New York: Cognizant Communication Corporation.

Azaryahu, M. 1996. The spontaneous formation of memorial space: the case of *Kikar Rabin*, Tel Aviv. *Area*, 28(4), 501–13.

Azaryahu, M. 1999. McDonald's or Golani Junction? A case of a contested place in Israel. *Professional Geographer,* 51(4), 481–92.

Bass, J. and Nelson, J. 1999. *The Orangeburg Massacre*. Macon, GA: Mercer University Press.

Bell, C. 1992. *Ritual Theory, Ritual Practice*. New York: Oxford University Press.

Bell, C. 1997. *Ritual: Perspectives and Dimensions*. New York: Oxford University Press.

Bills, S.L. (editor). 1982. *Kent State/May 4: Echoes through a Decade*. Kent, OH: Kent State University Press.

Carnegie Foundation for the Advancement of Teaching. 2009. Classifications data file (last updated 19 June 2009). [Online]. Available at: http://www.carnegiefoundation.org/classifications/index.asp?key=809 [accessed: 1 August 2009].

Copeland, T. 2006. *The Immortal Ten: The Definitive Account of the 1927 Tragedy and Its Legacy at Baylor University*. Waco, TX: Baylor University Press.

Cosgrove, D.E. and Daniels, S. (editors). 1988. *The Iconography of Landscape: Essays on the Symbolic Representation, Design, and Use of Past Environments.* Cambridge: Cambridge University Press.

Cram, B. and Richardson, J. Directors/Producers. 2009. *Scarred Justice: The Orangeburg Massacre 1968.* Northern Lights Productions.

Doss, E. 2002. Death, art and memory in the public sphere: the visual and material culture of grief in contemporary America. *Mortality*, 7(1), 63–82.

Doss, E. 2008. *The Emotional Life of Contemporary Public Memorials: Towards a Theory of Temporary Memorials.* Amsterdam: Amsterdam University Press.

Duling, J.A. 2007. Healing and hope: community redefined. *Bluffton* 4(3), 4–9.

Dwyer, O.J. 2004. Symbolic accretion and commemoration. *Social and Cultural Geography*, 5(3), 419–435.

Dwyer, O.J. and Alderman, D.H. 2008a. Memorial landscapes: analytic questions and metaphors. *GeoJournal,* 73, 165–78.

Dwyer, O.J. and Alderman, D.H. 2008b. *Civil Rights Memorials and the Geography of Memory.* Chicago: Center for American Places at Columbia College Chicago and University of Georgia Press.

Everett, H.J. 2002. *Roadside Crosses in Contemporary Memorial Culture.* Denton: University of North Texas Press.

Foote, K.E. 2003. *Shadowed Ground: America's Landscapes of Violence and Tragedy.* Revised Edition. Austin: University of Texas Press.

Foote, K.E. 2010. Shadowed ground, sacred place: Reflections on violence, tragedy, memorials and public commemorative rituals, in *Holy Ground: Re-inventing Ritual Space in Modern Western Culture*, edited by P. Post and A.L. Molendijk. Leuven: Peeters, 93–117.

Foote, K.E. and Azaryahu, M. 2007. Toward a geography of memory: geographical dimensions of public memory and commemoration. *Journal of Political and Military Sociology,* 35(1), 125–44.

Graham, M.W. 2006. Memorializing May 4, 1970 at Kent State University: Reflections on collective memory, public art and religious criticism. *Literature and Theology*, 20(4), 424–37.

Grider, S. 2000. The archaeology of grief: Texas A&M's sad study in modern mourning. *Discovering Archaeology*, 2(3), 68–74.

Grider, S. 2001. Spontaneous shrines: a modern response to tragedy and disaster. *New Directions in Folklore.* [Online]. (5 October). Available at: http://www.temple.edu/isllc/newfolk/shrines.html) [accessed: July 13, 2007].

Grider, S. 2005. Vernacular memorialization of school tragedies: a chronological study. *Australian Folklore*, 20, 80–98.

Grider, S. 2006. Spontaneous shrines and public memorialization, in *Death and Religion in a Changing World*, edited by K. Garces-Foley. Armonk, NY: M.E. Sharpe, 246–64.

Grider, S. 2007. Public grief and the politics of memorial: Contesting the memory of 'the shooters' at Columbine High School. *Anthropology Today*, 23(3), 3–7.

Grimes, R. 2000. *Deeply into the Bone: Re-inventing Rites of Passage.* Berkeley: University of California Press.

Hobsbawm, E. and Ranger, T. (editors). 1983. *The Invention of Tradition.* Cambridge: Cambridge University Press.

Lavergne, G. 1997. *A Sniper in the Tower: The Charles Whitman Murders.* Denton: University of North Texas Press.

Lennon, J. and Foley, M. 2000. *Dark Tourism: The Attraction of Death and Disaster.* New York: Continuum.

Loffreda, B. 2001. *Losing Matt Shepard.* New York: Columbia University Press.

Lukken, G. 2005. *Rituals in Abundance: Critical Reflections on the Place, Form and Identity of Christian Ritual in Our Culture.* Leuven: Peeters.

Margry, P.J. and Sánchez-Carretero, C. 2007. Memorializing traumatic death. *Anthropology Today*, 23(3), 1–2.

Margry, P.J. and Sánchez-Carretero, C. (editors). In press. *Grassroots Memorials: The Politics of Memorializing Traumatic Death.* New York: Berghahn Books.

Nichol, J.M. (McG) Director. 2006. *We Are Marshall.* Warner Brothers.

Norkunas, M. 1993. *The Politics of Public Memory: Tourism, History and Ethnicity in Monterey, California.* Albany: State University of New York Press.

Post, P, Grimes, R.L., Nugteren, A., Pettersson, P., and Zondag, H. 2003. *Disaster Ritual: Explorations of an Emerging Ritual Repertoire.* Leuven: Peeters.

Santino, J. 2004. Performative commemoratives, the personal, and the public: spontaneous shrines, emergent ritual, and the field of folklore. *Journal of American Folklore,* 117(466), 363–72.

Santino, J. (editor). 2006. *Spontaneous Shrines and the Public Memorialization of Death.* New York: Palgrave Macmillan.

Schwartz, B. 1982. The social context of commemoration: a study in collective memory. *Social Forces*, 82, 374–402.

Smith, J.M. 2007. The Texas Aggie Bonfire: A conservative reading of regional narratives, traditional practices, and a paradoxical place. *Annals of the Association of American Geographers*, 97(1), 182–201.

Spofford, T. 1988. *Lynch Street: The May 1970 Slayings at Jackson State College.* Kent, OH: Kent State University Press.

Sturken, M. 2008. *Tourists of History: Memory, Kitsch, and Consumerism from Oklahoma City to Ground Zero.* Durham, NC: Duke University Press.

Texas A&M University. 2009a. Aggie traditions: Muster. [Online]. Available at: http://aggietraditions.tamu.edu/muster.shtml [accessed: 1 August 2009].

Texas A&M University. 2009b. Aggie traditions: Silver taps. [Online]. Available at: http://aggietraditions.tamu.edu/silvertaps.shtml [accessed: 1 August 2009].

Walter, T. 1991. The mourning after Hillsborough. *Sociological Review,* 39(3), 599–625.

Walter, T. 1994. *The Revival of Death.* London: Routledge.

Walter, T. (editor). 1999. *The Mourning for Diana.* Oxford: Berg.

Weiss, K.J. 2008. *The Kent State Memorial to the Slain Vietnam War Protestors: Interpeting the Site and Visitors' Responses*. Lewiston, NY: Edwin Mellen Press.

Williams, C. 2008. *Orangeburg 1968: A Place and Time Remembered*. Orangeburg, SC: Cecil Williams Publishing.

Williams, P.H. 2007. *Memorial Museums: The Global Rush to Commemorate Atrocities*. Oxford: Berg.

Chapter 11

Private Spaces for the Dead: Remembrance and Continuing Relationships at Home Memorials in the Netherlands

Joanna Wojtkowiak and Eric Venbrux

'There is no place like home' and 'home sweet home' are common expressions indicating the significance attributed to the home as a very special place like no other. Our home is not only a physical space but also has great emotional value, a place where we transfer social and cultural norms to the private sphere and where we communicate our personal way of life to family, friends and other visitors (Lawrence 1987). Theorists conceive of the home as part of the self or as a symbol of personal identity (Francis et al. 2005, Lawrence 1987). William James describes the home as an extension of our material self: 'its scenes are part of our life; its aspects awaken the tenderest feelings of affection (James 1891: 292). Lawrence (1987) underlines that besides its psychological and emotional value, cultural, socio-demographic and economic factors also come into play. The occupants' economic background, for instance, affects the size of the house, allowing for more or less private space restricted to the family and shielded against access by the outside world. In this chapter we focus on home memorials for the dead in the Netherlands as examples of private spaces for remembrance practices.

The study is based on two surveys conducted in the Netherlands in 2005 and 2007, which yielded significant numbers of respondents (n = 1212; n = 508). The questionnaires included open and closed questions about home memorials and interpretations of death. The actual data will be discussed below in greater detail, but at this stage we can say that these spaces for the dead have become increasingly popular: In 2005 over 30 per cent of Dutch respondents reported having such a memorial space (Wojtkowiak and Venbrux 2009: 147). This chapter will therefore discuss the meaning of private space in contemporary mourning rituals. We argue that a private space at home is of great importance in the re-invention of individual remembrance rituals and continuing bonds with the deceased in the relatively secular society of the Netherlands. If there is no place like home, we want to show that there is no place like home for mourning.

Re-invention of Mourning Rituals

Death rituals in contemporary Western society have been described as highly personalised and privatised (Garces-Foley 2003, 2005, Laderman 2003, Venbrux et al. 2009). The bereaved do much to shape these rituals, such as the ceremony, the speeches, the music and even the clothes visitors should wear at the funeral. 'Please come dressed in bright colours' is a common request in Dutch death notices. Dying people are also becoming increasingly preoccupied with planning their own funerals (Wojtkowiak 2009). This preoccupation often takes the form of a codicil stipulating the funeral arrangements and signed by the person who is going to be buried or cremated. Ronald Grimes (2002) has drawn attention to the 're-invention' of rites of passage in contemporary Western culture. 'Re-invention' in his understanding is the re-shaping and re-using of traditional rituals or ritual elements to fulfil the need for personal expression. Such a ritual is said to be more efficacious and easier to identify with, at least for those who are not satisfied with traditional rituals. While these ritual re-inventions can be effective, they may fail if certain ritual elements are lacking or personal symbols are not chosen carefully. Whether current re-inventions of death rituals are satisfactory to convey the intended meaning is a very complex question that has been discussed elsewhere (Wils 2008). The appearance of home memorials in the Netherlands in the early twenty-first century may be considered in light of the re-invention of death rituals and the quest for new rituals of mourning (Wouters 2000). In the past, the religious version of the home memorial or house shrine was found predominantly in Catholic households in the Netherlands. Catholic house shrines used to be common but have mostly disappeared from Dutch homes (Margry 2003). By focusing on the concepts of ritual and continuing bonds with the dead in the private space of the home we want to unravel the role of private space in the re-invention of ritual in the bereavement process.

A discussion of ritual almost inevitably deals with religion or religiosity. The Netherlands is relatively secularised; church attendance has declined and Christian beliefs are also decreasing (Dekker et al. 2007, De Graaf and Te Grotenhuis 2008). Although 58 per cent of the Dutch population claim to be members of a church or religious group, a mere 11 per cent say they regularly attend church or mosque worship (CBS 2009: 14 and 42). Institutionalised religion is losing its former impact on everyday life; some state that religion has become a private matter in many Western countries (Cochran 1990). The influence of traditional Christian belief systems and rituals is dwindling in Dutch society, but historically traditional Catholic house shrines featured prominently, particularly in the southern Netherlands.[1] In traditional Catholic homes house shrines consist of sacred objects such as crucifixes and statues of Mary, often combined with candles (Margry 2003).

---

1 For a wonderful overview of all sorts of shrines around the globe we recommend: Martin, J.H. 2002. *Altäre. Kunst zum Niederknien*. Hatje Cantz Verlag.

Since Dutch society is highly secularised, it may be assumed that the home memorials in this study do not necessarily signify membership of a religious group. That is also what we found in our first study: most people with a home memorial are not church members (Wojtkowiak and Venbrux 2009).[2] Furthermore, there was no discernible influence of former church membership or parental church membership on the home memorials. Other features of the home memorial underscore that we are dealing with an intensely personalised form of death ritual; it is 'used' by separate individuals in the context of their homes, together with personal objects belonging to the deceased. We therefore prefer the term 'home memorial' to 'house shrine'. The difference between a shrine and a memorial is that the former is more closely associated with traditional religions, whereas 'home memorial' stresses the importance of memorialising the dead, rather than religious faith. Furthermore, home memorials mostly consist of one or more pictures and personal belongings or small objects associated with the deceased (see Figure 11.1). Besides these items most people have candles and flowers on their memorials. One person arranges the memorial for one or more deceased. Most home memorials were for deceased parents or partners. The place of the memorial is often in the 'living area' of the house: the living room or hallway.

Figure 11.1   Home memorial consisting of photograph, candles, mourning card and other personal objects

Source: Photograph by the author.

---

2   Secular people form the biggest group (in absolute numbers), but the tendency to have such a memorial is strongest among Catholics (the percentage is higher).

Creating a Place for Remembrance: The Cemetery and the Home

A common place for remembrance after death is the cemetery. Visiting graves gives mourners both public and private spaces to remember the dead. The actual grave is the most private, followed by other graves and monuments in the cemetery that form a more anonymous background. Studies of 'cemetery behaviour' show the importance of the grave in maintaining emotional bonds with the deceased, self-reflection by the bereaved and the attribution of spiritual and religious meaning (Francis et al. 2000, 2005). The cemetery, therefore, plays a vital role in creating space for remembrance. The Netherlands has recently seen a re-emergence of collective services at the cemetery on occasions such as All Souls' Day, the traditional Catholic feast to remember the dead (Quartier et al. 2008). Visiting the grave or the cemetery in general creates ritual space and makes it special for multiple reasons. First of all, it is the place where the deceased is buried or the ashes are scattered. The body, the most direct material symbol of the deceased, has its last resting place in the graveyard. It is permeated with vivid memories of the last goodbye. Second, the trip to the cemetery requires mental and physical preparation; you have to make the choice to go to the cemetery, decide on a day and a time for the visit, make the trip by car, bus or bike, and perhaps even buy flowers or some other grave gift. The nature of the cemetery obviously makes it a good place for remembrance rituals.

The home, by contrast, is a place of everyday life where the distinction between ritual space and everyday space is less clear. The home is not the place where memories of the funeral are captured; on the other hand, it is where memories of the life of the deceased are more vivid and less 'monumentalised' than at the cemetery. Home is where one shared meals, conversations and other private interactions. The home is less 'special' than the grave, but it offers the most personal place for remembering the deceased. The home represents the most private moments and memories of life with deceased loved ones. Moreover, at home people feel freer to express their feelings without a sense of being watched. Whereas public commemoration constructs the political identity of the deceased (Stengs 2009), private commemoration maintains the more intimate, social identity (Wojtkowiak and Venbrux 2009). Home memorials create space to express bereavement in a safe environment: one would probably not put grandpa's treasured watch on his grave.

Jonathan Smith's theory of ritual space is a significant contribution to ritual studies (Smith 1987). He argues that space is the most important determinant of the meaning given to ritual elements; for example, in everyday life water and blood have different meanings than in ritual space. Smith speaks of different 'nows' in ritual (Smith 1987: 110). In ordinary life water can spread impurity, but ritually it is used for cleansing us of impurity. A careful reader might observe that the distinction is not clear-cut. To give an example, water is also used to clean things in everyday life. Hence place alone does not determine the meaning of ritual or ritual elements. Grimes's (1999) comment on Smith's theory underlines that place is not the distinctive element of ritual. Whereas Smith pays most attention to space, Grimes attaches, if anything, more importance to action. The symbolic values underlying ritual actions make it possible to express things that are sometimes difficult to articulate. Smith writes that 'ritual is, first and foremost, a mode of paying attention […] place directs attention' (Smith 1987: 103). In addition he cites the example of ritual place as an obvious hallmark of ritual: 'the temple serves as a focusing lens' (Smith 1987: 104). The temple manifestly marks an event as a ritual, since you go to the temple for it. Based on what we said about cemeteries, following Smith, the graveyard, or more specifically the graves, can be seen as a focusing lens for certain ritual actions and moments of mourning. In the case of home rituals, the place as such does not immediately mark them as rituals. Nevertheless a home memorial creates a ritual space for mourning and remembering the deceased. Hence, adapting Smith's theory, we speak of (private) ritual *space* rather than place. We want to indicate that a home memorial serves as a focusing lens for ritual space and continuing bonds with the dead.

Reflecting on the grave and the home as places for remembrance rituals raises the question of the roles of collectivity and individuality in personal remembrance (Quartier 2009). Given the added dimensions of private and public, one might expect the home to be the setting for more individualised rituals as opposed to the cemetery, which is a public space with both collective and individual dimensions. Domestic memorials for more than one deceased, sometimes even 'for all the dead', have a stronger collective dimension. But having a home memorial does not mean that visits to the grave or public commemoration are not conducted. The home becomes a place for mourning in everyday life, whereas the grave is visited on special occasions or special dates. Sometimes the grave is also visited daily, especially in the period immediately following a person's death. Collective commemorations, such as re-inventions of All Soul's Day, have a strong togetherness dimension: 'I am not alone in my grief'. Home memorials, by contrast, have a marked 'I want to be alone with my grief' connotation; the private relationship with the deceased is expressed in the privacy of the home.

Religio-spiritual and Psychosocial Explanation of Home Memorials

Having discussed the role of place in private and public remembrance rituals, we now analyse the home memorial phenomenon in the Netherlands in greater detail. To this end we discuss two major explanations of the home memorials phenomenon: a religio-spiritual and a psychosocial explanation. The religious explanation focuses on religious dimensions of home memorials and is based on questions about (a) afterlife beliefs, (b) belief in a higher power, and (c) ritual activity associated with the memorial. The psychosocial explanation adds contextual dimensions such as (d) emotions accompanying the memorial, and (e) the relationship with the deceased. This distinction between 'religious' and 'psychosocial' is not an antithesis; it does not imply religious versus psychosocial, but clarifies the importance of both dimensions in the phenomenon of memorials for the dead.

The following discussion is based on the results from two large-scale surveys conducted in the Netherlands in 2005 and 2007. The first sample consisted of 568 males (46.9 per cent) and 644 females (53.1 per cent) recruited from the Dutch speaking population based on postal code areas. Participants ranged in age from 20 to 72 (Mean = 50 years, Standard Deviation (SD) = 14). Various educational backgrounds were represented in the sample, ranging from lower general secondary school to university educated. The percentage of religiously unaffiliated participants was 60 per cent. Of the 40 per cent who identified themselves as Christians, 20 per cent were Catholics and 15 per cent Protestants. Finally, 5 per cent were unspecified. We use additional information from our follow-up study of home memorials in 2007 (Wojtkowiak and Venbrux 2009). The quotations later in this chapter are from the follow-up study.

We found that 34 per cent of the Dutch population has or had a private memorial in their homes. In addition 80 per cent of people who have had a home memorial maintained it for longer than a year. This shows that memorials for the dead are not a temporary phenomenon briefly after the death of a loved one, but are longer-term memorial spaces. Home memorials are found in all age groups, among both males and females, all educational backgrounds and both religiously unaffiliated and affiliated people. A detailed profile of home memorial keepers is given elsewhere (Wojtkowiak and Venbrux 2009). In the follow-up study we asked people when they installed the memorial, which yielded a spread from 1954 to 2007. The frequencies showed an increase in home memorials from the late 1990s (around 1998) onwards. The increase in the number of home memorials in the 1990s is not surprising, as they are in line with more general developments in Dutch death culture. This period is often characterised in terms of increasing professionalism, education and an increasing openness towards death. We now turn to the focus of this chapter: the role of private space in remembrance rituals.

Private Versus Public Mourning?

> What a man does and says in public is but a fraction of him. There is what he does in private, and the reasons he gives for doing it. But even this is not enough. Beyond what he says is what he will not say but knows, and finally, what he does not know. (Murray 1940: 159)

In this quotation Murray highlights the complexities of private and public space in relation to self-perceptions. The 'public' in a society defines what is 'private', and vice versa. The home in the Netherlands is primarily a private space, but can become public when we have visitors with whom we are not familiar. Certain aspects of home memorials make them more private than graveside or public commemorations of famous figures. Home memorials accentuate the private and social identity of the deceased and the bereaved. In terms of the tripartite structure of rites of passage, the related behaviour more closely resembles an integration rite than a separation or transition rite (Van Gennep 1960). Integration takes places on two levels of identity: on the one hand the bereaved are reconstructing their self as divided from the deceased (who is physically dead); on the other hand they are constructing their relationship with the dead, and therefore reconstructing the identity of the deceased, the latter's postself, by maintaining a home memorial. The expression of the relationship between the bereaved and the deceased is largely, but not entirely, private. The home memorial is set up in private space, but it is not invisible. This means that visitors recognise the memorial as such, for example because of the picture of the deceased that forms the centrepiece of almost all memorials. Another point is that most home memorials are set up in the living room, which of all rooms in the house is most public.

'What is private, what is public changes rapidly' in a society, states Cochran (1990: 24). Besides, what is considered private depends on various factors: culture, history, educational and economic background, family history, and family traditions. Cochran points out that it makes no sense to define private and public in terms of certain domains of life, and the same is true of places of remembrance. Consider, for example, attending the public commemoration of a famous person at the central market of a big city. On first sight, defined by place, this would seem a very public domain of remembrance. But now imagine that you came here, not only to show compassion, but primarily to memorialise your deceased loved one who died in the same circumstances as the famous person, which is also the reason why you were so touched by the latter's death. Every time you hear about the death of the famous person, you remember the pain and sorrow of losing your loved one in the same way. Public and private places do not necessarily define private and public remembrance. Cochran usefully distinguishes between private and public matters in terms of social inclusion and exclusion (Cochran 1990), but both are important in private and public life.

Home memorials include the views of family, friends and sometimes strangers, but, in contrast to a grave, exclude general public views. Furthermore, home memorials include aspects of everyday life in an everyday living space, but are excluded from visits to the cemetery as part of remembrance preparation. The home memorial also excludes the influence of nature and public laws, which very much affects the cemetery. Home memorials exclude social identification with a bigger group, such as during a public commemoration. You might also say that home memorials exclude large-scale togetherness, but include private togetherness between the memorial keeper and the deceased. There are social conventions that operate in public but not in the home (Cladis 2003). Talking to the dead may come more easily at home than in the graveyard, but people do talk to the dead in cemeteries, at specific graves and in private conversation. This private space seems to be associated with a homeliness that is also expressed in graveyards. In *Secret Cemetery* the authors argue that 'just as the home is the material symbol of the self and the family, so the tomb, as home, comes to embody the personhood and meaning of the deceased' (Francis et al. 2005: 82). Descriptions by home memorial keepers show that people sometimes feel observed in a graveyard and therefore do not always dare to talk to the deceased aloud.

Religiosity in Home Memorials

In this section we deal with religious dimensions of home memorials. Analysis of the empirical data highlights important aspects of home memorials and, specifically, tells us more about communication with the deceased. Correlations were calculated between the variable of whether and for how long people have had a home memorial[3] and statements about the meaning of life and death. Statements about the meaning of life and death can be divided into two categories: religious meaning (*death only has meaning if you believe in God*) and denial of meaning (*death has no meaning at all*).[4] A broadly formulated question about respondents' prayer life was included: '*Prayer can be understood in very different ways. Even if you are not a member of a church or religious group, you may have the feeling once in a while that you are praying. Do you pray sometimes?*'[5]

The results show a strong, significant, positive correlation (p < .001) between duration of keeping the memorial and prayer (r = .48). This means that the more people pray, the longer they have had a home memorial. In other words, a home memorial stimulates people to pray in a broader sense, but prayer also stimulates people to keep a memorial. Our research design excludes statements about causal relationships; correlations stand for an influence in both directions. Second, there is a significant positive correlation (p < .001) with religious interpretations of life and death (r = .31). This means that the more people agree with religious interpretations of life and death (such as 'death is not the end but a gateway to another life'), the longer they keep a home memorial. Third, there is a significant but negative correlation with denial of the meaning of life and death and duration of keeping a home memorial (r = -.21). The denial category consists in statements that strongly disagree with the possibility of any life after death. This underlines the previous correlation with religious meaning, namely that the more people believe in life after death, the longer they have kept the memorial.

The results show that people with a home memorial have a strong tendency towards prayer/meditation/communication with someone or something and belief that death is not the end. Moreover, a significant number or home memorial keepers deny that there is no existence after death. At first sight, then, the religio-spiritual explanation applies; people who perform actions that have a religious root such as prayer are more inclined to keep home memorials. Besides, people

---

3   Answers ranging from 1 = no, never to 4 = yes, longer than a year.

4   Answers ranging from 1 = I totally disagree to 5 = I totally agree.

5   Answers ranging from 1 = no, never, 2 = sometimes, 3 = yes, regularly to 4 = yes, often.

who believe in an afterlife tend to have home memorials. These traditional beliefs may be triggered by a home memorial, or more generally by the death of a loved one. Interestingly, the religious backgrounds of parents had no influence on maintaining a home memorial. The design of this study does not allow us to make statements about causality. Nevertheless, it is worthwhile to reflect briefly on the relation between mourning and afterlife beliefs. We return to this issue below, but let us first proceed with the research and our findings.

Relationships with the Dead

Among the characteristics of mourning in contemporary Western society are two important factors that help us understand the emergence of home memorials: (1) in the mourning process the bereaved are not trying to move on without the deceased, but rather to find 'a place for them' (Walter 1996: 20), and (2) bereavement is part of someone's (auto)biography or self-narrative, which on an individual level means re-shaping and re-constructing your identity without the deceased loved one (Walter 1996). Home memorials are an obvious place for reshaping the bereaved's identity. One's personal identity and the ongoing relationship with the dead are also based on self-narratives and communication with the dead. In the following responses we analyse how the bereaved maintain bonds with the deceased and what religious and psychosocial elements can be found. The first examples represent religiously unaffiliated respondents (subsequent quotations are from religiously affiliated persons).

Our 2007 survey showed that half the respondents experience the deceased's presence at the memorial (n = 61). For example, a 50-year-old woman describes how she planted a tree in the garden for her deceased husband ten years ago and surrounded it with flowers, candles and a signboard with his name. She writes about experiencing him after burning paintings made by their children near the memorial tree in the garden:

> Yes, indirect, [I experience him] when I have problems with the children; then I think about it before I go to sleep and ask my deceased husband for help. I often have the solution the next morning.

This description has different layers of meaning. The fact that she thinks about a problem before going to sleep and has the solution the next day is not uncommon – it is even advised in many studies of decision making and the unconscious (Dijksterhuis 2007). But there is one part of the sentence that suggests an afterlife belief: 'I ask my deceased husband'. This widow attributes her finding the solution to asking her husband, and furthermore believes that her deceased partner can influence her thoughts. This is what we refer to as a religio-spiritual explanation of the experience.

Another quotation is from a 55-year-old man who lost his father 'too early', as he says: 'Yes, that he looks over my shoulder and that he approves of it all'. The son believes that his deceased father watches him and judges his actions positively. This quotation reflects belief in the father's after-death existence. A 59-year-old woman describes how she experiences the dead:

> Especially my parents, I experience their presence, feel, experience with my eyes closed and see them mostly and can talk to them (especially mother; she was also more spiritual than father). Sometimes they take the initiative or I ask them to come, this can happen anywhere.

This quotation gives different explanations for experiencing the dead. Firstly, she says that she feels and experiences them with her eyes closed. Psycho-cognitive processes can explain this; if you think about someone with your eyes closed, your senses are focused on your thoughts, not on your sight, and you feel more intensely what you are thinking about. Then she writes that she can see them and talk to them. Here the question is whether she is having audio-visual hallucinations or really having a spiritual experience. She also says that the dead take the initiative in contacting her; this clearly indicates agency by the dead. The quotation becomes even more complicated when we add her description of her religious affiliation. She writes that she is not a member of a church but 'I believe in the cosmos, God, love and spirituality'. This list of beliefs includes traditional religious terms (God) but also modern personal interpretations (love, spirituality).

The next two quotations are from self-reported religiously affiliated participants. A 55-year-old Catholic woman asks her deceased father 'for support, to think about us and be with us when we have a difficult time or see to it that everything goes well with births'. This personal request specifies the nature of the dead person's agency; she asks her father to think about them, which refers to a cognitive dimension of the dead agent. A Greek Orthodox woman (63 years old), who has a memorial for her parents, grandparents and other deceased loved ones, writes:

> In my dreams the clock stops and starts again on its own, flowers that open and close when I ask if there is a good soul present, cold air behind me (everything is closed and no draft possible), candlelight moves and flickers.

This description has some natural elements that we might expect from 'new' spirituality beliefs (flowers, cold air), but also traditional religious terms (soul). The quotation reveals a variety of sources that people draw on in their experiences of the dead. Some are everyday 'coincidences' (the clock stops), others invoke a spiritual layer in nature (flowers open and close).

Two general characteristics are interesting in this small set of quotations: all participants are between 50 and 65 years of age and most of them write about experiencing their parents. This can be explained by the fact that this is the age when most people have a home memorial (usually for one or both parents). But there is more; we discern a clear tendency to ask for help and support, which suggests the kind of relationship one has with one's parents during their lifetime. Home memorial keepers continue this relationship after the parents' death by asking the dead for support. The same applies to people who have lost a partner, brother or close friend. In almost all descriptions the bereaved think intensely about and ask something from the deceased.

The most important conclusion from this section is that people explain the experience of the dead differently: (1) in religio-spiritual terms varying from traditional to more recent, popular concepts, and (2) in psychosocial terms, such as thinking, dreaming and feeling the presence of the dead. The person's relationship with the deceased is continued through experiences with the dead, asking for support and asking if the dead are present. Religious affiliation does not influence the descriptions of these experiences. What is more, we find religiously affiliated people using personal, psychological terms and religiously unaffiliated ones speaking about a spiritual agency. These results underline that continuation of the relationship with the deceased is a personal experience based on personal beliefs, and not on a belief system imposed by a religious authority.

Religious and Psychological Explanation of Continuing Bonds

In Benoire and Park's (2004) review of literature on the relation between mourning and afterlife beliefs, they identify two major concepts in beliefs about death: belief in an afterlife and continued attachment. 'Belief in an afterlife' is the general term for all kinds of afterlife constructions (e.g. resurrection, life of the soul, reincarnation). Continued attachment refers to belief in an ongoing relationship with the deceased that can assume different forms, both symbolic and concrete. The main point of their model of death-related coping is that belief in an afterlife and ongoing attachment are both part of an individual's global meaning system and therefore not a direct result of the death of a loved one. Nevertheless belief in an afterlife and continuing bonds give meaning to the death of a loved one. Both forms of belief are involved in the mourning process, but are not a consequence of it. This means that people believe in an afterlife and continuing bonds without having personally experienced the death of a significant other. The death of a loved one makes these beliefs central in people's lives. The last point, 'central in peoples lives' (i.e. in people's everyday lives), is what we observe in respect of home memorials. The memorial space has a place in everyday life, but the ritual actions around it (for example lighting candles, changing flowers, actual communication or prayer) create moments for remembrance and continuing bonds. The memorial space stimulates mourners to perform symbolic acts of commemoration (such as arranging flowers, lighting candles) and to think of or communicate with the deceased.

Apart from the religious dimensions of actions and attitudes prompted by home memorials, there are psychosocial mechanisms. In an earlier article we pointed out that the memorial preserves social and material aspects of the deceased's identity (Wojtkowiak and Venbrux 2009). The memorial is central in the continuing relationship with the deceased. Earlier we analysed the religious dimension of an afterlife. Stroebe (2004) points out that there are forms of continued attachment that are strictly secular, such as telling stories about the deceased, naming children after them or keeping objects. Continued attachment is possible without religious beliefs. Our qualitative analysis of descriptions of experience of the deceased showed that most descriptions include a mix of religio-spiritual and secular elements. Stroebe focuses on the importance of attachment in the mourning process and less on the place of beliefs in an afterlife in a person's meaning system. Our study of home memorials shows that many memorials are set up for deceased parents. Here the role of attachment is of great importance, especially with a view to asking for support and help from the deceased.

We also asked what kind of feelings people experience when they are around or thinking about the memorial.[6] Firstly, 60 per cent of the participants describe positive feelings; 20 per cent have negative feelings; and 20 per cent named memories prompted by the memorial, although strictly speaking they were asked to describe feelings. In more detail, people named the following feelings, in order of frequency: peace and restfulness (Dutch: *rust*), warmth, love, comfort, gratitude and respect. Negative feelings were restricted to sadness and loss (Dutch: *gemis*). The feelings show a tendency to have positive associations with the memorial, maybe because of its comforting character. People described how happy they were to have the memorial, and therefore the deceased, close to them in the house. By talking to or 'honouring' the deceased with flowers and candles the memorial keeper creates little private moments of memory, ritual and communication. Feelings of mourning are focused on the memorial and accompanied by these ritual actions, which give them support and comfort in dealing with the loss. The private setting of the home offers numerous possible moments to perform ritual gestures and express feelings out of the public eye.

Park and Benore (2004) criticise categorising afterlife beliefs into religious and nonreligious; they range on a continuum from religious to nonreligious beliefs about death. In this study we, too, opt for a continuum approach. But there are different dimensional levels of religiosity, since there was no clear evidence that church membership motivated the appearance of home memorials, but Christian interpretations of an afterlife did. We cannot identify the memorials as either a religious or secular phenomenon; instead we study various secular and religious dimensions of the memorials, such as afterlife interpretations or relation with the deceased. Apart from the religio-spiritual explanation of home memorials we traced a significant psychosocial impact of and on the memorial. Continuing bonds are maintained at the memorial and the memorial keepers share positive feelings.

---

6   n = 111.

Conclusions: The Role of Home Memorials in Remembrance Rituals

In this study of home memorials in the Netherlands we focused on the role of private space and religio-spiritual and psychosocial elements of mourning rituals. Our results indicate that continuing bonds with the deceased, based on the participants' own psychosocial and religio-spiritual explanations and related to afterlife beliefs and prayer, play a key role in maintaining a home memorial. Hence we can say that continuing the relationship with the deceased by way of communication (literal and symbolic) accords perfectly with the private setting of the home. There are several arguments for the role of private spaces in these remembrance rituals.

The home memorial serves as a focusing lens for private remembrance rituals. Our survey indicated that home memorial keepers tend towards transcendent interpretations of death, but invest these with personal meaning based on their relationship with the deceased, for example by asking deceased parents for support. Church affiliation appeared to have no significance in this respect. Most home memorial keepers are secular, but echo Christian interpretations of life and death (religio-spiritual explanation). The relationship with the deceased is the focus of the memorial and is maintained by experiencing the presence of the dead and communicating with them. Sometimes the communication is felt to be mutual, including the agency of the dead. Furthermore, the memorial is associated with mostly positive and comforting feelings.

In general our results indicate that the impact of secularisation is uneven. Although church membership in the Netherlands is decreasing, traditional Christian interpretations of life and death are strongly associated with memorials. A methodological weakness of this study is that there is no representation of other religious interpretations of life and death such as Asian religions, which have gained popularity in the Netherlands, at least as a source of inspiration (Poorthuis and Salemink 2009). Traditionally the house shrine was a sacred place, where one communicated with a higher power (e.g. God or the saints); home memorials relate to beliefs in a transcendent world, but the difference between religious house shrines and home memorials for the dead is that God's existence is not crucial. A continuing relationship with the dead forms the common ground for beliefs and practices in regard to house shrines. The home has become a common place for mourning rituals in which a private relationship with a deceased loved one is continued. The dead are integrated with the everyday lives of the living in a very private place.

References

Benore, E.R. and Park, C.L. 2004. Death-specific religious beliefs and bereavement: belief in an afterlife and continued attachment. *The International Journal for the Psychology of Religion*, 14(1), 1–22.

Centraal Bureau voor de Statistiek (CBS). 2009. *Religie aan het begin van de 21ste eeuw.* The Hague: Centraal Bureau voor de Statistiek.

Cladis, M.S. 2003. *Public Vision, Private Lives. Rousseau, Religion and 21st Century Democracy.* Oxford: Oxford University Press.

Cochran, C.E. 1990. *Religion in Public and Private Life.* London: Routledge.

De Graaf, N.D. and Te Grotenhuis, M. 2008. Traditional Christian belief and belief in the supernatural: diverging trends in the Netherlands between 1979 and 2005? *Journal for the Scientific Study of Religion,* 47(4), 585–98.

Dekker, G., De Hart, J. and Bernts, T. 2007. *God in Nederland 1996–2006.* Kampen: Ten Have.

Dijksterhuis, A. 2007. *Het Slimme Onbewuste.* Amsterdam: Bakker.

Francis, D., Kellaher, L. and Neophytou, G. 2000. Sustaining cemeteries: the user perspective. *Mortality,* 5(1), 34–52.

Francis, D., Kellaher, L. and Neophytou, G. 2005. *The Secret Cemetery.* New York: Berg.

Graces-Foley, K. 2003. Funerals of the unaffiliated. *OMEGA,* 46(4), 287–302.

Graces-Foley, K. (editor) 2005. *Death and Religion in a Changing World.* London: Sharpe.

Grimes, R.L. 1999. Jonathan Z. Smith's theory of ritual space. *Religion,* 29, 261–73.

Grimes, R.L. 2002. *Deeply into the Bone. Re-invention of Rites of Passage.* London: University of California Press.

James, W. 1891. *The Principles of Psychology.* London: Macmillan.

Laderman, G. and Leon, L. 2003. *Religion and American Cultures: An Encyclopedia of Traditions, Diversity and Popular Expressions.* Santa Barbara: ABC Clio.

Lawrence, R.J. 1987. What makes a house a home? *Environment and Behaviour,* 19(2), 154–68.

Margry, P.J. 2003. Persoonlijke altaren en private heiligdommen. In *Materieel Christendom: Religie en Materiele Cultuur in West-Europa.,* edited by A.L. Molendijk. Hilversum: Verloren.

Murray, H. 1940. What should psychologists do about psychoanalysis? *Journal of Abnormal and Social Psychology,* 35, 150–75.

Park, C.L. and Benore, E.R. 2004. 'You're still there': beliefs in continued relationships with the deceased as unique religious beliefs that may influence coping adjustment. *The International Journal for the Psychology of Religion,* 14(1), 37–46.

Poorthuis, M. and Salemink, T. 2009. *Lotus in de Lage Landen. De Geschiedenis van het Boeddhisme in Nederland.* Almere: Parthenon.

Quartier, T. 2009. Rituelle pendelbewegungen. neue trauerrituale im Niederländischen kontext. *Jaarboek voor Liturgieonderzoek,* 25, 185–207.

Quartier, T., Wojtkowiak, J., Venbrux, E. and De Maaker, E. 2008. Kreatives totengedenken. *Jaarboek voor Liturgieonderzoek,* 24, 155–76.

Smith, J.Z. 1987. *To Take Place: Toward Theory in Ritual.* Chicago: University of Chicago Press.

Stengs, I. 2009. Death and disposal of the people's singer: the body and bodily practices in commemorative ritual. *Mortality*, 14(2), 102–18.

Stroebe, M.S. 2004. Religion in coping with bereavement: confidence of convictions or scientific scrutiny? *The International Journal for the Psychology of Religion*, 14(1), 23–36.

Van Gennep, A. 1960 [1909]. *The Rites of Passage*. Chicago: The University of Chicago Press.

Venbrux, E., Peelen, J. and Altena, M. 2009. Going Dutch: individualisation, secularisation and changes in death rites. *Mortality*, 14(2), 97–101.

Walter, T. 1996. A new model of grief: bereavement and biography. *Mortality*, 1, 7–25.

Wils, J.-P. 2008. Rituelen op drift. Een korte fundamentele reflectie in rituele creativiteit. In *Actuele veranderingen in de uitvaart- en rouwcultuur in Nederland*, edited by E. Venbrux, M. Heessels and S. Bolt. Zoetermeer: Meinema, 121–33.

Wojtkowiak, J. 2009. The postself and the body in the process of dying. In Proceedings of the dying and death in 18th–21st centuries Europe. International Conference, Alba Iluia, Romania. Edited by M. Rotar and M. Sozzi. Cluj-Napoca: Accent.

Wojtkowiak, J. and Venbux, E. 2009. From soul to postself: home memorials in the Netherlands. *Mortality*, 14(2), 147–58.

PART IV
Art and Design in Service of Remembrance and Mourning

Chapter 12

Living to Living, Living to Dead: Communication and Political Rivalry in Roman Tomb Design

Penelope J.E. Davies

Disposal of the dead was a relatively simple matter in early Republican Rome. In the fifth and first half of the fourth centuries Before the Common Era (BCE), stone- or tile-lined pits carved out of the living tufo sufficed to contain the deceased (Barbera et al. 2005, Lanciani 1875, Taloni 1973, Albertoni 1983, Bartoloni 1987, Colonna 1996). No grand memorials marked burial spots, perhaps because sumptuary laws imposed limits on funerary display (Cicero, De Legibus 2.23.58–59).[1] The mid-fourth and third centuries saw the individualisation of elite tombs as display practices changed, and by the end of the second century, tombs could be substantial monuments in the urban landscape. By the first century their numbers had increased dramatically, and they came in a wide range of shapes and sizes, vying for attention through external appearance (von Hesberg 1992). Two well-known tombs stand out for their sheer magnificence (not to mention their relatively good state of preservation): the Tomb of the Scipios and the Tomb of Caecilia Metella. Both are prominently located on the Via Appia of 312, the earliest major road leading out of Rome; both are monumental in scale, and judging by extant archaeological evidence, both incorporate features that seem designed to augment and control a viewer's experience. These features are the main focus of inquiry here: this chapter suggests a cultural framework for their deployment. After a brief description of the two monuments, I trace the evolution of a concept of theatricality in Roman architecture and society, and propose that the patrons of these tombs harnessed this concept in the service of political self-promotion. In a culture that controlled public commissions tightly, both men seem to have recognised the peculiar advantage of funerary art, as later Romans would also: these privately-funded, privately-conceived monuments were nevertheless part of public display, and offered unusual propagandistic potential and freedom of self-representation.

At some point in the early third century BCE, L. Cornelius Scipio Barbatus, consul of 298 (or his son, L. Cornelius Scipio, consul in 259) constructed a sizeable

---

1 The only remaining evidence for these sumptuary laws stresses expense and behaviour at the funeral; it does not address the issue of tombs.

tomb for the Scipio clan. Like contemporaneous chamber tombs on the Esquiline and beneath San Stefano Rotondo on the Caelian hill, the Tomb of the Scipios was excavated horizontally into an escarpment of tufo (Santa Maria Scrinari 1972). Yet even in this initial phase, it was unusually impressive. Inside, instead of the customary single cella there were four subterranean galleries in a quadrangular plan, with two additional intersecting corridors at the centre. The façade on the escarpment was monumental, at approximately 2 metres (m) in height and 8 m in length. Moreover, on cleaning its surface in 1926 and again in the 1970s, conservators discovered frescoes of military subjects that decorated its entire surface, over-painted in at least five superimposed strata dating from the beginning of the third century to the second century (La Rocca 1977, 1990, Talamo 2008). The tomb eventually contained as many as 33 sarcophagi for members of the Scipio family, of which the first, and the most impressive, was a peperino sarcophagus, now in the Vatican Museums. Inscriptions on its lid and chest identify it as that of L. Cornelius Scipio Barbatus, and it stood at the end of the central corridor, on axis with an arched entrance framed with voussoirs of Aniene tufo (Saladino 1970, Coarelli 1972).

A systemic renovation occurred at the tomb in the second half of the second century. At this point a new gallery was excavated to the right of the existing network, to accommodate additional burials. The earliest sarcophagus in this gallery, and the most prominently placed, was that of P. Cornelius Scipio Aemilianus, which makes him the likely instigator of the renovations, some time between 150 and 135. As part of the remodelling, he also extended the front of the tomb, and transformed it from an autonomous façade into a massive podium, with an entrance to the right of the existing doorway and a matching blind arch to the left (Figure 12.1).

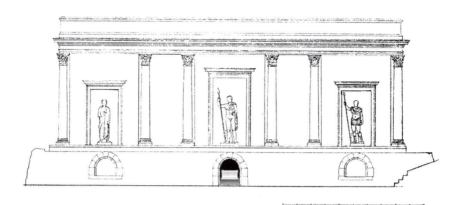

Figure 12.1   Reconstruction of the facade of the Tomb of the Scipios, ca. 150 BCE

Source: Illustration by the author, adapted from F. Coarelli

Above the podium he erected a substantial architectonic screen, which Filippo Coarelli reconstructs with six fluted half-columns on attic bases, supporting an entablature and framing *aediculae* above the entrances. In these *aediculae* he places three marble statues that ancient writers saw at the tomb and identified (albeit with some reservations) as Cornelius Scipio Africanus, L. Scipio Asiaticus, and the poet Ennius, who had extolled Scipio Africanus' memory with a poem on his exploits (e.g. Livy 38.56.3–5; Coarelli 1988).

From the start, the Tomb of the Scipios presents evidence for a fledgling interest in providing for a visitor. For one thing, the axial placement of Scipio Barbatus' sarcophagus presupposes a viewer, and the inscription demands a reader. Secondly, the design of the sarcophagus – with Ionic volutes on the lid, and a Doric frieze with rosettes on the upper chest – sets it apart from the existing Etrusco-Latial tradition and finds its closest parallels in a series of small altars produced in Sicily and a monumental altar to Hieron II in Syracuse (Zevi 1973). Design and placement of the sarcophagus, in fact, evoke a Hellenistic *heroön*, where subsequent generations might not just remember but venerate Scipio Barbatus (Humm 1996: 742). It is the second century phase of the tomb, however, that suggests a heightened interest in viewer experience. As Coarelli notes, with its superposed tiers, columns and *aediculae*, the façade closely resembled the *scaenae frons* or stage set of a Roman theatre (Coarelli 1996), and like a *scaenae frons*, the façade drew the eye's attention: sculptures provided visual interest on its surface, encouraging the eye to linger, and in the deepest recess, visually framed by the triple door arrangement, was the sarcophagus of the founding Scipio, inviting entry, interest and finally reverence.

The Tomb of Caecilia Metella consists of a circular drum resting on top of a high square podium, both built of concrete faced with travertine (Figure 12.2).

Encircling the drum at cornice level is a Pentelic marble frieze of bucrania and garlands, interrupted on the west side, facing the Via Appia, by a sculptural relief. Of this only a fragment survives, depicting a trophy flanked by bound prisoners. A second trophy probably mirrored the first on the right, and it is likely that, together, they flanked a larger relief panel. Beneath, a dedicatory inscription on a marble slab (perhaps pavonazzetto) names the tomb's honouree: *Caeciliae Q(uinti) Cretici f(iliae) Metellae Crassi* ('To Caecilia Metella, daughter of Quintus Metellus Creticus, wife of Crassus') (Gerding 2002: 38–9). With a podium once measuring approximately 100 Roman feet on a side, and a drum with a 100 Roman foot diameter, the tomb stands apart from other circular tombs of the late Republic for its vast scale; crowning the crest of a lava ridge, it overshadowed other smaller tombs in the region. Just as striking as the tomb's size is the arrangement of its interior. Generally speaking, in circular tombs of the late Republic, if there was an inner room it was a simple rectilinear (or more rarely circular) chamber, to which a radial corridor granted access from the exterior (Eisner 1986, Colvin 1991, Gerding 2002). Rare divergences from this design can usually be attributed to ritual usage. In the vast tomb of L. Munatius Plancus at Gaeta, for instance, an annular corridor around the chamber probably accommodated circumambulation

Figure 12.2   Section of the Tomb of Caecilia Metella

Source: Illustration reproduced with the permission of Rita Paris

rituals enacted before and after entry into the chamber (Winfeld-Hansen 1962, Davies 2000). In the Tomb of Caecilia Metella an arch in the south side of the podium, once framed with travertine voussoirs, opens onto a spacious passageway or dromos. About two-thirds of the way along the dromos, a substantial Gabine stone door-case indicates that access was once regulated (Gerding 2002). At the end of the passage, and at the centre of the tomb, is a circular cella, measuring 6.61 m in diameter (close to 30 Roman feet). Its walls – an early example of brick facing on concrete (Lancaster 2005) – soar upward, tapering to converge on an oculus, which allows air into the tomb, and a dramatic shaft of light. The oculus may or may not have been part of the ancient design; if it was, it was smaller than its present size (Gerding 2002). Where the dromos intersects with the cella, the floor level drops by about 3 m, to a surface of *opus signinum* or brick conglomerate. At 0.5–0.6 m above this floor surface is a ring of Gabine stone blocks, set into the walls, and cuttings in the surface of the blocks suggest that they supported a grille of metal bars (Gerding 2002). Immediately beneath the dromos (running parallel to it when viewed in section) is a barrel-vaulted chamber, connected to the cella by a low passage, barrel-vaulted at the level of the stone blocks. In the north wall of this chamber, above the low passage, is a niche. Unusual as it is, the interior of the tomb seems to provide an elevated viewing gallery into the cella.

Despite their differences in design, in both the Tomb of the Scipios and the Tomb of Caecilia Metella there seems to be a measure of interest in crafting a viewer's experience and, more specifically, constructing views. It is possible that the general monumentalisation and the unusual design features of both these tombs are manifestations of increased reverence for the dead and growing attention to funerary rituals. Scholars know a fair amount about these rituals: Romans visited tombs on personal anniversaries of the dead and during the *Parentalia* (13-21 February), when family and friends would make offerings of grain, salt, wine-soaked bread and violets to the dead, and dine in their company (see bibliography in Davies 2000, Erasmo 2008). Yet at least two other factors – pan-Mediterranean trends in architecture, and political competition – seem also to have merged together to inform the design of elite (and eventually non-elite) tombs in these years.

Scholars generally concur that the Hellenistic period in the Mediterranean was a time of escalating 'theatricality' in architecture. This term, first coined by Jerome J. Pollitt (1986), described a literal and a metaphorical change in architectural practice after the classical period. From the late fourth century, and increasingly by the second, architects focused mounting attention on theatres. Construction began on the renowned Theatre in the Sanctuary of Aesculapius at Epidauros in the early third century, and the uppermost tier of seating was completed in the second. At Pergamon, architects set a theater onto the steep western slope of the citadel in the third century, and it reached its most splendid form in the first half of the second, under Eumenes II (197–159). Spectacular views from the seating area stretched beyond the stage and the sanctuary of Dionysos to embrace the wide expanse of the Selinus River plain (Akurgal 2001). At the same time, architects began to design buildings that provided visitors with dramatic and personal experiences, often culminating in some kind of visual revelation. The most renowned instances of this trend are sacred: the sanctuaries of Athena Lindaia at Lindos on the island of Rhodes, and Aesculapios on Cos, and the great temple of Apollo at Didyma, where architectural form constantly surprised and confounded a visitor; but entire cities could be planned for similar effect. A visitor's path through the city of Pergamon, for instance, as expanded under Eumenes II, moved from profane to ever more sacred areas, and offered dramatic and surprising vistas as it wound between buildings (Pollitt 1986). In all of these cases, architects controlled paths of access to ensure views and sensations that surprised and/or awed a visitor, making for a personal (rather than a collective) experience of the site. Pollitt associates this trend at least in part with a contemporaneous dissolution of the individual's sense of involvement with the state.

The propensity toward theatricality in architecture also developed in Republican Rome. As in Greece, Roman politicians started building theatres. The first stone theatre in Rome was Pompey's great theatre-portico complex of 55 BCE

(Packer et al. 2007), but the tradition of temporary theatre construction went back to an earlier date. By Livy's account, *ludi scaenici* (scenic games–probably short plays, mimes and Atellane farces) began in Rome in 364, possibly in connection with the *ludi romani* (Roman Games) (Livy 7.2.1–3; Gruen 1992). By ca. 240 Livius Andronicus had introduced the city to fully-fledged drama, which enjoyed tremendous popularity by the end of the third century. Magistrates let contracts for theatrical entertainments as a component of religious celebrations; victorious generals put on plays to complement their triumphs, and some individuals sponsored funerary *ludi* to accompany elite funerary ceremonies (Beacham 1999, Bodel 1999, Gruen 1992). For some games, such as the *ludi Apollinaris* at the Temple of Bellona, or the *ludi Megalenses* at the Temple of Magna Mater on the Palatine, temple steps provided seating; but from about the second century, for most theatrical productions an aedile or a censor let a contract for the construction of a temporary stage and a limited seating area out of wood (Gruen 1992). Laura Klar hypothesises that the form of the *scaenae frons*, with its tiered aediculae, developed out of its use for housing booty that the sponsoring general had looted on campaign. As she puts it: 'The desire of second century generals to showcase plundered statues in the temporary theaters erected for their votive games mandated a structure with ample space for sculptural display … The abundant use of columns and entablatures that characterises the Roman *scaenae frons* may have begun to develop at this time as a means of framing plundered Greek statues, thus placing them in an appropriate architectural setting … ' (Klar 2006). If correct, this would suggest that a theatre's design was distinctly tied to the individual who commissioned it, and its motivation was in large part self-promotional.

In 154, precisely when Scipio Aemilianus was active in Rome, theatre construction became a loaded political issue. Ancient notices record that the censors of that year, C. Cassius Longinus and M. Valerius Messalla, let a contract for the first stone theatre of Rome, near the Lupercal on the south-western slope of the Palatine. According to Livy, it was close to completion when P. Scipio Nasica (censor of 159, and first cousin once removed of Scipio Aemilianus) incited a *senatus consultum* for its dismantling and the sale of its various components at auction (*Periocha Liviana* 48).[2] His stated reason was the danger to public morals and national character, but scholars disagree on a possible subtext to his opposition. He may have objected to the political edge it would have offered the censors. For Erich Gruen, a permanent theatre would have represented a relaxation of senatorial authority, since the ritual of erecting a temporary theatre was a reminder of the ruling class' authority in the realm of the arts; moreover, senatorial control promoted highbrow drama, such as Terence's comedy, rather than the more popular buffoonery of Plautus and gladiatorial games (Gruen 1992). For M. Dauster and Gary Forsythe, the phrasing of Scipio Nasica's remonstration suggests that elite

---

2   In alternative narratives, a consul, Caepio, blocked the theatre's construction. See Valerius Maximus 2.4.2; Augustine *De Civitate Dei* 1.31; Orosius 4.21.4; Velleius Paterculus 1.15.2; Appian *Bella Civilia* 1.28.125; Sordi (1988), La Rocca (1990): 384.

hegemony as a whole was threatened, not simply their control of culture, since a stone theatre might result in citizens assembling on their own initiative, without a magistrate, to conduct public business (Dauster 2003, Forsythe 1994, Gruen 1992, La Rocca 1990, Beacham 1999, Davies forthcoming). Whatever the reasons for dismantling the theatre, they seem to have been political; and the initiative to build a permanent theatre, successful or not, reveals that members of the Roman elite were growing accustomed to constructing both spectacle and spectatorship.

As in Hellenistic Greece, theatres went hand in hand with attempts to control views and the viewing process in other types of building. In Latium and Campania the development of high quality concrete amplified the propensity, freeing architects from the constraints of post-and-lintel construction and allowing them gradually to experiment with new forms. The *locus classicus* for visitor manipulation through concrete construction is not in Rome but at the oracular Sanctuary of Fortuna Primigenia at Praeneste (Palestrina), in the foothills of the Apennines, dated to the late second century BCE. Architects moulded structures over the entire surface of the hillside, crafting spaces that controlled and heightened a visitor's experience in the service of religion (and possibly local politics). Light alternated with darkness, exposed stairways contrasted with confined ramps, curved forms animated straight lines. A visitor had to overcome physical exhaustion and a sense of vulnerability in order to reach the familiar forms of a colonnaded court at the summit, and a small theatre where religious performances took place against the magnificent backdrop of the countryside below, which stretched away to blend with the far-off horizon (MacDonald 1982).

Though it is most evident in public building, the tendency to manipulate experience is just as prevalent in domestic architecture, as John Clarke and others have shown (Clarke 1991). Designers often crafted decorative ensembles of paintings, mosaics and sculptures to guide a visitor through late Republican houses in Pompeii and Herculaneum, and set up doors and windows serially to frame long vistas from one end of a residence to another. In the House of the Menander at Pompeii, for instance, Heinrich Drerup saw a series of framing devices creating a central axis from the entrance all the way through to a peristyle at the back (Drerup 1959). This was more than simply an aesthetic choice: looking in from the street, even a casual passerby could glimpse signs of the wealth and status of the owner.

By the first century in Rome, the attempt to emulate both aspects of a built theatre – a controlled viewing station and a constructed view – seems more determined, in public and private architecture. In the substructure known as the Tabularium on the edge of the Capitoline hill, of circa (ca.) 78, the consul Q. Lutatius Catulus incorporated a large viewing gallery.[3] Huge arched openings pierced one entire wall of the gallery to create spectacular views over the entire Forum, which was undergoing repaving contemporaneously (Giuliani and Verducchi 1987, Purcell 1995). Close at hand, a viewer would see the Curia

---

3   For new interpretations of the building's function, see Tucci (2005), Coarelli (2008), Davies (forthcoming).

Hostilia, newly enlarged to accommodate Sulla's expanded Senate of 600 – a hallmark of his systemic attempt to restructure and regain control of the flailing Republic (Coarelli 1993). Not long thereafter, in 55, Pompey succeeded in erecting the first stone theatre in Rome. Built out of concrete on the flat plain of the Campus Martius, the massive structure rivalled the very hills of Rome; to this building the consul brought the rituals of Rome, where they were intimately tied to his person and his name. By 52, the penchant for theatrical architecture had turned itself into spectacle: G. Curio commissioned a two theatres made of wood, set back to back on a pulley system so that they could revolve to transform into a single amphitheatre. Years later, Pliny comments on the precariousness of the contraption, which served to enthrall and fascinate (Pliny *Natural History* 36.118–119, Carey 2003, also Pollard 1999).

Viewing stations developed simultaneously on elite estates. One of them, built in the Late Republic, achieves a small measure of fame from Suetonius' account of the great fire of Rome in 64, when 'Nero watched the conflagration from the Tower of Maecenas, enraptured by what he called 'the beauty of the flames'; then put on his tragedian's costume and sang *The Sack of Ilium* from beginning to end' (Suetonius, *Nero* 38.2). The tower must have been in Maecenas' garden estate on the Esquiline; Antonio Colini locates it in the vicinity of the Sette Sale (Colini 1979). Orosius later describes it as *altissima*, 'very tall' (Orosius *adv. Paganos* 7.7.6.), and on analogy with a second-century of the Common Era (CE) structure on a villa close to Anguillara Sabazia, Colini reconstructs it as a three-storey structure about 18 m tall, with windows piercing all four sides and a terrace on top. Inspired, perhaps, by lighthouse architecture, viewing stations such as this were the precursors of towers at the villa of Pliny the Younger at Laurentina, and the East and West Belvederes at Hadrian's second-century villa at Tivoli (as well as Trajan's Column of 113) (Pliny the Younger, *Ep.* 2.17.12, Colini 1979, MacDonald and Pinto 1995, Davies 1997). In Horace *Odes* 3.29.5–10, the poet invites Maecenas to enjoy a view over the hills of Tivoli, Aefulae and Tusculum, and Pseudo-Acron glosses the lines to explain that this was the view from Maecenas' tower (also Porphyrion on lines 6–8, Colini 1979). Private belvederes such as his offered their owners a chance to see beyond the confines of their estates; yet they also provided guests with panoramic vistas across the owner's estate – and it would be a mistake, probably, to suppose that such views were simply pleasing panoramas; they guided viewers to bear witness to ownership and status.

What these examples suggest is that in late Republican Rome members of the elite were exploring ways to construct environments in which the act of viewing, and that which was to be viewed, were closely manipulated and controlled. Moreover, the view achieved often framed a political message. This tendency in architecture began in the second century, and heightened in the first. The Tomb of the Scipios belongs in the early phase of this development; that of Caecilia Metella in the later stage. In both, elements of 'theatrical' design can be discerned; and both, there may have been a political agenda.

The set-up of the Republic was inherently competitive: elective magistracies diminished in number the higher one climbed the rungs of power. Still, the Tomb of the Scipios in its restored phase and the Tomb of Caecilia Metella belong to periods of dramatic escalation in competition for political office (Davies forthcoming).

The conquest (or 'liberation') of Macedonia and Greece in the early second century had the effect of introducing and spreading unprecedented wealth in Rome, with the result that more men than ever could entertain the possibility of embarking on a political career. Moreover, territorial gains necessitated adjustments to existing command structures, so in 197 the number of praetors (second in command to the consuls) was increased from four to six (then reduced to four in alternate years after 181), and this further heightened tension over consular elections (Lintott 1999). For the first time electoral bribery became a serious issue, and the senate called for repeated laws to curb it, and to control the lavish entertaining that won clients (Lintott 1990). The established elite hoped these laws would prevent men with new wealth from destabilising existing client-patron networks (and thus their stranglehold on the magistracies) by buying electoral support. As newcomers to politics experimented with aggressive, innovative tactics to attract clients, families with long-standing histories in politics emphasised their point of advantage: an illustrious pedigree.

Within this atmosphere of general rivalry, competition was particularly intense for Scipio Aemilianus. Adoptive son of P. Cornelius Scipio and grandson of Cornelius Scipio Africanus, he was a celebrity in Rome. As consul in 147 while still under the minimum age of 42, he led Roman forces against the Carthaginians, and in 146 razed Carthage to the ground, thereby definitively concluding hostilities with Rome's long-standing enemy. For this he earned a triumph and the right to use his adoptive *agnomen,* Africanus. Yet as huge as his accomplishment was, by a misfortune of timing he had company in his success. In the same year, Q. Caecilius Metellus triumphed for a resounding victory over Macedonia, and took the *agnomen* Macedonicus, first used by Scipio Aemilianus' own natural father, L. Aemilius Paullus Macedonicus; the following year, L. Mummius celebrated the second triumph of his career, over Achaea and Corinth, and became the first 'new man' (*novus homo*) to receive an *agnomen* – Achaicus – for military services (Ziolkowski 1988). All these men had an eye on future office, and the contest for self-glorification was acute.

Architecture served as the natural outlet for self-promotion. Yet there were long-standing constraints built into the constitution: only magistrates in office could vow, dedicate or let contracts for public buildings, and the brevity of magisterial terms (one year for all except the censorship, held for 18 months at five-year intervals) imposed limits on magistrates' ambitions for their building projects. Among the elite, there was also a consensus, albeit fluid, on appropriate public behaviour (e.g. Hölkeskamp (1993). The senate's willingness to intervene forcefully in projects it deemed inappropriate was precisely exemplified by the destruction of the censors' theatre of 154. Still, there were ways to work within and around the system, and Scipio Aemilianus, Caecilius Metellus and Mummius all schemed to make a mark in the urban landscape (Davies forthcoming). Scipio Aemilianus and Mummius exploited the *triumphator*'s right to erect a votive temple on behalf of the state with competing temples to Hercules, the mythological hero *par excellence* (Coarelli 1988, Ziolkowski 1988). Caecilius Metellus designated a portion of his spoils to construct a portico where he could display the pride of his booty from Macedonia: a crowd of bronze equestrian statues by the famed Greek artist Lysippos, representing Alexander, the historical model of military prowess, with his companions at the battle of Granikos (Coarelli 1997). All three projects were legitimate, and must have been sanctioned by the Senate.

The beauty of a tomb was that it needed no such sanction. Like a house, it was a private building, funded by private means, and in its construction a patron had great freedom. As long as it stood outside the city walls, as required by the laws of the Twelve Tables, the senate had no power of veto over it (Davies forthcoming). In the heat of competition with two plebeian men, the patrician Scipio Aemilianus redesigned his family tomb to extol his greatest asset, his pedigree. The Greek historian Polybius explains that elite funerals were a principal venue for recalling, re-enacting and reinforcing lineage:

[An elite man's] body is carried with every kind of honour into the Forum to the so-called Rostra ... The whole mass of the people stand around to watch, and his son, if he has left one of adult age who can be present, or if not some other relative, then mounts the Rostra and delivers an address which recounts the virtues and the successes achieved by the dead man during his lifetime. By these means the whole populace ... is involved in the ceremony... Then after the burial of the body and the performance of the customary ceremonies, they place the image of the dead man in the most conspicuous position in the house, where it is enclosed in a wooden shrine ... And when any distinguished member of the family dies, the masks are taken to the funeral, and are there worn by men who are considered to bear the closest resemblance to the original both in height and in their general appearance and bearing ... dressed according to the rank of the deceased ... They all ride in chariots with the fasces, axes and other insignia carried before them, according to the dignity of the offices of state which the dead man had held in his lifetime, and when they arrive at the Rostra they all seat themselves in a row upon chairs of ivory. It would be hard to imagine a more impressive scene for a young man who aspires to win fame and to practice virtue. For who could remain unmoved at the sight of the images of all these men who have won renown in their time, now gathered together as if alive and breathing? What spectacle could be more glorious than this?

Moreover, [when] the speaker ... has delivered his tribute, [he] goes on to relate the successes and achievements of all the others whose images are displayed there, beginning with the oldest. By this constant renewal of the good report of brave men, the fame of those who have performed any noble deed is made immortal, and the renown of those who have served their country well becomes a matter of common knowledge and a heritage for posterity. But the most important consequence of the ceremony is that it inspires young men to endure the extremes of suffering for the common good in the hope of winning the glory that waits upon the brave. (Polybius 6.53–54, trans. Ian Scot-Kilvert)

Polybius is clear that while these rituals honoured the dead, they were about the living also: by displaying ancestors as *exempla* for the living, they projected the deceased's family into the future. The tomb was the culmination of these rituals. It functioned as a resting place and memorial for the dead, but also provided a controlled historical setting that promoted the living. At his remodelled family sepulchre Scipio Aemilianus exploited architectural trends of the time to turn the fleeting performance of pedigree into a permanent narrative display. He seems to have accomplished in the unregulated space of his tomb what the censors of 154 had been unable to achieve in the public arena: a *scaenae frons* in stone. The façade could serve as a permanent backdrop for privately-funded plays in honour of the dead, or, at the least, as a perpetual evocation of the *fabulae praetextae* that were sometimes performed as part of elite funerary celebrations (Bodel 1999). Set within that theatrical façade, statues of two illustrious ancestors and the poet whose writings were a lasting testament to Scipio Africanus' reputation kept the ancestral glory of the Scipios alive. The new façade framed a view of the sarcophagus of Scipio Barbatus, founder of the tomb and the ancestor responsible for bringing the family to the forefront of Roman politics; heroön-like, this altar-sarcophagus became the dramatic denouement for the funeral celebrations, honouring Scipio Barbatus, to be sure, but also emphasising lineage for political ends.

Dating the Tomb of Caecilia Metella and determining its patron have proved arduous tasks for archaeologists; yet on the basis of construction techniques and materials, as well as iconographic, epigraphic and prosopographic analysis, most now concur that it was Caecilia Metella's son, M. Licinius Crassus (iii), who commissioned the monument in circa 30–25 BCE (Paris 2000, Gerding 2002). This places it at another moment of acute political rivalry. By this date the Republic had been in jeopardy for about half a century, as first Sulla then Pompey and Julius Caesar manoeuvred into positions of supreme power, threatening the entire collegial basis of government. A period of mob violence that had begun with the murder of Ti. Sempronius Gracchus in 133 came to a peak with Caesar's assassination in 44, when a group of senators revealed their intolerance for and readiness to curb autocratic behaviour. M. Licinius Crassus (iii) and his family had lived this history: his grandfather, M. Licinius Crassus (i), known for his extraordinary wealth and for subduing the Spartacan revolt in 72, had been consul with Pompey in 70 and in 55; he was also one of the first triumvirs in 59 with Pompey and Julius Caesar, and served as proconsul in Syria in 54 before dying in the battle of Carrhae in 53. M. Licinius Crassus (ii), Licinius Crassus' father and husband of Caecilia Metella, had fought with Caesar in Gaul in 57–51. By the time of the tomb's construction, Caesar's adoptive son Octavian had bested M. Antonius at the battle of Actium in 31, where they had faced off for the prize of sole power. Licinius Crassus (iii) had first sided with Antonius, and then switched his allegiance to Octavian, perhaps in 31 (Gerding 2002). Consul with Octavian in 30, he became governor of Macedonia and Achaea, and engaged in two successful campaigns against tribes in Thracia and Moesia, during which he killed Deldo, the enemy king of the Bastarnae, in single combat, and retrieved from the Getae the Roman standards that C. Antonius had lost in 63 (Cassius Dio 51.24.4, Gerding 2002). By rights, the former feat entitled him to donate Deldo's armour at the Temple of Jupiter Feretrius on the Capitoline. This honour, known as the *spolia opima*, was the only award higher than a triumph, and Roman tradition named only three men who had earned it: Romulus, A. Cornelius Cossus (in 437) and M. Claudius Marcellus (in 222). Returning to Rome in late 28, Licinius Crassus was a real and formidable threat to Octavian – who immediately machinated to ensure that the award could not take place. Shortly thereafter, in January of 27, Octavian claimed to restore the Republic, and accepted the title of Augustus from the Senate. After Licinius Crassus' triumph on 4 July, 27, he vanishes from history (Gerding 2002, Bastien 2007).

Exactly when Licinius Crassus commissioned the tomb is not known; yet two factors were constant throughout this sequence of events. First, it was not yet evident that monarchy was inevitable, or who a possible monarch might be; that is, it was still possible to be a contender for power. Second, both the Senate and Octavian were forces to be reckoned with. Vivid memories of Caesar's assassination and the First Triumvirate's proscriptions must have lingered, and a savvy politician would not want to offend. Licinius Crassus' monument seems to answer both these points. Vast, highly visible, ornamented with signs of his military accomplishments, the monument evoked battlefield trophies and *heroa* as well as an established tradition of circular tombs (Gerding 2002). Yet at the same time as it promoted him, it specifically denied that it honoured him (being a tomb for his mother).[4] The modest but beautifully incised inscription that ostensibly celebrates his mother but reminded a viewer that Licinius Crassus was the scion of two prominent families: the fame of the Metelli went back to the First Punic War (264–241), and thereafter they were involved in major conflicts in Macedonia (Caecilius Metellus as conqueror in 148), against Jugurtha in North Africa in 109, and against Catiline in 62. Consul in 69, Caecilia Metella's own father had subsequently dispensed with a pirate fleet and subordinated Crete (in 68–65), earning a triumph and the *agnomen* Creticus used in the inscription (Paris 2000).

How the interior affected those who gained access to it is harder to grasp; yet it is clear that a visitor's experience was closely controlled, and dramatic. At the least, having proceeded along the dark dromos, perhaps toward a bright shaft of light, he was supposed to stop and view, because the corridor leads to a spot from which he could not easily advance. Perhaps the architect meant him to see something displayed in a niche in the wall opposite the dromos, where there is now

---

4   Just as Augustus' Mausoleum would be a victory monument while concealing that fact: Davies (2000): 49–74.

a large cavity, measuring approximately 2.7 m by 1.6 m (Gerding 2002: 41, 102).[5]
If there was a niche there in Roman times, it must have been substantially smaller
than the present cavity to allow for framing and lining brickwork. Alternatively,
a grille resting on the stone ring may have supported an object above floor level.
In either case speculation might fall on a cinerary urn, a sculpture, or even the
armour of King Deldo. Beyond simply viewing, there is also the possibility that
the interior design of the tomb accommodated a visitor's interaction with the
deceased. A channel cut beneath the damaged doorsill in the dromos leads down
to the chamber beneath, and may have been intended for offerings to the dead
interred there; in fact, Henrik Gerding would place the ashes of Caecilia Metella in
the niche on the chamber's north wall (Gerding 2002). More than that, he looks to
a variety of architectural models for the design of the tomb: the cave-like aspect of
mystery cult buildings; *bothroi* or sacred pits for offerings to underworld deities,
like those at the Asklepieion at Athens or the Arsineion at Samothrace, identified
with *mundi* (entrances to the underworld) in Rome; and Greek oracular tombs
such as the grave of Trophonios at Lebadeia and the Nekyomanteion at Ephyra,
which incorporated passages to the Underworld (Gerding 2002). Considering
ancient concepts about the whereabouts of the spirits of the dead, in the general
vicinity of the tomb or beyond a body of water, he speculates that the Tomb of
Caecilia Metella was a shrine to one or more of the infernal gods, with an entrance
to the underworld, or an oracle of the dead (where the concave floor contained
a pool of water collected through the oculus, which separated the living in the
upper corridor from the dead in the lower chamber), or a mystery cult building, for
initiation rites or banquets (Gerding 2002).

The exterior of the Tomb of Caecilia Metella was a masterpiece of self-
promotional insinuation, responding to the complex political atmosphere of late
Republican Rome. However the interior functioned, it seems that an advanced
propensity for theatricality was at play – probably at least in part in the interests of
Licinius Metellus' career. Whatever a visitor witnessed from the end of the dromos,
it appeared as a sort of revelation. If it honoured Licinius Crassus, the advantage is
self-evident; if it honoured Caecilia Metella, and the architecture of revelation was
deployed to mystify her memory, it may have been a carefully-calibrated response
to Octavian's widely-diffused claim to be the son of the deified Caesar (Weinstock
1971). Octavian would spell out this very claim, which enhanced his charisma
and buttressed his constitutional supremacy, in the placement of his own tomb
on the Campus Martius, location of Caesar's burial, and in its close formal and
topographical relationship with the Pantheon, where Caesar's statue stood side by
side with those of the gods (Davies 2000).

The Tombs of the Scipios and Caecilia Metella differ vastly in form. Yet
unusual features in each of them suggest that their respective architects employed
contemporary design tactics to control experience and viewing processes. Their
choices may reflect a heightened attention to the cult of the dead; but they appear

---

5   Gerding believes this cavity was made in the Middle Ages at the earliest.

also to serve the political agendas of their patrons, who took advantage of the ambiguous place of funerary art – lodged between private and public – in Roman culture. It is no exaggeration to say that it was by exploiting ambiguities such as these that Roman politicians wore away at the state's constraints on visual self-representation and promotion; and that the erosion of those constraints contributed to the final demise of the Republic.

References

Akurgal, Ekrem. 2001. *Ancient Civilizations and Ruins of Turkey*. 9th edition. Istanbul: Net Turistik Yayinlar San. Tic. A.S.

Albertoni, M. 1983. La necropolis Esquilina arcaica e repubblicana, in *L'Archeologia in Roma Capitale tra Sterro e Scavo*. Venice: Marsilio Editori, 140–55.

Barbera, M.R., Pentiricci, M., Schingo, G., Asor Rosa, L. and Munzi, M. 2005. Ritrovamenti archeologici in piazza Vittorio Emanuele II. *Bullettino della Commissione Archeologica Comunale di Roma* 106, 302–37.

Bartoloni, G. 1987. Esibizione di ricchezza a Roma nel VI e V secolo: doni votivi e corredi funerari. *Scienze dell'Antichità* I, 143–59.

Bastien, Jean-Luc. 2007. *Le Triomphe Romain et son Utilisation Politique à Rome aux Trois Derniers Siècles de la République*. Rome: École Française de Rome.

Beacham, Richard. 1999. *Spectacle Entertainments of Early Imperial Rome*. New Haven and London: Yale University Press.

Bodel, John. 1999. Death and display: looking at Roman funerals, in *The Art of Ancient Spectacle*, edited by Bettina Bergmann and Christine Kondoleon. New Haven and London: Yale University Press.

Carey, Sorcha. 2003. *Pliny's Catalogue of Culture: Art and Empire in the Natural History*. Oxford: Oxford University Press.

Clarke, John R. 1991. *The Houses of Roman Italy, 100 BC–AD 250: Ritual, Space, and Decoration*. Berkeley: University of California Press.

Coarelli, Filippo. 1972. Il Sepolcro degli Scipioni. *Dialoghi di Archeologia* 6, 36–106.

Coarelli, Filippo. 1988. *Il Sepolcro degli Scipioni a Roma*. Rome: Fratelli Palomba Editori.

Coarelli, Filippo. 1993. Curia hostilia, in *Lexicon Topographicum Urbis Romae*, volume 1, edited by Margareta Steinby. Rome: Edizioni Quasar, 331–32.

Coarelli, Filippo. 1996. *Revixit Ars: Arte e Ideologia a Roma. Dai Modelli Ellenistici alla Tradizione Repubblicana*. Rome: Edizioni Quasar, 179–238.

Corelli, Filippo. 1997. *Il Campo Marzio, dalle Origini alla Fine della Repubblica*. Rome: Edizioni Quasar.

Coarelli, Filippo. 2008. *Substructio et Tabularium*. Paper delivered at the British School at Rome.

Colini, Antonio M. 1979. La torre di Mecenate. *Atti dell'Accademia Nazionale dei Lincei. Classe di Scienze Morali, Storiche e Filologiche. Rendiconti* 34, 239–50.

Colonna, G. 1996. Roma arcaica, i suoi sepolcreti e le vie per i Colli Albani. *Albalonga, mito Storia e Archeologia.* Rome: 335–54.

Colvin, Howard. 1991. *Architecture and the Afterlife.* New Haven and London: Yale University Press.

Dauster, M. 2003. Roman Republican sumptuary legislation, 182–102, in *Studies in Latin Literature and Roman History*, edited by Carl Deroux. Brussels: Latomus, 65–93.

Davies, Penelope J.E. 1997. Politics and perpetuation: Trajan's Column and the art of commemoration. *American Journal of Archaeology* 101, 41–65

Davies, Penelope J.E. 2000. *Death and the Emperor: The Funerary Monuments of the Roman Emperors from Augustus to Marcus Aurelius.* Cambridge: Cambridge University Press.

Davies, Penelope J.E. (forthcoming). *Architecture, Art and Politics in Republican Rome.* Cambridge: Cambridge University Press.

Drerup, Heinrich. 1959. Bildraum und realraum in der römischen architektur. *Römische Mitteilungen* 66, 145–74.

Eisner, Michel. 1986. *Zur Typologie der Grabbauten in Suburbium Roms.* Mainz: P. von Zabern.

Erasmo, Mario. 2008. *Reading Death in Ancient Rome.* Columbus: Ohio State University Press.

Forsythe, Gary. 1994. Review of Gruen, Erich S. 1992. *Culture and National Identity in Republican Rome.* Bryn Mawr: Classical Review.

Gerding, Henrik. 2002. *The Tomb of Caecilia Metella. Tumulus, Tropaeum and Thymele.* Lund: Lund University.

Gruen, Erich S. 1992. *Culture and National Identity in Republican Rome.* Ithaca: Cornell University Press.

Giuliani, Cairoli Fulvio and Patrizia Verducci (1987). L'area centrale del Foro Romano. Florence: Leo S. Olschki Editore.

Hölkeskamp, Karl J. 1993. 'Conquest, competition and consensus: Roman expansion in Italy and the rise of the nobilitas'. *Historia* 42, 12–38.

Humm, Michel. 1996. Appius Claudius Caecus et la construction de la via Appia. *Mélanges de l'École Française de Rome: Antiquité* 108, 693–749.

Klar, Laura S. 2006. The origins of the Roman scaenae frons and the architecture of triumphal games in the second century B.C., in *Representations of Warfare in Ancient Rome*, edited by Sheila Dillon and Katherine E. Welch. Cambridge: Cambridge University Press, 162–83.

La Rocca, Eugenio. 1977. Cicli pittorici al sepolcro degli Scipioni. *Roma Comune* 1 Supplement to no. 6–7, 14ff.

La Rocca, Eugenio. 1990. Linguaggio artistico e ideologia politica a Roma in età repubblicana, in *Roma e l'Italia: Radices Imperii.* Milan: Libri Schweiwiller, 289–498.

Lancaster, Lynn C. 2005. *Concrete Vaulted Construction in Imperial Rome: Innovations in Context.* Cambridge: Cambridge University Press.

Lanciani, Rodolfo. 1875. Le antichissime sepolture Esquiline. *Bullettino della Commissione Archeologica Comunale di Roma*, 41–56.

Lintott, Andrew. 1990. Electoral Bribery in the Roman Republic. *Journal of Roman Studies.* 80, 1–16.

Lintott, Andrew. 1999. *The Constitution of the Roman Republic.* Oxford: Clarendon Press.

MacDonald, William L. 1982. *Architecture of the Roman Empire 1. An Introductory Study.* New Haven and London: Yale University Press.

MacDonald, William L. and Pinto, John S. 1995. *Hadrian's Villa and its Legacy.* New Haven and London: Yale University Press.

Packer, James E., Burge, John and Gagliardo, Maria C. 2007. Looking again at Pompey's Theater: The 2005 excavation season. *American Journal of Archaeology* 111, 505–22.

Paris, Rita. 2000. *Via Appia: The Tomb of Cecilia Metella and the Castrum Caetani* Milan: Electa.

Paris, Rita. 2001. Mausoleo di Caecilia Metella e Castrum Caetani, in *Archeologia e giubileo: Gli interventi a Roma e nel Lazio nel piano per il grande giubileo del 2000*, edited by F. Filippi. Naples: Electa Napoli, 316–21.

Pollard, N. 1999. Theatrum Scaurum, in *Lexicon Topographicum Urbis Romae* volume 5, edited by Margareta Steinby. Rome: Edizioni Quasar, 38–39.

Pollitt, Jerome J. 1986. *Art in the Hellenistic Age.* Cambridge: Cambridge University Press.

Purcell, Nicholas. 1995. Forum Romanum (The Republican Period), in *Lexicon Topographicum Urbis Romae*, volume 2, edited by Margareta Steinby. Rome: Edizioni Quasar, 325–36.

Saladino, V. 1970. *Der Sarkophag des Lucius Scipio Barbatus*, Beiträge zur Archäologie I. Würzburg.

Santa Maria Scrinari, Valnea. 1968 [1972]. Tombe a camera sotto via S. Stefano Rotondo presso l'Ospedale di S. Giovanni in Laterano. *Bullettino della Commissione Archeologica Comunale di Roma*, 17–23.

Scott-Kilvert, Ian (trans.) 1979. *Polybius The Rise of the Roman Empire.* London: Penguin.

Sordi, Marta. 1988. La decadenza della Repubblica e il teatro del 154 a.C. *Invigilata Lucernis* 10, 327–41.

Talamo, E. 2008. La fronte depinta del sepolcro degli Scipioni, in *Trionfi Romani*, edited by Eugenio La Rocca and Stefano Tortorella. Rome: Electa, 119.

Taloni, M. 1973. La necropolis Esquilina, in *Roma Medio Repubblicana: Aspetti Culturali di Roma e del Lazio nei Secoli IV e III a. C.* Rome: Assessorato antichità, belle arti e problemi della cultura, 188–200.

Tucci, Pier Luigi. 2005. 'Where high Moneta leads her steps sublime': the 'Tabularium' and the Temple of Juno Moneta. *Journal of Roman Archaeology* 18, 6–34.

Von Hesberg, Henner. 1992. *Römische Grabbauten*. Darmstadt.

Weinstock, Stefan. 1971. *Divus Julius*. Oxford: Clarendon Press.

Winfeld-Hansen, H. 1962. Les couloirs annulaires dans l'architecture funéraire antique. *Acta Instituti Romani Norvegiae* 2, 35–63.

Zevi, Fausto. 1973. Il Sepolcro degli Scipioni. In *Roma Medio Repubblicana: Aspetti culturali di Roma e del Lazio nei secoli IV e II a. C.* Rome: Assessorato Antichità, Belle Arti e Problemi della Cultura, 234–39.

Ziolkowski, Adam. 1988. Mummius' Temple of Hercules Victor and the Round Temple on the Tiber. *Phoenix* 42, 309–33.

Chapter 13

Maxwell Fry and the 'Anatomy of Mourning': Coychurch Crematorium, Bridgend, Glamorgan, South Wales

Hilary J. Grainger

Coychurch Crematorium in Bridgend, Glamorgan opened in 1970 and enjoys a special status on two counts. First, coincidentally together with two other crematoria close in date and location, Margam (1969)[1] and Llywdcoed, Aberdare,[2] Coychurch is one of only a handful of British examples to warrant mention in Pevsner's *Buildings of Wales* as being of architectural interest. Second, it was designed by a British modernist of international reputation, Edwin Maxwell Fry (1899–1987), distinguished by having worked with Walter Gropius in the 1930s and with Le Corbusier in the 1950s. Architectural historian John Newman described Coychurch as a 'major work' by Fry, Drew & Partners and one 'strongly influenced by late Le Corbusier'. Furthermore it was in his view, 'the most important recent display of stained glass in the country' (Newman 2004: 334).

This chapter explores the ways in which Fry's ideas on crematorium design and what he called 'an anatomy of mourning' (Fry 1968: 43) were embodied at Coychurch by means of architectural space, the formalist language of Modernism and the elements of *delay* and *distance* in order 'to make people participate more closely in the cremation service through the design of the building and its approaches' (Newman 2004: 334).

In order to evaluate Fry's remarkable achievements it is first necessary to understand something of both the history of cremation and the development of crematorium design in Britain and more particularly, in Wales. Although these are well documented (White 2002, 2003, Jupp 2006, Grainger 2005, 2009) a brief outline is useful here. Cremation was revived in the late nineteenth century as an alternative to burial. But despite a rich pre-Christian heritage dating back to the Bronze Age, it did not claim widespread support in the UK until the second half of the twentieth century, on account of its long and sustained struggle against resistance on religious grounds and strongly rooted conservatism. The movement in Britain was historically secular, informed by concerns over hygiene, overcrowded burial

---

1   Designed by F.G. Williamson & Associates, see also their work at Parc Gwyn, Narberth, Pembrokeshire (1968).

2   Designed by H.M.R. Burgess & Partners, opened in 1970.

grounds and cemeteries, and supported by advances in late-Victorian technology. For very nearly 2,000 years Christian influences throughout Europe had fostered burial as the traditional form of funeral. A belief in the doctrine of the resurrection of the body further reinforced the importance and symbolism of burial. However, as cremation slowly gained acceptance in Britain, notably by most Protestant churches, this progress was reflected both in its architectural expression and planning and as theologian and social anthropologist Douglas Davies points out, each crematorium can therefore be seen as 'a symbol of social change' (Davies 1995: 1).

By the end of the twentieth century 72 per cent of people in Britain were choosing burial over cremation.[3] The crematorium building increased in significance as a public building in the latter part of the twentieth century, perhaps often replacing the church as the main focus for the important function of saying farewell to loved ones. Thirty years ago the crematorium was a place for cremation and a brief committal ceremony. Now the ritual, the function and the remembrance are centred on the crematorium. As a result, Britain's crematoria have emerged as a significant element within contemporary life worthy of scholarly attention and deserving of a new critical reading, since they offer an architectural form that reflects the values and social life of a modern, urban and increasingly socially and geographically mobile society.

In contrast to the Hindu and other traditions of cremation, the practice in Britain and Europe was developed from the outset as an indoor disposal activity and as such, one that called for a new building type, by definition, without architectural precedent. The crematorium presented architects with the challenge of creating new spaces to serve very specific functional, emotional and symbolic needs. The first crematorium in the UK opened in Woking, Surrey in 1889 and the most recent in West Lothian in 2008.

Cremation came late to Wales. Indeed of Britain's 253 crematoria, only thirteen are in Wales.[4] Although initial efforts to establish a crematorium date from 1919, it was not until 1924 that Pontypridd, the first in Wales, finally opened. It was not a new building, but a conversion of one of two cemetery chapels in Glyntaff Cemetery, coincidentally only a few miles from where Dr William Price had committed the offence of attempting to burn the body of his infant son in a large cask containing half a barrel of paraffin oil. This act gave rise to the landmark trial in Cardiff in 1884, in which Judge Stephen cleared the Welsh eccentric, concluding that no criminal act was committed by burning a dead body provided that it did not 'amount to a public nuisance at common law'. However the concerns expressed over the practical issues raised drew attention to the need for dedicated buildings. Wales was to wait another 29 years for its second crematorium at Thornhill, Cardiff (1953) where Cardiff Architects' Department engaged in a comparatively adventurous adoption of contemporary European forms, primarily Scandinavian.

---

3   Information supplied by The Cremation Society of Great Britain.
4   See Gazetteer in Grainger 2005: 363–502.

Swansea (1956) and Colwyn Bay (1957) followed, but were deemed somewhat undistinguished.

However, post-war, the Welsh began to lead the field by turning to modernism for a solution to the vexed problem of crematorium design, their endeavours culminating in the acclaimed designs for Margam, first mooted in the 1950s, Coychurch and Llywdcoed.

Developments in Wales therefore beg two questions; why did cremation come late, and why was modernism espoused so enthusiastically in Glamorgan, leading as it did, to the commissioning of Fry? Keith Denison argues that despite the long desired objective of disestablishment and disendowment of the English Church achieved finally in 1920 with the constitution of the Church of Wales, historic Welsh Nonconformity has been in long-term decline. He maintains that there is still in rural Wales a 'residual attachment to "the chapel", but the challenge of increasing secularisation is very real in Wales and is reflected in the growing acceptance of cremation' (Denison 2008: 238). But despite the fact that more people are cremated than buried in Wales, it remains a smaller proportion in Wales than in England (Denison 2008: 241). For many living in rural areas, journeys to the nearest crematorium are inevitably long and time-consuming and there remains a strong attachment to burial. Indeed for millennia the Welsh buried their dead, processing on foot to the local burial ground. Although a growing percentage of modern Welsh society chooses to travel, sometimes not inconsiderable distances, to cremate its dead, the echoes of the imagery of burial are not too hard to find. At cremation coffins tend, in the main, to descend in imitation of burial and the Constitution of the Church in Wales insists that adherents' ashes be buried, although no such restriction is imposed on Nonconformist or secular funeral celebrants.

As far as the espousal of Modernism is concerned, post-war architecture in Glamorgan can be divided into two halves, which Newman terms: 'Modernist and rather than Post-Modernist – Pluralist. For 30 years, from 1945 to the mid 1970s, a Modernist idiom established itself as the *lingua franca*' (Newman 2004: 116). This came about first 'through the intervention of a small number of English architects' and from the mid 1970s onwards, when the impact of the newly founded Welsh School of Architecture at University of Wales Institute of Science and Technology (UWIST) began to be felt. Notable was George Pace of York, who worked at Llandaff Cathedral during the 1950s, where he showed his debt to Le Corbusier's church at Ronchamp (1950–55) and second, 'through the conversion to Modernism of local authorities architects' departments and the leading private practices (Newman 2004: 116). Here the outstanding names are the Cardiff City Architects' Department, under John Dryburgh from 1957–74, and Sir Percy Thomas & Son, (from 1964, Percy Thomas & Partners, and from 1971, the Percy Thomas Partnership).

Although F. Buckley of Percy Thomas & Son was the architect responsible for Gwent Crematorium in 1960, Sir Percy Thomas is reported to have taken a personal interest in the project, since it was the firm's first crematorium. Designed in a dignified domestic style, the materials used externally and internally were of the highest quality and the site of the building was given careful consideration. Thereafter, designs in Wales became increasingly modern, beginning with Wrexham (1966) by Sanger and Rothwell – architects of the acclaimed conversion of Oldham Crematorium in 1953 and winners of the competition for Kirkaldy Crematorium in 1954. F.D.Williamson & Associates of Porthcawl, a practice beginning to make a name for progressive work in Glamorgan, was responsible for the design of Narberth Crematorium, Pembrokeshire (1968). Compact and simple in its configuration of two blocks with asymmetrical roofs linked by a cloister, its modern design was mitigated by the use of traditional local materials, which lessened its impact on the surrounding countryside. Narberth was a prelude to Margam, opened in 1969, where Williamson & Associates were to make an uncompromising statement about the modernity of cremation.

Arguably the most dramatic crematorium design in Britain, Margam is quite unprecedented in its frank expression of materials and bold configuration of geometric forms drawn from European Modernism. The architects however stopped short of what might have been the inappropriate rawness of Brutalism, choosing instead to execute the crematorium in white concrete. The chimney was conceived as the focal point and celebrated its function in a refreshingly uncompromising way.

Perhaps because of its location, adjacent to the British Steel Works at Port Talbot, the architects chose to celebrate the modernity of cremation in their choice of industrially produced materials and forms inspired by the machine age. While the crematorium itself could be read as a paradigm of modernity, the plain cross at the entrance to the site speaks of a universal truth for Christians and is a direct reference to Gunnar Asplund's Woodland Cemetery in Stockholm (1935–40), much admired by contemporary architects. The cross, a feature also employed by the architects at Narberth, heralds a solemn approach to the crematorium by motor vehicle.

The Welsh authorities had undoubtedly benefitted from observing the increasing dissatisfaction with post-war local authority crematoria playing out in England and this perhaps goes some way towards explaining the leaning towards progressive architectural style in Glamorgan. In England cemetery chapel conversions had found favour during the 1950s, being cheaper and easier to secure planning permission. The architectural response in Wales was significantly different; only one of Wales' thirteen crematoria – Pontypridd, was a conversion, all the others were new – and with a couple of exceptions, situated in magnificent locations.

By the 1960s the minutes of the Joint Committees of Margam, Coychurch and Llywdcoed reveal that the need for crematoria in Glamorgan had become more apparent on account of the increasing practice of cremation and the worsening of traffic conditions, coupled with the need to conserve land 'for many necessary purposes' (Minutes of the Mid-Glamorgan Joint Committee, Coychurch Crematorium). All three crematoria are important in that they challenged the strong tradition of burial in Wales by confidently – indeed, defiantly, adopting a Modernist idiom, generally eschewed in England. However, it was Fry's involvement that was to prove particularly significant, given his international profile, but perhaps more importantly because the design and thinking behind Coychurch emphasised Fry's deeply held conviction over 'how great a part emotion should play in design' (Fry 1969: 259). This was particularly pertinent in the case of the inherently emotional space of a place of departure. His designs for Coychurch would further expose the mediocrity of much British crematorium design of the 1960s and in so doing, provide a benchmark of excellence for the future.

In June 1964 The Cremation Society of Great Britain invited Fry to address their annual conference in Bournemouth on the *Design of Modern Crematoria* (Fry 1964). Quite why he was approached to speak is not documented, but it is reasonable to assume that it was in an attempt on the part of the Society whose primary purpose was, and still remains, the advocacy of cremation, to bring a fresh eye to an unresolved building type clearly attracting adverse public comment. Paradoxically, despite the growing popularity of cremation, those using crematoria found (and indeed often still find) many of them unsatisfactory – their design uninspiring, banal and inconsequential. It must, however, be borne in mind that the crematorium presented a series of unprecedented challenges to the architect, given the nature of its function, a building frequented by a large cross-section of religious, secular and ideological movements, all with different, but overlapping needs. From the outset the lack of a shared and clear expectation of what was required from a crematorium had given rise to the cultural and arguably intentional, ambivalence lying at the heart of many designs. Not surprisingly therefore, architectural responses had been often ambiguous and evasive. Furthermore, given that very few architects were in the position of having designed more than one crematorium, there was little accumulated wisdom on the subject.

At once utilitarian and symbolic, religious and secular, crematoria are fraught with complexity. Two very distinct spaces are required: the functional and the symbolic, linked by a transitional space through which the coffin passes from the chapel or meeting hall to the cremator. This space acts both as a barrier and as a threshold between the 'death' and 'life' sections of the crematorium, the former being the utilitarian space devoted to the disposal of the dead body and the latter, the space occupied by the celebrant and the mourners during the funeral service. While the utilitarian purpose – that of reducing a dead body at high temperature to vapour and ashes has remained unequivocal, the search for symbolic architectural forms to create meaningful spaces in which to accommodate the range of emotional, spiritual or religious needs of the mourners, has proved highly problematical.

From the outset concerns had abounded over form, ritual and symbolism, but by 1960, criticisms were emerging over the low order of design and architectural expression of crematoria, articulated by the public, death care professionals and architectural commentators alike (Grainger 2005). At best, the new post-war crematoria were 'solemn, sentimental and modestly pious', on the one hand and 'jaunty, efficient, hygienic and civic-minded on the other' (Miller 2003). At worst, they represented the formulaic dreariness of much municipal architecture and were not always to the public eye, humane. The Cinderellas of local authority provision, they were almost invariably denied the architectural embellishment reserved for prestigious civic buildings. The Cremation Society was keen to change this.

Fry had never designed a crematorium, but he enjoyed a reputation as a forward thinker and promoter of social values in architecture. His professional status, endorsed by having just received the Royal Institute of British Architects' Queen's Royal Gold Medal, would have recommended him to a Society that had long recognised the promotional value of architecture evidenced by their publications and conference papers (Grainger 2005).

In his introduction to Fry's lecture, the chair, landscape architect and Society Council Member, Leslie Milner White, made reference to the previous speaker from Holland, who having identified the crematorium as a 'challenge to architecture' hoped and wished that the creation of *new* crematoria would call for 'a real cultural and artistic contribution' on the part of Dutch architects (Fry 1964: 39). The question of whether or not the challenge had been met in Britain was to be the subject of Fry's discourse. The number of deaths resulting in cremation had risen dramatically after the Second World War; in 1945 the figure stood at 7.2 per cent, rising to 43.2 per cent in 1964, the year of Fry's paper, and to 55.6 per cent in 1970 the year that Coychurch opened. The demand for crematoria had been met by a building boom between 1950 and 1970 from which period three-fifths of Britain's total provision dates.

Fry later explained that in preparing his address, he had 'made a survey of existing crematoria', tackling first the issues of architectural expression and economy, arguing:

> Though there existed crematoria the design and treatment of which aroused feelings appropriate to the assuagement of grief and bereavement, too many were of a character unfitted to the task in this country owing to an unawareness of the solemnity of the final rites and pitiable need for true consolation; coupled, it must be said, with a general understanding that crematoria should cost about £70,000 to £80,000 where loan sanction is concerned. (Fry 1968: 258)

In contrast, Fry pointed to two Scandinavian examples, which in his view represented meaningful architectural expressions of the complex human emotions surrounding cremation. The first, Gunnar Asplund's Stockholm Woodlands Crematorium (1935–40), Fry argued 'owes much of its reputation as a highly devotional building to a simple cross placed on the skyline of the approach to it, but proving to me so conclusively how great a part emotion should play in design' (Fry 1968: 258–9) The other was Gävel, in Sweden, designed by Engström, Landberg, Larsson and Töreman (Kidder Smith 1964: 233) where Fry believed that 'by its utter simplicity and the way in which it collaborates with the forest landscape so dear to its inhabitants, performs the same true function with a sensitiveness appropriate to its purpose' (Fry 1968: 259). Fry concluded:

There are crematoria of various kinds, which in one way or another, emphatically, demonstrably, anciently and modernly, simply and often dully, convey some part of the emotions they were built to evoke. But too many of the British examples gave me the impression of insufficiency, of a ritual becoming truncated, of work done to a formula, down to a cost, with materials of poor quality, as though crematoria could be as much run of the mill design as any other buildings valued beneath town halls. (Fry 1968: 259)

Fry took exception to the emotional emptiness of both the buildings and the 'ritual', believing that what was lacking was any sense of participation in the ceremony of cremation. He recalled that in Chandigargh in India, where he and his wife Jane Drew had worked with Le Corbusier in the 1950s, he had attended the funeral of the father of one of his architects and had been struck particularly both by the immediacy of the event and by the active involvement of the mourners in the ceremony. The burning of the body on the riverside pyre he found both direct and moving in the same way that services of marriage and death in the Church of England, with their 'nearly bitingly significant phrases' (Fry 1968: 257), could be equally unambiguous and pertinent to the occasion. This directness led to a shared purpose and understanding. Fry was convinced that architectural language played a central role in articulating emotion and belief:

> The monuments of an age of faith when everything that was concerned with ceremony was part of the life of those who took part in it were always extremely direct and the feelings significant, and the monuments with which death was memorialised both in the middle ages and later were the work of the highest talent available. They were objects of great beauty and everything connected with them was done at the highest level available. (Fry 1964: 41)

Fry cited medieval York as a near perfect example of this integration of architecture, art and design, 'from the Minster down to any one of a dozen parish Churches' (Fry 1964: 41). The population would at that time only have been fifteen or twenty thousand adding testament to the significant role that these buildings played in the community.

In contrast he remembered his mother's funeral conducted at an outer London crematorium and recalled

> [...] how low I felt, and how little uplifted or comforted by the ceremony. Indeed, the burdens of grief were immediately taken away from us by the undertakers. We drove to the ceremony in motor cars, and did nothing more than assist. The music, I remember was mechanical and the coffin jerked away mechanically out of sight to its final disposal. (Fry 1964: 41)

He was left feeling 'frustrated and uncomforted' and 'emotionally cheated' (Fry 1964: 41) on his homeward journey and concluded that what was required was an architecture and liturgy commensurate with the significance of its purpose.

Fry identified the lack of definition surrounding the concept of the building type to be at the root of the problem. The 'secular state' was not sure how to deal with death and was, as a consequence, unable to provide the architect with a clear brief. Above all, he felt keenly the 'denigration of true feeling, the loss of simplicity and the absence of a really significant ceremony' (Fry 1964: 42).

This attenuation of feeling could be attributed to our reliance upon 'machinery and mechanistic effects in the conduct of our lives' (Fry 1968: 257). The issue was fresh in his mind, having just returned from the USA, where he felt that the social significance of what was being done was of 'a very low order'. He saw the problem as being rooted in the 'preponderance of mechanism, industrialisation and commercialism traducing and falsifying feelings that should be kept simple and direct' (Fry 1964: 41). Fry maintained that the quickening pace of secularisation in Britain had led to the 'replacement of clergy in the ceremony of cremation by the undertakers' who were 'by definition and practice, commercial' and the question to debate was how to 'raise the profoundly important offices of the dead to the dignity and meaningfulness that a civilised people stand in need of, and lacking which, will lose its hold on life itself' (Fry 1968: 257–8). Similar concerns had been articulated in somewhat more rebarbative terms, by Jessica Mitford in *The American Way of Death*, originally published the previous year, 'Gradually, almost imperceptibly, over the years the funeral men have constructed their own grotesque cloud-cuckoo-land where the trappings of Gracious Living are transformed, as in a nightmare, into the trappings of the Gracious Dying' (Mitford 1998: 14–15). 'Good building' argued Fry could not 'come forward in an atmosphere of commerce or of second-class feeling' (Fry 1964: 42).

Fry threw down the gauntlet by suggesting that the Cremation Society ought to address the question of 'the place that cremation takes in a modern secular world', asking:

> How should it come that a ceremony that should have quite naturally and properly belonged to the religion to which the deceased had belonged in his life should take place in a building that is specially made by the secular power, the cremation authority – and serves indifferently for several religions. The sense of property in belonging is absent. (Fry 1964: 42)

He ended with the plea that:

> We need a clarification of feelings, and this Society of yours should, I feel, try to ask fundamental questions in order to arrive at the programme of a crematorium which is at the highest level possible, freed from all the considerations, the commercial considerations especially, that have gathered round it in the course of its history. (Fry 1964: 42)

He later implored an audience at the Royal Society of Arts to:

> [...] consider the true purpose of a crematorium serving the region in which it was to be built, the need for a ceremony so much more than perfunctory that it ministered directly to grief, the avoidance of whatever served to introduce emotions at variance with the ends in view, whether through mechanistic short-cuts or duplicities of any kind. (Fry 1968: 259)

In suggesting an absence of a meaningful relationship between spatial configuration, architectural language and inner condition, Fry had moved the argument away from its pragmatic base, which hitherto had placed emphasis on planning and provision, towards an acknowledgement of the psychological impact of the buildings. There had been a time when there was no question or doubt as to the ceremonies and feelings appropriate to the fact of death, but 'today' he argued, 'it seems to me more than elsewhere that in the cult of cremation there are doubts and confusion that require some clarification' (Fry 1964: 39–40).

Fry's exhortation to those at the Cremation Society conference fell on receptive ears, for in the audience were Mansell Matthews, the Engineer & Surveyor to Penybont Rural District Council and representatives of a group of small rural authorities, Ogmore and Garw Urban District Council, Bridgend Urban District

Figure 13.1   Coychurch Crematorium
Source: Photograph by the author.

Council, Cowbridge Borough Council and Cowbridge Rural District Council, who together formed the Joint Mid Glamorgan Crematorium Committee. Fry later recalled that 'a Welshman came to me saying that he agreed with every word I said, and that if only I would come to South Wales he would do all that he could to see that these ideas were carried out' (Fry 1968: 259). The commission that ensued early in 1965 would allow Fry to make the leap from theory into practice.

The Joint Committee had secured a rather beautiful site on the outskirts of Bridgend, near the crest of a hill sloping down into a valley, to which was later added a 20 acre wood, providing a scenic background, with low hills in the distance. Fry, Drew & Partners began work in February 1965 reporting to Matthews that they were 'attempting to correlate the figures now in their possession', but had not yet 'fully completed' their 'basic research' (Letter from Peter B. Bond to Mansell Matthews 19 February 1965, Coychurch Crematorium). It was agreed that Fry would meet with representatives of religious bodies and the first scheme, with one chapel, was approved in June 1965 at a projected cost in June of £140,445. There followed some attempts to reduce the cost before seeking approval from the Welsh Office, but Fry was anxious not to compromise the crematorium or Chapel of Remembrance, but was instead prepared to look again at the external works. The final estimate appeared to have fallen to £126,500 by July 1965. The following year Fry was in a position to articulate his ideas to the Welsh Office – largely drawn from his address to the Cremation Society. The design was finally approved in March 1967, by which time the cost had risen to £142,723. In 1968, when the commission was well under way, Fry was invited to deliver The Alfred Bossom Lecture on *The Design of Crematoria,* to the Royal Society of Arts in which he developed his ideas further and explained the ways in which these were coming to fruition at Coychurch. He explained that foremost in his mind were two things:

> [...] first, purity and clarity in the functions of what was to take place, and secondly, the need everywhere for what would comfort and console, in large elements and small. And hovering over these the need to connect it all with history, to embed it into the region as part of the language and story of it. (Fry 1968: 260)

Fry believed that the impediments to mourning were 'abruptness, obviousness, peremptoriness and banality':

> [...] what mourning requires is first of all dissociation from the busy life of the streets the cortège must pass through, followed by a sympathetic ceremony. All surroundings, forms and concerted actions conspiring to revive, receive and finally to compose private grief by dignifying it in association with the past history of life and death in the region, and thereby joining the crematorium to the history and life of the churches and chapels equally with that of a secular life. (Fry 1969: 43)

In addressing the Welsh Office in June 1966, Fry had proposed that:

> This description of the interior explains the exterior which in the first view of it will establish the contact with mourners that serves to give true significance and comfort. The external forms are in general rounded which is in itself both solemn and comforting, helped by the beautiful texture of old re-used stone. The curved drums of the approach cloister accent the processional ceremony, the tall window marks the last few steps to the catafalque, while the aspiring, convex and receiving form of the catafalque niche prolonged to a pointed arch termination speaks for a finality not without hope. The long and heavy canopies act as a restraint and provide deep shadows.
>
> The chapel of remembrance precedes the main buildings on the line of entry, joined to the entrance by a line of yew trees. This chapel contains not only a book of remembrance but wall plaques and commemorative window panes and can be considered as a place of prayer in addition to its office as containing the Book of Remembrance. (Fry 1966: 2)

Architecturally, the final design represents a subtle synthesis of elements of European modernism. Some of the municipal crematoria of the late 1950s and 1960s owed a debt to the humane modernism of both the Festival of Britain and the Scandinavian Welfare State. Fry invoked this most noticeably in the main chapel interior, but he built upon this formula by adding the use of indigenous building materials, giving what Keith Miller calls 'a mannerist play of textures straight out of Frank Lloyd Wright – the whole place looks crazy-paved, admittedly a favoured vernacular technique in the district' (Miller 2003). The most overt references to iconic modernism are taken from Le Corbusier's pilgrimage church at Ronchamps in the Vosges (1950–55) and are clearly apparent in the concrete cowl surrounding the cross, the overhanging 'cushiony' (Miller 2003) concrete roofs and the simplicity of the interiors. The regrettable removal of the cowl in 1993 altered the visual balance of the building, by excising the vertical counterpoint to the elements in the foreground. The concrete entrance canopy and the stone-faced drums of the cloister together with the enlarged chimney shaft now serve to dominate a predominantly horizontal composition.

Fry recognised the need to connect the crematorium with history, to embed it into the region 'as part of the language and story' of it and this he achieved through, amongst other things, the recycling of old local stone (Fry 1969: 260). In November 1969 Matthews was applying to Bridgend Urban District Council to remind them of their agreement some two years earlier 'together with the Surveyors of the other constituent authorities, that you could arrange for any suitable stone which would become available in your area, as a result of the demolition of old buildings, to be reserved for the crematorium building' (Letter from Mansell Matthews to Fry, 16 November 1967, Coychurch Crematorium). The smaller, second chapel, Capel

Coity was built in cedar woods, with a copper roof to distinguish it externally from the main chapel.

Fry, himself a committed Christian, first resolved that there should be a Christian chapel, 'unequivocal and belonging' and so he designed Capel Crallo to seat 130 mourners, with a manual organ, for Christian worship and Capel Coity, for those of other faiths and beliefs, smaller and intended to seat about 70, but of the same quality. Fry would later have to defend his use of a cross, when at the Royal Society of Arts lecture in 1968 funeral director Ivor Leverton challenged him by asking 'Is cremation meant to be universal or is it just meant to be on sufferance for the people to whom a cross is an insult?' maintaining there should be no external cross. But Fry was resolute, arguing that 'a Christian community should be buried in its faith' and that 'to take away the rites and rituals associated with a particular faith seems to me to take away too much' (Fry 1968: 266). Having just returned from Iran, he invoked the 'curious spectacle' of pilgrims coming:

> [...] miles and miles and miles as a pure act of faith to go through the Moslem shrine in Mashhad. Why they do this is entirely illogical, and one cannot make the ceremony too logical. For that reason, I would in South Wales still maintain the Christian chapel and have a secondary one rather than make it all non-denominational. (Fry 1968: 266–7)

He was adamant that:

> Any degree of falseness or equivocation, such as the temporary removal of crosses or other tell-tale symbols like the reproduction of religious-seeming music, is more destructive of feeling than architectural ineptitude ... (Fry 1969: 43)

Central to Fry's design and planning was the reinstatement of an involvement in ritual, he believing that the procession of mourners through the grounds and the crematorium could in itself offer spiritual significance. This was expressed architecturally by the privileging of two elements – *delay* and *distance* and these underscored the design and planning, thereby enriching the ceremony 'so that both it and our own lives thereby become significant' (Fry 1968: 262).

This he orchestrated by means of the positioning of the building on the site and by the route that mourners take through the grounds and the crematorium building, both in motor vehicles and on foot.

In his presentation to The Welsh Office in March 1967, Fry outlined the practicalities. The rural character of the site was to be preserved by his proposal for an informal road plan leaving the major part of the space in rough grass, although some intervening hedges would have to be removed. The woodland would be maintained 'with a measure of good husbandry only'. Though informal in general appearance, the woodland landscape had been:

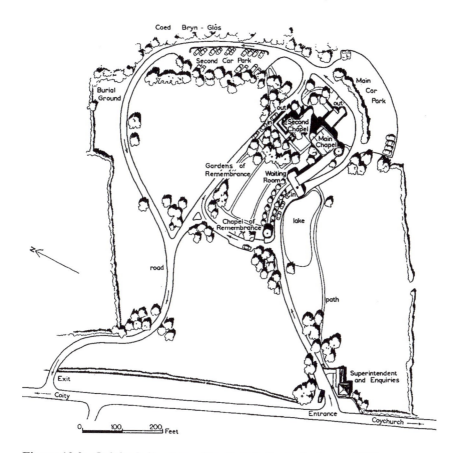

Figure 13.2   Original site plan – Coychurch Crematorium c. 1970

Source: Illustration published by kind permission of Coychurch Crematorium.

[…] very carefully planned to bring mourners into the right sort of contact with the crematorium buildings as they enter the site, to follow a determined flow that takes account of the timing of cortèges approaching and leaving the building, the visual separation of different cortèges, and the final separation of cortèges entering and leaving the site. The routing of visitors has also been considered. (Fry 1967: 1)

In addition to a central car park for 40 cars on the downward slope of Coed Bryglas, there were to be small lay-bys for funeral directors' cars near the dispersal points of both the large and the small chapels, near the Remembrance Chapel and at the office by the main entrance.

Fry pointed out that, though beautiful, the site sloped away from the road at an inclined hollow, falling towards Coychurch and into the wood and was not therefore 'without its difficulties'. He had concluded that a slight shelf of rock emerging from the wood would provide a level base and was therefore the obvious position for the building, 'The road falls therefore from the entrance in a direct line, keeping close to the contours, until it swings with the line of the valley and allows the front of the buildings to be seen over a shallow lake that skirts the wood' (Fry 1967: 1). In so positioning the building, Fry was able to exploit the natural contours of the landscape in order to match the functional requirements, while at the same time affirming and enhancing the symbolic significance of the route.

Furthermore, Fry recognised the role that motor vehicles must necessarily play in funerals resulting in cremation. All crematoria must conform to the 1902 Cremation Act by being 200 yards from the nearest dwelling and 50 yards from any public highway. As a consequence they were often confined to the margins of towns and cities and dependent on transport. Vehicular access had determined historically, to a large extent, both the layout of crematoria sites and the design of the buildings themselves, and over time had played a significant role in the ritual and ceremony of cremation in contrast to the procession to a grave on foot. Fry determined that vehicular access would predicate the layout of the site. Although he believed that 'as early as possible in the circulation of a crematorium, the cars should be banished out of sight', he nevertheless dealt with their necessity with sensitivity. He held that 'the hearse having come to a place appropriate to the Chapel that lies beyond should at that point be given some significance, but after that there should be if possible, some form of procession reformed with the coffin or the litter carried for this distance' (Fry 1964: 43) in order to regain something of the solemnity of the proceedings that might have been 'dispersed in the passage through the traffic of the street to the crematorium' (Fry 1964: 43).

The entrance to the site is marked by a tall stone pylon, which sounds the emotional note. Mourners experience a change of state as they move from the busy road into the tranquil confines of the crematorium passing first the Superintendent's House and Office, which Fry kept deliberately low and made to give the appearance of a long stone wall. Fry, in common with Le Corbusier, recognised the importance of the processional approach to a building, especially by motorcar, which accorded with the tenets of modernism. Fry reported to his clients that the view of the crematorium would be at first hidden by groups of trees so that it would be realised properly as mourners reached the lower ground. The building would be seen through the windows of the vehicles across a small lake that followed the curving line of the road. Fry explained that by planting trees and shrubs:

> First to conceal and then to reveal, reflected in a small lake, the forms of the crematorium itself – a climax is announced to which everything else will contribute up to the final moments of dispersal and return. (Fry 1968: 260)

He claimed that the purpose of these arrangements was to provide an atmosphere of calm and solemnity before the crematorium is entered after the journey through the 'busy heedless traffic of the town' (Fry 1968: 260). The fact that the crematorium building lay some distance into the site allowed Fry to exploit the emotional value of *delay*, thereby affording mourners the time to compose themselves.

The road then sets mourners down under a wide canopy and proceeds around the building to a further lay-by for funeral directors' cars, the entrance to a car park hidden on a sloping ground among trees, and the point of dispersal for those attending the funeral. It goes on to circle the building, serving a service yard on the way – is joined by the exit to the car park, and rejoins the main road some distance away from the entrance. The mourners meanwhile take up a carefully orchestrated procession on foot through the cloister to the catafalque, which is invested with quiet drama, by the introduction of the temporal element of shafts of light from above, moving around the space with the sun.

This was to be a ceremonial procession that is:

> […] repeated in smaller scale on the second chapel occurring at the other side of the cremation area and approached by its own road, so that no two cortèges will overlap to make disturbing confusion, or provoke a feeling of hurry. (Fry 1968: 260)

By privileging the approaches in this way, Fry wanted to obviate 'our disassociation from the ceremony and our non-involvement with grief' by allowing mourners to participate in ritual (Fry 1969: 43).

The emotional resonance of procession is further enhanced by the position of the Chapel of Remembrance, designed to announce:

> The slow rhythm of the stone drums that terminate in the circular vestry and mark, with the window in the chapel, the turn to the last few steps to the catafalque, playing a part in emphasising and prolonging the ceremony that ends with the committal. (Fry 1968: 260)

The form and character of the buildings are therefore determined, argued Fry, by their purpose. Both chapels are approached on foot along a cloister so that 'on the cortège arriving, there should be a pause before entering' (Fry 1968: 260). At the entrance canopy, Corbusian in its prominence, the procession takes up its load of the encoffined body – the approach cloister, with its semi-circular recesses encrusted with the memoria of past mourners, beating out a slow and comforting rhythm to the gradual slope where the officiating priest of whatever denomination, with or without choir, awaits the procession to lead the last few steps, still upward – indeed the whole of Capel Crallo is inclined towards the catafalque. Externally, Fry deliberately contrived the forms of the main chapel and its approaches to be 'convex, rounded, comforting but undemonstrative, the chapel in a warm-textured concrete and the lower walls of a series of urn-like bays that form niches within, in a local limestone reused from old buildings' (Fry 1968: 260) Around the central group of buildings were further stone walls disguising the cremation area, echoing and therefore emphasising the circular forms of the chapel.

The climax of the sequence is the committal of the body at the moment when it is imperceptibly lowered from the catafalque into the cremator area below, Fry arguing that 'I can consider no other or further disclosure of the process that would not transgress beyond the limits of normal human emotions under the stress of, in some cases, the acutest manifestations of grief' (Fry 1967: 2). This climax is expressed architecturally, in the form of a semi-circular niche in which the catafalque is placed, carried upwards into an open eye that allows light to fall onto the beautiful curved surface of the niche and to be slowly dimmed at the moment of committal. This almost Baroque feature allows the sunlight to mark the passage of time as the service progresses:

> So, in items large or small, the true nature of the rite is made implicit; but there remains the continuing memory of what took place, fading with time, for nothing is eternal; but adding to the history of the region stretching backward in time still to come. (Fry 1969: 261)

The dispersal follows the line of a covered way unobtrusively backed by cloakrooms, a small room for pall bearers, and a small flower room, with flower display outside the line of canopy, which draws quite explicitly in its detail from Le Corbusier, served by a small pool. Gentle steps take the mourners to their waiting cars.

Fry then turned his attention to the issue of memorialisation. One of the intended outcomes of the cremation movement was the creation of a 'new landscape for mourning' (Grainger 2005: 261). From the outset cremationists had been anxious to abandon the concept of the mournful, elegiac Victorian cemetery, calling instead for a new setting for remembrance, one in which a balance might be struck between individual commemoration and a more collective response to the shared human experiences of loss and memory. The jostling rows of competing memorials and the emblems of those memorialised, but long forgotten, were considered dreary, depressing and redolent of decay.

During the first half of the twentieth century new and varied forms of memorialisation would be instrumental in determining and shaping the design and landscape of crematorium gardens (Grainger 2005: 261–304). Furthermore, as Douglas Davies argues the 'crematorium chapel and garden of remembrance are likely to be invested with a sense of the sacred when they become closely associated with repeated memorial visits in honour of the dead' (Davies 1996: 87–8).

So far as the crematorium was intended to take the place of churches and chapels, where formerly the dead were buried and their memory sustained by tombstones and windows, it should, if it is to perform its true office to the community, argued Fry, provide the means of commemorating the dead in a number of ways (Fry 1967: 2). The secondary purpose of the Approach Cloister and the Chapel of

Remembrance at Coychurch was therefore to act as the most beautiful means of commemorating the dead – in plaques of stone or metal of standard size inserted into the niches, in small windows inscribed panes, again of prescribed size in the windows, in stained glass windows or inscribed into slate slab pavements. In the Garden of Remembrance, Fry proposed that stone slabs be incorporated into both the brick paving wall of the shelter. Generally within the grounds and woodland, he suggested the placing of seats, the insertion of stone slabs along the sides of the paths and the planting of bushes and other trees.

Fry believed that these memorials, which needed only a measure of uniformity, could initially be the work of local craftsmen drawn from the schools until a level was established, which would thereafter be maintained through a Faculty with appropriate membership 'to stand guardian over its future, and to indicate, perhaps, the composition of an organisation that may be the means of raising the function to the place it should occupy in our lives, to become one firmly established in the region' (Fry 1968: 261).

In the cloister pupils of Swansea School of Art were 'given their head' to produce stained glass windows intended to incorporate memorial tablets. There are four contrasting south facing sections, one between each drum of the cloister and to the north, a continuous composition between concrete mullions.

The circular Chapel of Remembrance was afforded careful attention. Built from local stone, with a copper roof, it was situated at the end of a tree-lined route and could be approached by car without disturbing entering funeral cortèges and had its own lay-by. Fry was not himself a devotee of Books of Remembrance (Fry 1968: 261) but nevertheless ensured that the specially designed book in sheepskin vellum was of the highest order. Stained glass in the Chapel of Remembrance was by Swansea tutor, Timothy Lewis and is abstract, highly coloured and in the best tradition of John Piper, the artist responsible for the tapestry in Coventry Cathedral. Fry was anxious that the Chapel of Remembrance be a space for private prayer, 'such a room being a sort of extension of the chapel and in no way a sight-seeing item of the whole' (Fry 1967: 2).

These devices would leave a variety of memoria in the gardens and surrounding grounds, with further opportunities along the paths and steps of the acres of woodland. Other forms of commemoration were devised, in slabs and garden seats that extended the commemorative motif. The idea behind this commemoration was, argued Fry:

> [...] to attach the buildings to the local community by something much stronger than convenience and to make it finally as evident a piece of regional history and culture as the oldest building by which we set store. Only by doing this can crematoria gain the dignified place they should occupy in the life of the modern community. (Fry 1967: 3)

Fry nevertheless had residual concerns that 'however well it may be done' this 'removes the ceremony from the living centres of population where ideally it still should be'. Fry had ended his lecture in 1964 by suggesting somewhat controversially that crematoria should be located somewhere in the centre of the town and not away on the outskirts. They ought to be 'brought back into our life as much as the Church was the centre of villages' (Fry 1964: 43). He proposed cloisters, 'upon which we lavish the very best work that can be got in carving, painting and sculpture' to be situated near the centre of the town and if necessary separated from the crematorium itself. He invoked the Campo Santo, the medieval cloister lying at the heart of Pisa, with the Cathedral and Baptistery, where 'the deaths of the citizens of Pisa have been celebrated with the greatest art, with murals and monuments on the floor and walls' (Fry 1964: 42). Fry admired its directness:

> The fact of death has been made a commentary on life and not, as in our own country, something we turn away from or sentimentalise and do not feel directly and with feeling. (Fry 1964: 42)

Fry was not the first to be moved by this edifice. The late Victorian architect Ernest George had made a study of cloisters in Italy and the Campo Santo in particular, while preparing his designs for Golders Green Crematorium for the London Cremation Company in 1902. George's cloisters, completed in 1916, were the first to introduce what was to become a standard form of memorialisation in crematoria in Britain.

In 1969, Fry coined the phrase, an 'anatomy of mourning' in order to:

> [...] draw attention to what some personal reflection will confirm, namely that however much we gloss over or otherwise dismiss them from our consciousness, the fact of death and the necessity for the full expiation of grief as a communal act must be freely admitted for the good of our soul and the health of our spirit. (Fry 1969: 44)

In so doing, Fry invoked the sociologist Geoffrey Gorer, whose pioneering study, *Death, Grief and Mourning in Contemporary Britain* (1965) proposed that mourners who were subjected to a more ritualistic form of mourning seemed able to adapt themselves more readily to life afterwards. Fry's contention was that architecture had the responsibility of providing the spatial context in which these subtle human experiences were expected to take place.

At Coychurch Crematorium he succeeded in expressing the complex human emotions associated with death and grieving in an architectural form and landscape articulated in such a way as to ensure that tradition, ritual, history and sense of place and belonging are privileged. Coychurch emerges accordingly as one of the finest crematoria in Britain.

References

Davies, D.J. 1995. *British Crematoria in Public Profile*. Maidstone: Cremation Society of Great Britain.

Davies, D.J. 1996. The sacred crematorium. *Mortality*, 1(1), 83–94, 87–88.

Denison, K. 2008. Funeral ministry in Wales, in *Death Our Future*, edited by P.C. Jupp. London: Epworth, 234–43.

Fry, E. Maxwell. 1964. The design of modern crematoria. *Report of Proceedings of the Cremation Society Conference, Bournemouth, 23, 24 and 25 June, 1964*, 31–8.

Fry, E. Maxwell. 1966, an unpublished paper written on behalf of the Joint Mid-Glamorgan Crematorium Committee, for presentation to the Welsh Office in June 1966, in the possession of Coychurch Crematorium.

Fry, E. Maxwell. 1967. Unpublished paper written for the Joint Mid-Glamorgan Crematorium Committee, in the possession of Coychurch Crematorium.

Fry, E. Maxwell. 1968. The Design of Crematoria, The Alfred Bossom Lecture, delivered to the Royal Society of Arts, 11 December 1968, *Journal of the Royal Society of Arts*, 117, 1968–9: 256–68.

Fry, E. Maxwell, 1969, *Art in a Machine Age: A Critique of Contemporary Life through the Medium of Architecture*, quoted in *Fry Drew Knight Creamer Architecture*, edited by Stephen Hitchins. London: Lund Humphries 1978, 43–4.

Gorer, G. 1965. *Death, Grief and Mourning in Contemporary Britain*. London: Cresset Press.

Grainger, H.J. 2005. *Death Redesigned, British Crematoria: History, Architecture and Landscape*. Reading: Spire Books Limited in association with The Cremation Society of Great Britain.

Grainger, H.J. 2009. 'Crematoria to die for': Modernity in Mid-Glamorgan, *Pharos International*, 74(3), 6–11.

Jupp, P.C. 2006. *From Dust to Ashes: The Development of Cremation in Britain*. Basingstoke: Palgrave Macmillan.

Kidder Smith, G.E. 1964. *The New Architecture of Europe*. London: Penguin.

Miller, K. 2003. Making the grade, Coychurch Crematorium, Bridgend. *The Daily Telegraph*, 15 November 2003.

Mitford, J. 1998. *The American Way of Death Revisited*. London: Virago Press.

Newman, J. 2004. *The Buildings of Wales, Glamorgan*. New Haven and London: Yale University Press.

White, S.R.G. 2002. A burial ahead of its time? The Crookenden burial case and the sanctioning of cremation in England and Wales, *Mortality*, 7(2), July 2002, 171–90.

White, S.R.G. 2003. 'The Cremation Act 1902: From Private to Local to General', *Pharos International*, 69(1), Spring 2003, 14–18.

Chapter 14

The Living, The Dead and the Imagery of Emptiness and Re-appearance on the Battlefields of the Western Front

Paul Gough

Taking as its field of enquiry the trenches of the First World War, this chapter explores the processes of death, burial and exhumation on the Western Front. Deserted by daytime, yet crowded with action at night, the Great War battlefield was a lethal tract where death was often random and anonymous. However, the battlefield could also be a phantasmagoric, at times enchanted place, replete with myth, superstition and sublime moments of dread and fascination. By looking at the war through the eyes of a number of artists this chapter examines the role of painting and photography in appearing to bring the dead, the disappeared and the dying back to figurative life. Possibly the best known work of this kind is Stanley Spencer's vast panorama of post-battle exhumation *The Resurrection of the Soldiers*, a mural-scale panorama of earthly redemption which was painted in the 1920s at the same time as vast tracts of despoiled land in France and Belgium were being brought back from apparent extinction, and planted with thousands of military gravestones. While salvage parties recovered and re-buried thousands of corpses, Spencer and such artists as Will Dyson, Otto Dix, Max Beckmann and Will Longstaff were conjuring up images of barren and blighted landscapes populated by phantom soldiers emerging from shallow graves.

The chapter opens with an examination of how soldiers populated an apparently emptied landscape which was actually teeming with subterranean activity, how they died, how they were buried, and how they were made to 're-appear' through art, film, and poetry. Having examined the crowded emptiness of No Man's Land, the chapter briefly explores the complex processes and iconography of remembrance, including the ritual surrounding the exhumation and re-burial of the Unknown Warrior in Westminster Abbey. Focusing on Stanley Spencer and his fascination with the ideas of redemption and resurrection, the chapter explores how different artists created images that appeared to revive and resurrect the battle-dead. Finally, through a reflection on Jeff Wall's epic photographic battlescape 'Dead Troops Talk', the chapter connects Spencer's ontology of reconciliation with Wall's bleaker montage of debacle and death.

Dying

> One was tall, gaunt Tom Gunn, the Limber-gunner of F. Sub-section. As we
> stood by his corpse someone lifted the blanket that covered his face. It was
> emaciated and the colour of pale ivory. The other man had died from shell shock.
> He stood upright by the wheel of his gun unmarked but quite dead. Just a short
> while before he had invited some of us to share in a parcel of food he had just
> received from home, but the party had to be cancelled. (Roberts 1974: 13)

Presented statistically, the loss of young life in the First World War is quite
overwhelming. Even when broken down into smaller numbers the scale of loss is
numbing: in 1916, Richard Tawney went 'over the top' on the Somme battlefield
with 820 fellow Manchesters; 450 men died in the initial attack; after the second,
just 54 answered the roll-call (Tawney 1953: 78). During the same battle the 1st
battalion of the Newfoundland Regiment – 801 officers and men – were reduced
to 68 uninjured men after a single day's fighting (Gough 2004: 238). However, as
many historians warn, these raw and terrible statistics must be treated with some
care. Figures for the hardest hit units must not be projected onto the whole war,
nor should one battle be regarded as typical of the experience of every foot soldier.
While it is estimated that 4.5 per cent of British fighting soldiers died during the
Second World War and 5 per cent in the Boer War, some 10 per cent died during
the Great War. The daily attrition rate on the British stretch of the Western Front
was over 200 soldiers (Terraine 1980).

Death came in various guises. For many it came anonymously and suddenly.
Artillery was the most lethal killer. While front-line soldiers dreaded the prospect
of hand-to-hand fighting, it was the awesome power of cannon and mortar that was
the real killer (Sheffield 2001: 110). A wounded man was three times as likely to
die as a result of a shell wound to the chest as of a bullet wound. Not only could
distant guns pound a specific tract of earth for hours, sometimes days on end, but a
direct hit from a heavy metal shell would completely obliterate the body, reducing
a living being to little more than a putrid whiff of air (Conrad 1999: 215). To
troops under sustained heavy shelling, artillery destroyed not only the body, but
the mind, inducing new depths of fear by its random anonymity.

Death by bullet was equally hideous. Sniper fire nearly always targeted the
head. Machine-gun fire was less clinical, tearing capacious holes in the body.
Obscene and random, death was like black magic:

> [...] bodies continued walking after decapitation; shells burst and bodies simply
> vanished. Men's bodies 'shattered': their jaws dropped and out poured 'so much
> blood'. Aeroplane propellers sliced men in to pieces. (Bourke 1996: 213).

Death from gas brought an entire new realm of suffering. Chlorine gas acutely
irritates the lungs and bronchial tubes, causing vomiting, violent coughing and
breathing difficulties. Heavy doses would cause the lungs to deteriorate in seconds,

the victim would cough up blood and die in minutes, 'doubled up, fists clenched, in agony' (Slowe and Woods 1986: 28).

Many, however, died without any visible sign of death, perhaps from shell percussion or from a hidden wound, though experienced soldiers would be able to detect the 'tell-tale blood drips on lips, in ears or lungs' (Winter 1978: 206). Siegfried Sassoon came across one unmarked body which he lifted upright from its prone state in a ditch:

> Propped against the bank, his blond face was undisfigured except by the mud which I wiped from his eyes and tunic and mouth by my coat sleeve. He'd evidently been killed while digging, for his tunic was knotted loosely about his shoulders. He didn't look to be more than eighteen. Hoisting him a little higher, I thought what a gentle face he had. (Sassoon 1930: 112)

What happened next depended entirely on the ebb and flow of the battle, and the specific conditions pertaining at the moment of death. Individual soldiers killed by sniper fire or shell explosion whilst holding a front-line post or near a road behind the lines would be easily identified and buried in small or individual plots near the front or reserve lines. Artilleryman William Roberts' drawing *Burying the Dead After a Battle* captures the poignant scene of gunners, heads bowed gathered around their comrades grave, while in the near-distance a town burns, an aeroplane falls from the sky, and tanks pitch about under billows of shellfire. 'We buried our own dead', he wrote, 'together with some left over from the infantry's advance, shoulder to shoulder in a wide shallow grave, each in his blood-stained uniform and covered by a blanket. I noticed that some feet projected beyond the covering, showing that they had died with their boots on, in some cases with their spurs on too' (Roberts 1974: 14). During the static years of siege warfare on the Western Front graves were dug in advance, some regiments setting aside plots of land for their own dead, even barring others from 'trespassing'. Many of these regimental plots would later be retained by the Imperial War Graves Commission as small and compact cemeteries dedicated to particular units – Gordon's Cemetery near Mametz, for example. Most of these small and isolated plots would later be dug up; the bodies exhumed by Graves Registration Units and brought into one of the vast 'concentration' cemeteries, located nearer villages and roads so as to allow ease of access by visitors after the war (Longworth 1967: 14).

Not all of the dead would be buried intact. Souvenir hunters would strip a body of its every article, especially if the dead were the enemy, and the further from the front-line the more cleanly picked. As Guardsman Stephen Graham later recalled: 'Those (bodies) nearest our encampment at Noreuil all lay with the whites of their pockets turned out and their tunics and shirts undone by souvenir hunters.' He remembered in particular, a well clothed six feet three inches tall dead German, whose boots were taken first, then his tunic, 'A few days later he was lying in his pants' (Graham 1921: 67).

After a major set-piece battle, the work of the burial parties was unrelentingly grim. Those who had been killed during an attack or patrol might be less easily identifiable, having lain in No Man's Land or other parts of the battlefield exposed to enemy fire. Mass clearance was attempted even when a battle was in progress or where a front was still strewn with the recent dead. In the aftermath of the Battle of Loos in autumn 1915 Scots officer George Craike spent nights with groups of his men scurrying into No Man's Land to hastily cover the bodies of East Surrey soldiers who had died in large numbers a week earlier:

> We crawled out of the trenches with caution in small parties, and dealt with the dead simply by putting them into depressions in the earth, or into shell holes. This was not a pleasant task and occasionally the arms disengaged from the bodies. However, the bodies were placed as far as possible in these holes and covered over with a light layer of earth, this earth being brushed or dug in by the entrenching tools. All the work had to be done on all fours, for to stand erect was courting disaster. ... The work was slow, laborious and difficult. (Arthur 2002: 105)

Once the worst of the fighting had passed over, larger burial parties would be organised. Often consisting of soldiers from different units, motivated by the need to maintain morale, to achieve some modicum of hygiene and out of common humanity, they combed the former battlefield, their noses and mouths covered by fragments of gas masks, removing the identity discs if they could be recovered, the red disc destined for the orderly office, the green one left on the body to ensure accurate identification. Pockets would be searched to uncover paybooks and other personal effects. There was little time for niceties: 'you put them in a hole ready dug with boots and everything on. You put in about 10 or 15, whatever the grave will hold, throw about 2 feet of earth on them and stick a wooden cross on top' (Winter 1978: 206).

Fragments of British dead were collected in empty sandbags and buried in mass graves as quickly as possible; their grave markers often listing little more than 'an unknown soldier', or possibly some indication of the regimental or unit title. Sapper Richards remembered gathering the remnants of one ghastly bombing accident, rescuing 'bits from telegraph wires where they'd been blown at great velocity', and burying them in a common grave (Arthur 2002: 106).

Sometimes the interval between death and discovery was too long and bodies had literally fallen to pieces, ravaged by rats, weather, and biological processes:

> As you lifted a body by its arms and legs, they detached themselves from the torso, and this was not the worst thing. Each body was covered inches deep with a black fur of flies, which flew up into your face, into your mouth, eyes and nostrils as you approached. The bodies crawled with maggots. There had been a disaster here. An attack by green, badly led troops who had had too big a rum ration – some of them had not even fixed their bayonets – against a strong

position where the wire was still uncut. They hung like washing on the barbs, like scarecrows who scared no crows since they were edible. The birds disputed the bodies with us. This was a job for all ranks. No one could expect the men to handle the bodies unless the officers did their share. We stopped every now and then to vomit... the bodies had the consistency of Camembert cheese. I once fell and put my hand through the belly of a man. It was days before I got the smell out of my hands. (Cloete 1972: 121)

Under these extreme circumstances every effort at conventional decency was attempted. Burial parties tried to give 'these poor bleeding pieces of earth' a Christian burial by reading sections from the Book of Common Prayer. Even if a Minister, chaplain or priest could not attend, all combatants felt it important that 'you buried your comrades and saw to it that their graves were marked with a wooden cross and a name ...' (Carrington 1965: 127–28). The hope of creating a more decent burial improved with distance from the front; whereas a loose covering of earth might be all that was possible at the trench lines, 'some old sacking' was considered an adequate and appropriate covering further from the front, while canvas sheets were regarded as a suitable substitute for coffins in the military hospitals located in the rear zones (Bourke 1996: 215–16).

Enemy dead were often left until last to be cleared from a battlefield. As Charles Carrington (1965: 128) coolly noted they 'came last in priority, and more than once I have cleared a trench of its defunct tenants by throwing them over the parapet where someone might or might not find and bury them.' Experienced soldiers could estimate the date of death from the colour and condition of corpses left out in the open, Caucasians turned from yellow to grey to red, and then to black. In death, white soldiers turned black and black Senegalese soldiers turned white (MacDermott in Bourke, 1996: 214).

Official War Artist William Orpen was astonished at the weird colours of the enemy corpses he stumbled across while roaming the abandoned Somme battlefields in 1917, and made a number of precise drawings describing the polished skeletons of German soldiers 'bleached white and clean' by the fierce summer sun. As he wandered over the emptied downlands of the Somme everything shimmered in the heat; abandoned clothes were baked into strange combinations of colour, 'white, pale grey and pale gold. The only dark colours were the deep red bronze of the "wire", wild flowers sprouted everywhere... in the evening, everything golden in the sunlight' (Orpen 1923: 23–4). His only companions on these sojourns were distant burial parties who were diligently digging up, identifying, and re-burying thousands of scattered bodies in the larger concentration cemeteries. Orpen passed one such group near Thiepval Hill, resting from their unpleasant work, and trying to identify the dead from their meagre finds – a few coins, pocket knives, an occasional identity disc – garnered from their long labour. Perhaps it was only artists who, commissioned to seek out the novel and unique faces of war, sought the imagery of death *in extremis*:

Then suddenly round the bend in the trench I came to a great bay which was full of dead Germans, but they weren't a bit horrible. They had been dead for about six weeks and weather and rats and maggots and everything else had done their stuff. Now they were just shiny skeletons in their uniforms held together by the dry sinews, that wound round their bones ... It was a most weird and extraordinary picture and I was absolutely fascinated. (Talbot Kelly 1980: 5)

The Desert: Deserted but 'Populated'

Not far from where Orpen sat drawing the picked remains of soldiers in foul-smelling trenches, Charles Carrington scanned the scorched earth of the southern battlefield, its few remaining trees snapped short with splintered ends 'like monstrous shaving brushes', everywhere the smell of burnt and poisoned mud, every yard of ground 'ploughed up by shell-fire and ...tainted with high explosive, so that a chemical reek pervaded the air ... and through it one could distinguish a more biotic flavour – the stink of corrupting human flesh.' In fact, Carrington reckoned that the best part of 200,000 men had been killed in the last few months somewhere in the 30 square miles around his trench. Buried hastily in shallow graves, or buried and subsequently blown out of those graves, he estimated '7,000 corpses to the square mile [was] not much of an exaggeration, ten to the acre shall we say, and your nose told you where they lay thickest' (Carrington 1965: 127).

To the scrutinising eye the landscape may have seemed deserted but the dead lay just beneath its ruptured surface and the living led an ordered and disciplined existence in underground shelters and deep chambers (Redmond 1917: 39). It was one of the greatest contradictions of modern warfare, a landscape that gave the appearance by daylight of being empty but was emphatically not: it teemed with invisible life. Few paintings have captured the immensity of that void; even words failed to convey the intensity of its emptiness. Faced with the phantasmagoric lunar face of the Western Front, the imagination froze:

It seemed quite unthinkable that there was another trench over there a few yards away just like our own ... Not even the shells made that brooding watchfulness more easy to grasp; they only made it more grotesque. For everything was so paralysed in calm, so unnaturally innocent and bland and balmy. You simply could not take it in. (Farrer 1918: 113)

One writer who visited the Western Front – Reginald Farrer – suggested, that it was quite wrong to regard the 'huge, haunted solitude' of the modern battlefield as empty. 'It is more', he argued, 'full of emptiness... an emptiness that is not really empty at all' (Farrer 1918: 25). Paul Nash visualised this idea – borrowing Farrer's phrase the 'Void of War' and populating its emptinesses with latent violence. The very concept of space as an undifferentiated, homogeneous void which surrounded solid objects had already been challenged by contemporary artists; cinema was revolutionising the visual arrangement of time; the act of film editing fractured continuous events, reshaping and compressing story-lines into new patterns of narrative. Just as geographers were developing regional approaches on the interrelationship between people and their local environments, (Baker 1988) so scientific research pioneered by Einstein argued for a number of distinct spaces equal to the number of unstable reference systems. Braque and Picasso smashed forever the belief in a neat pictorial system based on the single static eye of one-point perspective. It was a period of extraordinary innovation, as if 'an earthquake had struck the precisely reticulated sidewalks of a Renaissance street scene' (Kern 1983: 179). War accelerated these changes: when Picasso saw trucks heading out of Paris towards the Front he is said to have pointed at their camouflage and exclaimed 'yes, it is we who made that, that is cubism' and to a degree he was right. Deceptive and disruptive camouflage is the perfect exposition of the new way that the world's spaces had to be seen, or to be more exact, not seen (Stein 1938: 11).

By contrast, the benighted battlescape was always busy as troops set to work repairing their entrenchments, reinforcing the wire, bringing forward fresh troops, food and provisions, or setting out into No Man's Land on raid or patrol. The tract of land between the trenches was a 'debatable', fluid and near-mythical zone that soldiers learned to fear, but which also exercised a dread fascination with many. The poet David Jones captured its liminal qualities, the threshold between two different existential spaces:

> The day by day in the wasteland, the sudden violences and long stillnesses, the sharp contours and unformed voids of that mysterious existence profoundly affected the imaginations of those who suffered it. It was a place of enchantment. (Jones 1937: x)

At the intersection of these two worlds – the dangerous emptiness of the daylight battlescape and the crowded busy-ness of the benighted No Man's Land – came one of the critical moments of any soldier's experience of war: the moment he left the relative safety of the front-line and stepped up into the danger zone. One soldier remembered it thus:

> The scene that followed was the most remarkable that I have ever witnessed. At one moment there was an intense and nerve shattering struggle with death screaming through the air. Then, as if with the wave of a magic wand, all was changed; all over 'No Man's Land' troops came out of the trenches, or rose from the ground where they had been lying. (Stuart Dolden 1980: 39)

Moving from the horizontal to the vertical, from subterranean security to maximal vulnerability, was an ultimate transformation for every combatant. It compounded the central tenet of militarised service; the transformation from civilian to soldier, from innocence to experience, and, in many cases, from youth to adult. Indeed, every level of the military experience seemed to be permeated by the rhetoric of transformation and conversion. One officer, for example, relieved from an exposed front-line outpost, described how marvellous it was to be out of the trenches: 'it is like being born again.' (Plowman 1927: 54) Another described those who survived one particular battle as 'not broken, but reborn' (Williamson 1988: 10). Throughout the memoirs of the Great War, (and perhaps maybe all wars) there is a common language of initiation, of 'baptisms of fire', of inner change and transmutation brought about by ecstatic experience, of 'immense exultation at having got through the barrage' (Owen in Fussell 1975: 115). Edmund Blunden, returning to his lines after a desperately dangerous patrol in No-Man's-Land, recalled how 'We were received as Lazarus was' (Blunden 1928: 172). When Siegfried Sassoon discovered that his friend Robert Graves was not dead and had in fact survived an artillery barrage, the news was celebrated as though he 'had risen again from the dead' (Sassoon 1930: 128). As Paul Fussell has written, it was this plethora of 'very un-modern superstitions, talismans, wonders, miracles, relics, legends and rumours that would help shape the dominant mythologies of the war.' It was a world of 'conversions, metamorphoses, and rebirths in a world of reinvigorated myth' (Fussell 1975: 115). The transformation of the body through moments of extreme tension was matched by the transformation of the pulverised landscape both during and after the war. Hampshire's officer Paul Nash saw how war wreaked its havoc, but was astonished that nature should prove so extraordinarily resilient. He wrote of walking through a wood, or at least what remained of it after the shelling, when it was just 'a place with an evil name, pitted and pocked with shells, the trees torn to shreds, often reeking with poison gas'. Two months later this 'most desolate ruinous place' was drastically changed. It was now 'a vivid green':

> [...] the most broken trees even had sprouted somewhere and in the midst, from the depth of the wood's bruised heart poured out the throbbing song of a nightingale. Ridiculous mad incongruity! One can't think which is the more absurd, the War or Nature ... (Nash 1949: 33)

Re-membering

In 1919 Paul Nash and his brother John, were provided with a truck load of shards from the Western Front – metal fragments, sheets of corrugated roofing, concrete blocks and other detritus – delivered to their studio in the Chilterns to jog their memories as they embarked on paintings commissioned by the British War Memorials scheme. Both were encouraged to revisit the old battlegrounds, but having served on the front-line they chose not to, accepting that its cruel complexion was impressed indelibly on them. How could they forget the state of northern France and western Belgium after years of siege warfare. Objective measurements attest to the utter scale of desolation across a great tract of northern

Europe where some 333 million cubic metres of trench had to be back-filled, barbed wire covered an estimated 375 million square metres, over 80,000 dwellings had been destroyed or damaged, as were 17,466 schools, public buildings and churches, and the population of the devastated regions had diminished by 60 per cent (Clout 1996). A map drawn up by the British League of Help for Devastated France superimposed the scale of war damage onto the Shires of England with the startling prediction that no fewer than 21 English counties would have been severely blighted by war – a swathe of destruction that reached from Kent to the north Midlands (Osborne 2001).

While the native populations in France and Belgium toiled to reconstruct their homes and land, pilgrims and veterans roamed the former battlegrounds to locate places that might contain the memory of significant events. Outwardly there was nothing to see; the landscape that drew them was an imaginary one. It was a place of projection and association, a space full of history, yet void of obvious topography, where physical markers had been obliterated but the land overwritten with an invisible emotional geography (Gough 1993). When the painter Stanley Spencer travelled to the Balkans in 1922 he was undertaking a journey made by thousands, indeed tens of thousands, of travellers who were uniting intense memories with places that no longer existed; indeed the wasted landscapes in France, Belgium, the Dardenelles and Macedonia were outwardly empty places 'you take your own story to', bereft of identifying landmarks except for painted signposts indicating where things once were – former villages, churches or farmsteads – and littered with war refuse and unspent ammunition (Shepheard 1997).

By 1920, some 4,000 men were daily engaged in combing the battlefields in the search for human remains. On the Western Front, the ground was divided into gridded areas, each searched at least six times, but even ten years later up to 40 bodies were being handed over each week to the French authorities (Middlebrook and Middlebrook 1991: 3). In France a ten franc bounty was given for each corpse returned to the authorities. A systematic method to identify graves and locate shallow burials had been put in place by the British as early as September 1914, although initial attempts to co-ordinate the burial and recording of the dead were somewhat haphazard. It was the zeal of Fabian Ware and his Graves Registration Unit that laid the foundations of a systematic audit of British and Empire dead and their place of burial (Longworth 1967). Once it had been decided that bodies would not be exhumed and repatriated, Ware began to establish a method for graves registration and a scheme for permanent burial sites. He also arranged that all graves should be photographed so that relatives might have an image and directions to the place of burial. By August 1915 an initial 2,000 negatives, each showing four grave markers, had been taken. Cards were sent in answer to individual requests, enclosing details that gave 'the best available indication as to the situation of the grave and, when it was in a cemetery, directions as to the nearest railway station which might be useful for those wishing to visit the country after the war' (Ware in Hurst 1929: vii). Nine months later Ware's makeshift organisation had registered over 50,000 graves, answered 5,000 enquiries, and supplied 2,500 photographs. Little over a year later the work to gather, re-inter and individually mark the fallen had become a state responsibility. The dead, as Heffernan states, were no longer allowed 'to pass unnoticed back into the private world of their families'. They were 'official property' to be accorded appropriate civic commemoration in 'solemn monuments of official remembrance' (Heffernan 1995: 302). Ware's band of searchers took to their work with zealous diligence. One described it as requiring the patience and skills of a detective 'to find the grave of some poor fellow who had been shot in some out of the way turnip field and hurriedly buried.' After the war, local people, especially young children, joined in the searches with sometimes grisly outcomes:

> It occasionally happens that the grave which we believe to contain the remains of a certain person is, in fact, a pit into which large numbers of dead bodies have been thrown by the enemy. When such a grave is opened we are able not only to identify the body for which we are searching, but also by their discs, the bodies of many others. One example – the latest – will suffice. The trench containing the bodies of Colonel _, Captain _, Lieutenants __ _, held also the bodies of 94 non-commissioned officers and men. Of these 66 still wore their discs, etc., and thus their deaths were certified, and their graves ascertained. The trench was then prolonged, the bodies laid side by side, and the burial service read over them. (Report in Longworth 1967: 4–5)

According to Longworth, a good registration officer quickly came to know intimately the ground allotted to him: he knew its recent military history, every raid, skirmish or significant action, the regiments involved and in which fields unmarked or unrecorded graves were likely to be located. Following up every scrap of information, sometimes gleaned from veterans who had served on that part of the front, he pieced together the scanty evidence so as to identify burials. Graves Registration staff had actually undertaken such work during the war, often within range of enemy gunfire, but after the death of one staff member working in an Ypres cemetery they were ordered back from the front-lines into the safer areas where battle had moved on. This necessary, but unfortunate, decision may explain 'the extraordinarily high proportion of unidentifiable graves when the count came after the war' (Longworth 1967: 10).

Known and Unknown

Between 1921 and 1928 some 30,000 corpses were dug-up from their last burial place, and re-interred. Each body was marked by a standard stone headstone, which carried a modicum of military detail, as much as could be gleaned from the corpse or from its first grave marker, usually name, rank, regimental number (except for officers) military unit, date of death, and age (if supplied by next of kin). Personal inscriptions paid for by the family of the dead man were allowed to a maximum of 66 characters, including the spaces between words (Batten 2009). It is reckoned that only a quarter could be identified because fibrous identity discs issued before 1916 had disintegrated. In those instances the headstone would simply state 'A Soldier/of the Great War/Known Unto God'. In some cases the inscription indicates that the body was known to have belonged to a particular unit, but could be identified in no greater detail, or that the body lies not directly beneath the stone but somewhere within the plot of the cemetery. Despite the occasional attempt to have an individual body brought home for private burial, the principle – approved by the Imperial Conference of 1918 and endorsed by the British government in May 1920 – that all bodies were to be buried near to where they fell was rigorously applied. There was, however, one notable exception – the exhumation and burial in Britain of an 'Unknown Warrior'.

Many individuals have been credited with the idea of exhuming the body of an unknown soldier and entombing it in the sacred centre of the British State, 'the Parish Church of the Empire', at Westminster Abbey. Most scholars agree, however, that the idea originated with a young army padre, the Reverend David Railton MC who wrote first to Sir Douglas Haig, and then to the Dean of Westminster, the Right Rev Herbert Ryle (Inglis 1993) in August 1920. *Our Empire* later explained his motives:

> He was worried that the great men of the time might be too busy to be interested in the concerns of a mere padre. He had also thought of writing to the King but was concerned that his advisors might suggest some open space like Trafalgar Square, Hyde Park etc …Then artists would come and no one could tell what weird structure they might devise for a shrine! (*Our Empire* in Gavagan 1995: 9)

The popular press railed against 'weird artists' and were aghast at the exhibitions of official war art that were being shown in London. Railton's letter, however, struck a popular chord, and the Dean soon gained the approval of the Prime Minister, who in turn convinced the War Office and (a rather reluctant) King. Cabinet established a Memorial Service Committee in October. It was hoped that the entombment would take place at the unveiling of the permanent Cenotaph in Whitehall that November.

Necessarily a sensitive act, the selection of a single British body was clouded in secrecy. Historians differ as to the number of bodies actually exhumed, whether four or six. (Wyatt 1939) Whichever, a number of unknown bodies were dug up from the areas of principle British military involvement in France and Belgium – the Somme, Aisne, Arras and Ypres. The digging parties had been firmly instructed to select a grave marked 'Unknown British Soldier', one who had been buried in the earlier part of the war so as to allow sufficient decomposition of the body. The party had to ensure the body was clad, or at least wrapped, in British khaki material (Inglis 1993).

Funeral cars delivered four bodies in sacks to a temporary chapel at military headquarters at St Pol where at midnight on 7th November 1920, Brigadier General Wyatt, officer commanding British forces in France and Flanders, selected one of the flag-draped figures (described later by Wyatt as 'mere bones') by simply stepping forward and touching one of them. Before this ultimate selection each sackload had been carefully picked through to confirm that they were British (or at least British Empire) remains and that no name tags, regimental insignia or any other means of means of identification remained.

While the single selected body was made ready to embark on its highly ritualised journey, the others were quietly reburied. Other countries followed suit: having chosen their 'Warrior', the Americans returned three bodies to the soil without ceremony; in France, at precisely the same moment that the single chosen body was being buried to great ceremony under the Arc de Triomphe in Paris, seven other bodies that been dug up but not chosen, were re-interred under a cross in a Verdun war cemetery.

After an extraordinary choreography of ceremony and ritual the coffin – freshly constructed by the British Undertaker's Association from an oak tree that had stood in the parks of Hampton Court Palace – reached Westminster Abbey. After the clamour of the crowds that lined the railway lines from Dover to Victoria Station, and the masses gathered on the streets and squares of central London, the Abbey was hushed, if not tense with anticipation. Here, as Geoff Dyer has observed 'the intensity of emotion was reinforced by numerical arrangement': one hundred winners of the Victoria Cross lined the route to the burial place; a thousand bereaved mothers and widows stood behind them (Dyer 1995). Lowered into a grave dug in the entrance of the abbey, the coffin was sprinkled with soil from Flanders. Later the earth in the six barrels would be added – 'making a part of the Abbey forever a part of a foreign field' – and the grave sealed with a large slab of Belgian marble.

On Armistice Day that November over a million people passed by the Cenotaph in Whitehall in the week between its official unveiling and the sealing of the tomb. By way of lending a sense of proportion to the nation's loss, it was estimated that if the Empire's dead could march four abreast down Whitehall it would take them over three days to pass the monument, a column stretching from London to Newcastle. Not long after, this incredible idea was rendered actual as endless columns of troops marched past memorials all over the country. *The Times* intoned: 'The dead lived again'. In these memorable images it seems as though the soldiers are the dead themselves 'marching back to receive the tribute of the living'. (Dyer 1995: 24) It is an insight that provokes memories of Eliot's lines in *The Waste Land*:

> A crowd flowed over Westminster Bridge. So many,
> I had not thought death had undone so many.

In post-war Britain it would have been almost impossible to avoid the intensity of remembrance. One authority declared it the greatest period of monument-building since Pharaonic Egypt, (Ware 1937) and Stanley Spencer's painting of the unveiling of Cookham war memorial captures an event that was repeated countless times as the nation sought to mourn the common man. Indeed the line of young men who crowd the foreground of Spencer's painting seem less concerned with paying homage to the dead as vicariously acting out their missing townsmen, a surrogate army of ghosts returned home.

The Dead Rising

In the decade after the war, the image of the dead rising from the tortured landscapes of the old battlefields became a familiar part of the iconography of remembering. During the war, artists had created occasional images of a ghostly figure wandering wraith-like across No-Man's-Land; poets and the popular press had played with the notion of guardian angels or spectred hosts. There were many legendary (and largely apocryphal) tales of 'mysterious Majors', or benevolent phantoms who return to help, warn or merely stand alongside comrades in the twilight hours of stand-to. Many combatants found this entirely understandable; sudden departures and unexplained absences were common experiences, in battle soldiers literally vanished into the air, dematerialised before their comrades' eyes, every trace gone. Sudden absences, emptiness and invisibility became the hallmarks of the war. Despite the scale of commemoration in stone, many of those who returned to the former battlefields craved some form of spiritual connection with their vanished loved ones. In part this explains the upsurge in séances and similar activities in the years after the war, (Winter 1995) and perhaps also the fascination with battlefield pilgrimage and the need to gather 'mementos' or relics from the same landscapes that had apparently swallowed whole the sons, brothers and fathers of the massed armies, and which persists today (Gough 1996).

In film, in painting and even in photography, however, the disappeared and the dead could be made to live again and images of the dead rising from the earth gained a wide currency. In 1927 the *Melbourne Herald* published a drawing – *A Voice from ANZAC* – by war artist Will Dyson, which depicted two Australian soldiers on the shores of Gallipoli, one of them asking, 'Funny thing, Bill. I keep thinking I hear men marching.' That year another Australian artist, Will Longstaff, had attended the unveiling of the Menin Gate at Ypres – with Plumer's rhetorical message 'the dead are not missing, they are here' – and in response had painted *Menin Gate at midnight* which depicts a host of ghostly soldiers emerging from the Flanders battlegrounds and walking, as one, towards the massive monument through fields strewn with red poppies. So struck had Longstaff been by the ceremony at Ypres that he later had a vision of 'steel-helmeted spirits rising from the moonlit cornfields around him'. He returned to London and, it is said, painted the canvas in a single session while still under 'psychic influence' (Gray 2006).

Reproduced in tens of thousands of copies the painting had an extraordinary reception. It was displayed in London, viewed by Royal Command, toured to Manchester and Glasgow and then sent to Australia where it is still exhibited in a darkened chapel-like room at the Canberra War Memorial. Its appeal was strong in part because spiritualism was in vogue, but mainly because those who wished to communicate with the war dead found some consolation in its pictorial verity. His work was championed by the likes of Sir Arthur Conan Doyle who endorsed the spiritualist message it evoked. Longstaffs' work carried none of the venomous acrimony of Siegfried Sassoon's post-war poetry, which by comparison was populated with 'scarred, eyeless figures deformed by the hell of battle ... supernatural figures of the macabre' whom he pitied for the loss of their youth (Dollar 2004: 235). The poets' bitter realism was perhaps more fully shared with the filmmaker Abel Gance, whose 1919 film *J'Accuse* ends when vast hordes of French soldiers – the unjustly dead – materialise out of the tortured earth intent on terrifying the complacency of those who could, if they wished, have ended the war (Gance 1937, Van Kelly 2000).

Having a studio in London, Longstaff may have been aware of Stanley Spencer's *Resurrection, Cookham* (which was on show during February 1927). Spencer had been 23 at the outbreak of war, a student-prodigy and an inspired innocent who would go on to become one of Britain's greatest twentieth century painters, famous (indeed infamous) for two things: the celebration of his home village of Cookham – his 'heaven on earth' as he lovingly called it, and the fusion in his paintings of sex and religion, love and dirt, the heavenly and the ordinary (Hauser 2001).

Spencer's visionary imagination was realised through many hundreds of paintings, endless drawings and thousands of letters, written to both the living and the deceased. They exposed a complicated reading of his world and an ability to transform the menial and the banal into intense images of joyous delight. Through his work Spencer transformed Cookham into a visionary paradise where his family and neighbours would daily rub shoulders with Old Testament figures; and where it seemed entirely appropriate that Christ would wander in the garden behind the local schoolyard.

In 1915 he had left his protective homestead to serve first as a medical orderly in a converted asylum in Bristol, then in a Field Ambulance on the Macedonia Front – a forgotten theatre of war, where a hybrid Allied force faced a strong Bulgarian army reinforced by German troops. Spencer served at the front until the Armistice, joining an infantry regiment in the latter stages of the war (Carline 1978). Some seven years later, as Longstaff was having his vision of exhumed troops marching on Menin Gate and German artists such as Otto Dix were revisiting their Flanders nightmares, Spencer translated his war experience into an extraordinary series of murals on the walls of a private memorial chapel in Hampshire, which are ostensibly about war, but where death is the 'absent referent' lingering in the wings, not even relegated to a walk-on part (Hauser 2001: 64).

The Sandham Memorial Chapel is perhaps the most complete memorial to recovery and redemption ever completed in the aftermath of the First World War. There is nothing like it anywhere in Europe. Spencer referred to it as his 'Holy Box', an affectionate reference to the Renaissance chapels in Padua and Florence, whose simple exteriors and busy interiors he revered. Commissioned by two patrons of the arts as a memorial to a relative who died of illnesses contacted on the Macedonian campaign, its panelled interior depicts Spencer's tedious chores as a medical orderly in Bristol; his field ambulance work near Salonika, and on a vast endwall – some 4 m wide by 7 m high – an epic panorama of recovery and redemption, the 'Resurrection of the Soldiers'. For nine months Spencer toiled on this endwall, his small, tweed-suited figure lost high in the scaffolding amongst the dozens of animated painted figures. After the chapel was opened in 1932 a visitor was heard to pronounce: 'My dear, the Resurrection is not in the least like that!' (Behrend 1967: 27). However, Spencer's idea of resurrection was not one of judgment, nor of the revival of the dead, or the re-appearance of Christ. Instead, it embraced the more holistic idea of the Resurrection of the body and the mind. Spencer's social background and his concern for the 'common man' (and woman) meant that his interest lay in an egalitarian and inclusive notion of the body, one that was indifferent to social hierarchy and ignorant of external trappings and trophies of wealth and position. The Oratory was not a resurrection solely of the dead, nor for that matter a resurrection merely of the soul, rather it was 'resurrections of his state of mind at different times' (Glew 2001: 11). For Spencer it was a time of 'release and change' whereby even the mules and the tortoises come in for some sort of redemption, or re-finding of themselves from their experiences. Even the soldiers under mosquito nets seem to be caught in an act of spiritual transubstantiation, altering from one state to another, and everywhere soldiers emerge from the earth to return their now-redundant crosses to the Christ-figure, just as they had, at the end of hostilities, returned their blankets and kit to the Quarter-Master. 'In the resurrection', said Spencer, 'they have even finished with that last piece of worldly impedimenta' (Carline 1978: 190–91).

In a complicated accumulation of ideas Spencer thought of Resurrection as a 'Last Day', a time of reconciliation, not judgment. It was without doubt momentous, but it was entirely peaceful and calm, with no need for clarion calls or lofty pronouncements. In his interpretation, Resurrection was a redemptory act, a re-finding of oneself freed of the burden of experiences, and a reconciliation of friends, lovers, peoples of all creed and colour, and of course, their belongings. One soldier, for example, takes a small red book from his pocket, identified by Spencer as 'a little red leather-covered Bible' that he had been given by his sister Florence but which he had lost. 'Being the Resurrection', he writes simply and matter-of-factly, 'I find it' (in Hauser 2001: 153). However, despite his protestations of innocence, there is an air of apocalypse about elements of the chapel, traceable in its sombre mood, inexplicable incidents and suppressed fears. Not easily could Spencer ignore the terrible past and the recent present, with tens of thousands of displaced and maimed veterans wandering the land, stranded by the fiscal gloom of the late 1920s (see Figure 14.1).

During the war Spencer had buried dozens of soldiers. He painted the chapel during the years when the former battlefields were being combed for the dead and concentrated into cemeteries. In the early 1920s, when he was originating the murals in Dorset, quarrymen in nearby Portland were hacking out vast slabs of the shelly, coarse white stone to be chiselled into tens of thousands of headstones bound for the battlefields. The Resurrection wall at Burghclere is a testament to those thousands of unknown soldiers who were blown into pieces and who are remembered only in their names carved on panoramic slabs of stone.

Spencer's figures emerge from the torn earth intact, unsullied and calm, almost beatific; very different from the homunculi embedded in the Flanders mud as devised by Otto Dix. In his apocalyptic canvas, *Flanders*, the dawn may be epic, but the demise of the small troupe of soldiers is tawdry and banal, their bodies enmeshed in a thicket of webbing, wire and waste. Far from emerging from the glutinous mud, the soldiers are immersed in the land, becoming a part of its subsoil, embedded in their *totendlandschaft* – the dead landscape – where there may be biological metamorphosis, but there is absolutely no hope of resurrection

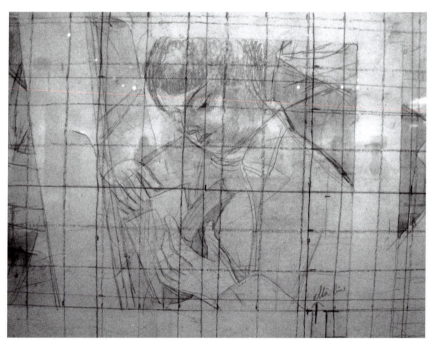

Figure 14.1   Stanley Spencer, drawing study for 'The Resurrection of the Soldiers', Sandham Memorial Chapel, Burghclere

Source: Thanks to the Trustees of the Bishop Otter Collection.

(Eberle 1985: 30). At least the skyscape holds an element of tentative promise, however ironic; in Jeff Walls vast panorama of an Afghanistan ambush even the redemptive possibility of a horizon is stripped out (Wall 1992, Chevrier 2006). Instead, in place of Spencer's serene and demilitarised figures, we find a platoon of traumatised soldiers with bulging eyes and contorted faces, tearing at each other, horsing around and stuffing their spilled entrails back into their soiled uniforms. Wall's dystopia shares more in common with Sassoon's bitter verse or Abel Gance's film in which the dead don't merely wander the earth, they are disgorged in rotting uniforms with mutilated bodies and torn faces, or Max Beckmann's savage panorama '*Resurrection*', which is dominated by a black-sphered sun. In their common scale, their subdued tonal range and their powerful sense of camaraderie there is some common ground between Wall and Spencer. But Wall's gurning and abandoned infantrymen appear to bear nothing in common with Spencer's mute and elegiac armies. Dead soldiers don't talk; but in Wall's visionary photo-piece they do. In fact it's hard to shut them up. His thirteen slaughtered soldiers cavort, play with strips of flesh, smile knowingly at each other, and chat from casual slouching positions. But their pain is palpable. How far is this from Spencer's notion of a reverential resurrection? In its unexpurgated depiction of pain it draws from Callot and Goya, whereas Spencer takes his inspiration from the Italian Primitives. Yet, like Spencer, there is no eye contact with us; no accusation outwards, no one turning into our world. As Susan Sontag says:

> There's no threat of protest. They are not about to yell at us to bring a halt to that abomination which is war. They haven't come back to life in order to stagger off to denounce the war-makers who sent them to kill and be killed. … Why should they seek our gaze? (Sontag 2003: 112)

Perhaps Wall, like Spencer and Dix before him, knows that we are unable to fully empathise with these wretched souls; we will never understand the dreadfulness of war. We can only peer in and glimpse these momentarily reprieved lives. However, where Wall re-imagines the Day of Judgment as something horribly Sisyphean, Spencer dreams a vision of reconciliation and arbitration, even though the figures in his haunted Macedonian hillside appear isolated, disengaged and rather sedated when compared to the livid lunacy of the doomed Russian platoon capering in their cruel crater.

Concluding Remarks

This chapter has addressed some of the key visual and phenomenological tropes of the British experience of the Western Front during the Great War. By focusing on the apparent emptiness of the face of the battlefield, I have been able to deconstruct the nature of absence, invisibility, and the void, suggesting instead that both during and after the conflict the battlefield was in fact a crowded emptiness, crowded with soldiers hidden in noisome labyrinths and 'occupied' for ever after by the bones and bodies of the dead. Quite literally there is many a corner of some foreign field saturated with their limbs and blood and these 'memoryscapes' became the figurative contexts for a succession of artists, photographers and film-makers who brought the battle dead back to 'life'. This approach to re-visualisation has many parallels in the literature of bereavement with its fascination for absences, presences and continuing bonds of place, body and tragic narratives. In re-visualising the dead, I have offered a brief background to the cycle of death, dying, disposal, and in some cases, the ritual exhumation for national causes. Many of those images still resonate today, even if the work of Jeff Wall is unrelentingly dark, the work of Stanley Spencer in particular is highly regarded as an unparalleled icon of redemption, recovery and reconciliation.

References

Arthur, M. 2002. *Forgotten Voices of the Great War*. London: Ebury Press.

Baker, S.J.K. 1988. Paul Vidal de la Blache: 1845–1917. *Geographers: Bibliographic Studies*, 12, 189–201.

Batten, S. 2009. Exploring a Language of Grief in First World War Headstones, in *Contested Objects: Material Memories of the Great War*, edited by P. Cornish and N. Saunders. London: Routledge, 163–77.

Behrend, G. 1967. *Stanley Spencer at Burghclere*. London: Macdonald.

Blunden, E. 1928. *Undertones of War*. London: Faber and Faber.

Bourke, J. 1996. *Dismembering the Male: Men's Bodies, Britain and the Great War*. London: Reaktion.

Carline, R. 1978. *Stanley Spencer at War*. London: Faber and Faber.

Carrington, C. 1965. *Soldier from the Wars Returning*. London: Hutchinson.

Chevrier, J.-F. 2006. *Jeff Wall: Catalogue Raisonne 1978–2004*. New York: Steidl & Partners.

Cloete, S. 1972. *A Victorian Son: An Autobiography*. London: Collins.

Clout, H. 1996. *After the Ruins: Restoring the Countryside of Northern France after the Great War*. Exeter: University Press.

Conrad, P. 1999. *Modern Times, Modern Place: Life and Art in the Twentieth Century*. London: Thames and Hudson.

Dollar, M. 2004. Ghost imagery in the war poems of Siegfried Sassoon. *War, Literature, and the Arts* 16(1–2), 235–45.

Dyer, G. 1995. *The Missing of the Somme*. London: Penguin.

Eberle, M. 1985. *World War 1 and the Weimar Artists*. New Haven and London: Yale University Press.

Farrer, R. 1918. *The Void of War: Letters from Three Fronts*. London: Constable.

Fussell, P. 1975. *The Great War and Modern Memory*. Oxford: Oxford University Press.

Gance, A. 1937. *J'Accuse*. Connoisseur 1937.

Gavagan, M. 1995. *The Story of the Unknown Warrior*. Preston: M & L Publications.

Glew, A. 2001. *Stanley Spencer: Letters and Writings*. London: Tate.

Gough, P. 1993. The empty battlefield: painters and the First World War. *Imperial War Museum Review*, 8, 38–47.

Gough, P. 1996. Conifers and commemoration: the politics and protocol of planting in military cemeteries. *Landscape Research*, 21(1), 73–87.

Gough, P. 2004. Sites in the imagination: the Beaumont Hamel Newfoundland Memorial on the Somme. *Cultural Geographies*, 11, 235–58.

Graham, S. 1921. *Challenge of the Dead*. London: Cassell.

Gray, A. 2006. *Will Longstaff*. [Online]. Available at: http://awm.gov.au/encyclopedia/menin/notes.htm. [accessed: 25 July 2006].

Hauser, K. 2001. *Stanley Spencer*. London: Tate.

Heffernan, Michael 1995. For ever England: the Western Front and the politics of remembrance in Britain. *Ecumene*, 2(3), 293–323.

Hurst, S. 1929. *The Silent Cities*. London: Methuen.

Inglis, K. 1993. Entombing unknown warriors: from London and Paris to Baghdad. *History and Memory*, 5, 7–31.

Jones, D. 1937. *In Parenthesis*. London: Faber and Faber.

Kern, S. 1983. *The Culture of Time and Place (1880–1918)*. Massachusetts: Harvard University Press.

Longworth, P. 1967. *The Unending Vigil,* London: Leo Cooper.

Nash, P. 1949. *Outline: An Autobiography and Other Writings*. London: Faber.

Orpen, W. 1923. *An Onlooker in France*. London, Williams and Norgate: Ernest Benn, 1923.

Osborne, B. 2001. In the shadows of monuments: the British League for the reconstruction of the devastated areas of France. *International Journal of Heritage Studies*, 7(1), 59–82.

Middlebrook, M. and Middlebrook, M. 1991. *The Somme Battlefields*. London: Viking.

Plowman, M. 1927. *A Subaltern on the Somme*. London: J.M. Dent.

Redmond, W. 1917. *Trench Pictures from France.* London: Andrew Melrose.

Roberts, W. 1974. *Memories of the War to End War 1914–18*. London: Canada Press.

Sassoon, S. 1930. *Memoirs of an Infantry Officer*. London: Faber and Faber.

Sheffield, G. 2001. *Forgotten Victory: The First World War Myths and Realities*. London: Headline.

Shepheard, P. 1997. *The Cultivated Wilderness: or, What is Landscape?* Cambridge: MIT Press.

Slowe, P. and Woods, R. 1986. *Fields of Death: Battle Scenes of the First World War*. London: Robert Hale.

Sontag, S. 2003. *Regarding the Pain of Others*. New York: Picador.

Stein, G. 1938. *Picasso*. London: Heinemann.

Stuart Dolden, A. 1980. *Cannon Fodder.* Blandford: Blandford Press.

Talbot Kelly, R. 1980. *A Subaltern's Odyssey: A Memoir of the Great War, 1915–1917*. London: William Kimber.

Tawney, R.H. 1953. *The Attack and Other Papers*. London: George Allen and Unwin.

Terraine, J. 1980. *The Smoke and the Fire: Myths and Anti-myths of War*. London: Sidgwick and Jackson.

Van Kelly, A. 2000. *The Ambiguity of Individual Gestures*. South Central Review, 17(3), 7–34.

Wall, J. 1992. *Dead Troops Talk (a Vision After an Ambush of a Red Army Patrol, near Moqor, Afghanistan, Winter 1986)*. Transparency in lightbox, 2290 × 4170 mm, Marian Goodman Gallery, New York.

Ware, F. 1937. *The Immortal Heritage*. Cambridge: Cambridge University Press.

Williamson, H. 1988. Foreword to *The Wipers Times*, editor Patrick Beaver. London: Papermac.

Winter, D. 1978. *Death's Men*. London: Allen Lane.

Winter, J. 1995. *Sites of Memory, Sites of Mourning: The Great War in European Cultural History*. Cambridge: Cambridge University Press.

Wyatt, L.J. 1939. The unknown warriors of 1920: how one was selected for Abbey burial. *Daily Telegraph*, November 8th, 23.

Chapter 15
Art and Mourning in an Antarctic Landscape

Polly Gould

How does the past engender the future in our personal lives and in art? I have approached this question as an artist. Or rather, as an artist, the problem of art's history, and my place in it, is of concern to me. It gives me something to think about and to worry over. Something else that has given me considerable pause for thought is the death of my father. This has also furnished me with something to think about.

Both of these situations have generated a comparable anxiety, although perhaps on differing registers of feeling. They have both connected me to the question of how the past generates the future. The question of what can otherwise be termed as the sometimes acute problem of inheritance. Not wishing to sound over-dramatic, but preferring to put it bluntly and in the banal vernacular, what I am talking about is the question of how to go on?

What follows is a recollection of a journey. It was undertaken in my imagination with regards to looking at artworks and making art, and it also took place in the outside world. It was a journey in search of a place to go to, an imaginary or psychological space to retreat to, in order to undertake the task that death leaves us with; the work of mourning as Freud termed it. Each of us uses whatever is at hand in undertaking this task. In my case I have the habit of looking at art, of thinking about it, of making it.

What I wish to put forward is the connection between mourning for a dead parent and a state of mourning to be identified in art and the burden of inheritance in both. I propose that it is pertinent to theorise from biography to history and possible to extrapolate from something particular to something general, and, from something that is individually true to something more widely the case. I also examine this question through the prism and perspective of my own work as an artist, and take it as a place from which to start.

Recollection is a feature of my aesthetic. My work re-combines fragments in the pin-board of paper ephemera, printmaking and the photographic print. I work with sequences of drawings, composite printed pieces and samples used in performances, video and audio works; the bits and bobs, flotsam and jetsam of everyday life. I seek to contrive narratives from a technique of collage, citation, montage and assemblage. Art can be a way of encountering and mediating absence and loss, a way to mourn through our engagement as viewers, collectors or makers.

This is the territory of art and its relation to time, to history, to the past, to memory and loss. This chapter is a search for transcendence in a time in which, for some (and for me), the afterlife is bankrupt as an ideal or as a realistic hope. Encountering the relentless cycle of losses which time's passage brings confronts one with the existential problem of how to be in a present that holds no promise of a redeeming future. Does the fantasy of another place become active in this context? Can there be a spatial response to this problem of time? Can imagining the future as another place (rather than a later time) offer some answer to the problem of time and mourning? Is there something particular about Antarctica that made me want to go there in order to undertake this work of mourning? This writing is a journey through those questions.

As my father lay dying in the sterile and mechanised space of the hospital, in that time of waiting with nothing to be done, I made a few drawings of him. I took out my journal and pencil that I always carry with me and passed some of that most interminable of time (time that one does not wish to end but which will end with a terrible certainty) by tracing out a few lines onto the empty pages. I took myself simultaneously in and out of that moment by looking at him and making a likeness of his features.

You can think that you know a familiar face until, that is, you try to see the person in your mind's eye. You can become tormented at how difficult it is to recall the impression of them, let alone the details. The faces of those that we love best become the hardest to retrieve because, perhaps, we hope to achieve a re-constitution and resurrection more materially convincing than casual recollection. We want to bring the remembered person within our grasp, to make them tangible or to bring them back to life.

Mantegna's *Dead Christ* (Figure 15.1), painted nearly 500 years ago, is 89 centimetres (cm) by 71 cm in dimension, just wide of a perfect square, and portrays Christ dead and laid out before us.

To the upper left corner of the painting two tearful, grieving faces look on in mourning at his form. This painting is odd and notable as Christ is shown to us lying down, his body foreshortened by perspective. The crucified soles of his feet are exposed to us at the bottom of the painting and his head, turned slightly to the right, away from his attendants, at the top of the painting. There is a pained expression on the face. He is set out before us in a vulnerable pose; not as a vertical portrait but boxed into a foreshortened view. It is a claustrophobic, entombed composition of the Renaissance, circa 1480. This is the dead Christ and not the risen Christ. The figure is made of earth and matter. The density of his body weighs it down, as heavy and stone-like as the stone slab upon which it rests. Christ as a corpse. We are invited to bear witness, to see for ourselves, joined with the mourners pictured at his side. 'Seeing with one's own eyes' is the meaning of the Greek word *autopsia* from which our contemporary term autopsy is derived. We are given the illusion of perspectival realism as if the scene were actually laid out before us. The perspective seems realistic but is actually distorted for the sake of vision. Reality can interfere with perfect visibility, so, although the

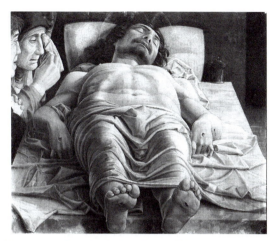

Figure 15.1 **Mantegna** *Dead Christ*, **89 × 71 cm, oil on canvas, fifteenth century, Pinocoteca di Brera, Milan. Reproduced under licence from the Italian Ministry for Culture Goods and Activities**

development of perspective in painting pursued the real and convincing depiction of the world, here the reality has been sacrificed to aesthetic requirements, to the demands of a good painting. Mantegna has reduced the proportion of Christ's feet in order to make the rest of the figure visible. A white shroud of fabric folds around the contours of his legs, groin and across the stone slab. The white cloth of the shroud holds blue shadows and warm yellow highlights in corrugated striations. His hands are folded over with their fingers bent underneath. The stigmata are exposed on the back of his hands. His skin is lined and veined like the markings in the marble that he lies upon, and the coloured veins in the onyx jar at his head. The stone holds more fleshy tones than Christ's pallid skin.

I sat at the foot of my father's bed viewing his body from a similar perspective as that shown in Mantegna's painting. I drew the lines representing his body against the white expanse of the bed sheets and filled them in with a watercolour wash; the pre-eminently portable medium of the landscape artist. I painted the folds and creases of bedclothes, the turquoise-green of the hospital gown breaking the white of the sheets, my father's pale skin and the wisps of fine grey hair, a few colours in the shadows and the chiaroscuro helping the likeness take form. The layers of fabric were folded over his body, which showed through like underlying geological strata. My father, dying, lay there exposed and unconscious, his un-seeing body vulnerable under my gaze. I exchanged the look of waiting relative for the artist's eye. I viewed him from the end of the bed along a line of perspective that shrunk his covered body. To look at someone with the eye of an artist takes that person outside of straightforward recognition. It makes them strange. The person retreats as you turn their skin and flesh into patterns of light and shade. In the process the distinction between what *is* and what *is not* them does not matter.

What matters is the illusion of form, the composition of line, the distribution of tone and the combination of colours.

I sat at the foot of his bed and drew a few pencil portraits. I sketched out and filled in a foreshortened view of his body lying under the bed sheet. The bare skin of his arm, elbow to wrist, crossed the bottom horizontal of the picture plane. I traced a view of the side of his face, his cheek bone and eye socket crowned by grey-white hair, against the pillow. The creases and curves and twists of white-grey-green bedclothes folded into the shapes of his arm and the side of his face. The watercolour paper was turned side-ways to a landscape format in which the horizontal plane dominates. In the portrait format the vertical axis dominates. To view a landscape we expect the horizontal axis to extend to the horizon giving us scope for wandering eyes to explore the sea, the mountains, the plains, to extend into the illusionary depth of the space and to roam across the picture plane.

My act of drawing passed the time. My drawing made a record. Drawing created a trace or a document. The drawing bore witness to what was there and what was happening, or was it a way of trying to hold on to what was imminently passing, an act of reparation in anticipation?

Art has an affinity with loss. You can only really make art out of dead things. In the philosophy of language put forward by Hegel, amongst others, we encounter the intimate bond between representation and absence. This 'linguistic turn,' as it is termed, describes how the *word* displaces the *thing*. Once a dog is named as 'dog,' the concept rises up in place of that particular hound, and dislodges the real in preference for the representation: it replaces the signified for the sign. So too, is the pencil-line founded on the absence of that which is described. Some theorists go further and even impute something murderous in the act of representation. Others see it as a process more akin to recuperation. From this perspective, absence and loss are inextricably bound up in the very possibility of conceptual thought. They are the preconditions for language and representation. In an apparent paradox, absence is the ground for thought.

Works of art also persist in the absence of what they portray. Over five hundred years later, Jan van Eyck's portrayal of a small dog (Figure 15.2) at the feet of the husband and wife in the *Portrait of Giovanni Arnolfini and his Wife* remains. Some commentator's interpret the dog as a sign of fidelity, others as a sign of lust. Either way, for them it signifies something to do with the couple's conjugal bond. It is the dog that gazes out at us and holds our eye, not the putative subjects of the portrait.

This painting has been interpreted as a marriage certificate, as a way to bare witness to their conjugal contract. In rather a sad contrast to this, the painting has also been interpreted as a memorial to a dead wife. Margaret Koster (2003) has proposed that this painting is a posthumous portrait, painted in memory of Constanza Arnolfini after her death. Koster argues that the dog is included not as a symbol of fidelity, but as a common symbol found at the feet of effigies of the time, indicating a companion in the after-world.

Figure 15.2 Jan van Eyck, a detail from *Portrait of Giovanni (?) Arnolfini and his Wife*, 82.2 × 60 cm, oil on oak, 1434, National Gallery, London. Reproduced with permission of the National Gallery Picture Library

In the historical reading, art in its pre-modern and classical embodiment was the cultural form for preserving what is known and transmitting this from generation to generation. The artist was consequently concerned with the eyewitness role of art. The formal developments in art consequently pursued techniques that would mimic reality; the use of perspective to create an illusion of space in the flat picture plane, or the skills of the artist to create the look of truth in their paintings, to make a likeness or a copy. The task was to make a portrait look like the person, or a drawing of a dog *look* like a dog. With the advent of the new technology of photography in the nineteenth century, this mimetic function of art was usurped.

In amongst the sequence of portraits by E.J. Bellocq (1873–1949), taken in New Orleans between 1911 and 1913 and known as the *Storyville Portraits*, there is a portrait of a young woman with a dog (Figure 15.3). She is wearing a shift dress that looks like a nightgown or her undergarments, and a pair of bloomers. She is sitting outside on a simple wooden chair on a brick paving. A paper screen is scrolled out behind her, creating a photographer's studio in the open air. The photo looks spontaneous, and she is smiling, holding the shiny black dog on her lap, leaning into its shoulder, its sharp little ears next to her cheek and its lolling panting tongue hanging out.

Every photograph is an intimation of mortality. With the advent of photography in the mid-nineteenth century traditional painting was usurped as conveyor of cultural and personal memory. Art was left to tag along behind the representational skill of the mechanical eye. Art was dethroned from its role of representing reality and left with the choice of imitating the photograph or turning back onto its own subjectivity. As this turn developed, the medium of the artwork rather than the subject matter pictured through the medium becomes the focus of artistic endeavour. This can be characterised as the closer consideration of the material process and the self-conscious mark of paint on the surface of a painting, which

Figure 15.3   E. J. Bellocq, (1873–1949) *Storyville Portrait,* **ca.1912, MoMA,
plate 31, image by E.J. Bellocq © Lee Friedlander, reproduced
courtesy of the Fraenkel Gallery, San Francisco**

increasingly produces melancholic objects, rather than images of mourning or
memorial.

Freud's essay *Mourning and Melancholia* provides a useful distinction
between these reactions to loss. Freud writes that mourning and melancholy are
comparable because of their shared similarities. Mourning is a reaction to the loss
of a love object. Sometimes the loss is of a more ideal object such as the loss of
a value like one's nation or one's career. The object hasn't actually died but has
been lost as an object of love. Mourning is time-limited and is in response to
some external loss that is commonly recognised as reason enough for the grieving
person's behaviour. Melancholia and mourning display similar symptoms; a loss
of vitality and interest in the world, a sense of hopelessness. Melancholia might be
better described as an ongoing state of mourning that might otherwise be known
as depression and which does not work itself through. This is usually because the
basis of the melancholia remains in some ways unknown to the sufferer. As Freud
puts it:

> He knows *whom* he has lost but not *what* he has lost. (Freud 1917: 254)

It is the task of psychoanalysis to turn depression or melancholia into ordinary
grief. Freud doesn't make claims for the ability of psychoanalysis to create
happiness, but rather aims for ordinary sadness. In melancholia, the grief is
directed upon the self:

The analogy with mourning led us to conclude that he had suffered a loss in regard to an object; what he tells us points to a loss in regard to his ego. (Freud 1917: 256)

In melancholia, there is confusion between the loss of self and the loss of the other. Freud applies some of his theories on loss, melancholia and mourning to the understanding of the creative person's constitution. Psychoanalytically speaking and on the level of personal motivations, Freud proposes that the imaginative faculty comes from dissatisfaction and this is the cause of the drives or ambitions that propel creative characters. At the root of it is an experience of loss. Artists seem compelled by a more or less knowing or unknowing sense of loss. It is as if all artists are harbouring an unconscious motivation that is aware that a loved and whole world has been ruined, and they are consequently compelled to engage in their creative making as a process of reparative acts, that re-members and reconstitutes that ruined world, to make it available to them, once more. Art implies destruction, loss and restoration. The creative act is a successful act of mourning. In the case of unsuccessful mourning, melancholia ensues.

When John Dwight's daughter died on 3rd March 1673, at the age of six, he commissioned his pottery factory in Fulham to make a figurine in her image (Figure 15.4).

Figure 15.4   John Dwight's Fulham Pottery, *Lydia Dwight: Figure of Lydia Dwight on her deathbed*, 25.5 × 20.5 × 11 cm, hand-modelled and salt-glazed stoneware, 1673, Victoria & Albert Museum, London

John Dwight's technical innovations in the production of stoneware in England took the medium of clay and fired it into something as durable as stone, thus taking into the potter's hands the potential for the longevity that a stone mason might know. He started with the production of jugs and pots in the 1670s and culminated in busts and figurines; stoneware portraiture. It was this technique that he was in the middle of developing when his daughter died.

Lydia Dwight on her Deathbed circa 1673 was made of hand-modelled and salt-glazed stoneware. This semi-recumbent bust, measuring 25 cm by 20 cm by 11 cm deep, is square in composition. The white-grey piece pictures Lydia with her small hands clasped around a bouquet of flowers and her eyes closed, her little mouth tight shut with turned down corners. The figure features her bust only. The folded top of the blanket, under her linked arms, marks the limit of the effigy. The material of her little shirt follows around her wrists in moulded creases and folds. The hood of her headdress is figured in a gentle twist of fabric, circling her head to around her chin. It is met there by an open daisy at the top of the posy, like a miniature reiteration of her little face petalled by the hooded shroud. The detailing is precise; the lace-work is featured as tiny patterns of five-petal flowers, made up of pinpricks. She is resting her head against a plumped-up stone pillow. The contrast of the soft flesh represented and hard stoneware that it is fashioned in, adds poignancy to the impression. It speaks of the manner in which the coldness and stillness of death drains the warmth, colour and movement from those, once living, limbs. The colour of the cream-white stoneware of the face and hands merges one into the other.

Like Mantegna's painting of the dead Christ, the stone colour of the skin and fabric all meld into one continuum. In contrast to the Mantegna portrait of Christ, which seems to portray the feeling of lost hope by showing us the vision of the pained cadaverous and all too human body of the Son of God, Dwight's portrayal of his daughter expresses symmetry and composure in death. The suggestion of faith in the afterlife, that Lydia Dwight on her Deathbed might intimate, is corroborated by the companion piece, *Lydia Resurrected.*

This full body figurine standing only 28 cm high, she is diminutive but whole. In this piece she is shown upright, vivid and alert, her eyes open and gazing heavenwards, as if to indicate her anticipated journey into the afterlife. She is draped or dressed in her shroud, which falls around her like a Grecian toga, hanging down from her shoulder and fastened near her waist. No dog here, she carelessly ignores a skull at her feet, and steps upon a scattering of petals. These two portraits show mourning and its consolation; the deathbed scene and the ascent to an afterlife in heaven. Here she remains, in these little figures of love petrified, holders of a private grief, tokens of memory and reflection.

In *Sustaining Loss: Art and Mournful Life*, Greg Horowitz (2001) says that to associate the advent of modernism with the appearance of images of death in art is clearly an untenable argument. He points out that Western art is full of images of the dead Christ. The difference is that these dead Christ figures are understood within the Christian narrative of death, sacrifice, redemption and resurrection. The two figures of Lydia Dwight exemplify this. They are images of death expressed in the wider narrative of faith in an afterlife. *Lydia Dwight on her Deathbed* does not remain alone, she is not imagined as fixed in her corpse, but is partnered with the figure of *Lydia Resurrected*. Horowitz distinguishes between these images of death and those found in the modernist image, as follows:

> Death does not stay put in the modernist image; it fatefully penetrates the very medium of art, turning the medium into the site of non-transcendence. Whereas the dead Christ was the very image of life eternal, Courbet's trout is the image of eternal death. (Horowitz 2001: 2)

Perhaps Mantegna's *Dead Christ* speaks to our modern sensibility of the lack of transcendence; our lack of belief means that we founder on the shores of the painted surface, and cannot use it as springboard into contemplating resurrection. We are stuck in the medium as 'the site of non-transcendence.' We have lost the eyes to see within that perspective of redemption, so remain with the dead.

Horowitz characterises the pre-modern function of art as being mnemonic, of making memories, of holding memories and passing them on:

> This idea of self-transmission of art can be grasped most easily through one of its most obvious instances: carved or incised stones used for funerary monuments. In older cemeteries one finds monuments etched with versions of the imperative 'forget-me-not.' Such stones are the most powerful testimonials to the profound forgetfulness of human beings. (Horowitz 2001: 13)

In a paradox, neatly put, he expresses the vulnerability of our memories to loss, and the need to make them concrete in order for them to survive the passage between generations and the passing of generations. It is pertinent that stone is chosen for this purpose:

> The stone is the medium in which the striving for transmission is carved because the stone can be counted on to transmit itself. (Horowitz 2001: 14)

The form for lasting memory is sort through the materiality of stone, the medium of brute nature, with its qualities of endurance. Stone evokes the scale of geological time, against which our human lives are rendered insignificant and transient in the extreme.

Horowitz is exploring what he calls the 'crisis of inheritance'. He discusses the transmissive nature of art, which started to falter after the demise of classical Greece. Inheritance becomes problematic in art once newness becomes the guiding principle and the criteria for judgement to a discerning viewer. To remain at the forefront of artistic modernism demands constant revolution and the over-turning of precedents by innovation and progress. Cultures at other times and places have been less concerned with values of newness; in fact, they might have identified inventions as mistakes and misapprehensions in the proper aim of the continuation of tradition. Modernism has not sought continuity in the renewal of tradition, but the sensation of the radical cut with the past that an absolute break might offer; the shock of the new, like the crashing down of tonnes of compressed ice, breaking away from the monolithic mass of a slow glacial continuum. But utter newness risks incomprehensibility. The sensation of newness is sought within recognisable forms, which are therefore accessible to our cognition.

John Dwight's dead daughter also provoked a crisis of inheritance. His child will not, in her turn, outlive him and go on to bear her own children, nor take a stab at fashioning her own name into an immortal inscription upon the history of art. What persists is immortality through art; the name of the artist, the skill of the maker who profits on the absent and represented body of loss, to make a beautiful portrait, the silky white sleeping face of a dead daughter in stoneware.

The 'crisis of inheritance' in the wider culture is traced by Horowitz back to the point at which the concerns of art and philosophy split. Up until the advent of modernism the questions of art and philosophy had been in keeping. He describes modernism as unable to inherit a past which it none the less 'ceaselessly disgorges'. The endeavours of modernism entail a continual return to a past that it cannot inherit but cannot discard. This might be characterised as a mourning that cannot see itself through, that gets stuck in a continual compulsion to return; a melancholic stasis of repetition without progression.

So, in contrast to Freud's a-historical and personal psychoanalytical perspective, Horowitz's reading is gained through an historical interpretation that distinguishes the social and cultural context for the production of art. He identifies different eras: the pre-modern, in which artists acted as the witnesses to culture and the makers of artefacts, the purpose of which was to transmit that culture to the future; and the modern, in which confidence in the progress of history has broken down and artists are now solitary individuals left to grapple with existential anxiety, caught in a melancholy dilemma, expected to endlessly invent themselves anew while also being obliged to inherit the past.

As a contemporary artist, making art here and now, this is really the simple question of what to do next in the wake of the modernist avant-garde of nineteenth and twentieth century Western art. Elkins describes it as an endgame scenario:

> In visual theory, endgame art is a postmodern condition in which little remains to be done, and yet it is unclear whether the 'game' of art can actually be ended. Endgame artists make minimal moves, trying to finesse the dying mechanisms of art a few more incremental steps. (Elkins 2002: 35–6)

We are like the eager faces of the students of anatomy gathered around the dissected corpse in Rembrandt's 1632 *The Anatomy Lesson of Dr. Nicolaes Tulp*. The group portrait of the Amsterdam surgeon's guild depicts the cadaver subjected to scientific inquiry, undergoing an autopsy. It pictures those men in the process of an attempt to transcend mortality in the application of medical knowledge, investigating the body of the corpse to give clues to the mechanisms of life in order to evade death. Contemporary artists are gathered around the corpse of art, wondering what it might reveal to them, and how it might give clues to what steps might now be taken.

So the philosophical argument, the psychoanalyst interpretation and the art historian's analysis all re-iterate art's affinity with loss. The question of how to inherit art's history, and how to make art anew reverberates with this personal drama of inheritance. After my father's death, after his body was gone, I wonder what is to be done with his corpus, *his life's work*. How does one go on making art in the post modern and how does one live on post mortem? How does one deal with a past that one cannot inherit and cannot discard?

Approaching the third anniversary of his death, I was still perplexed by the hiatus, this halt in temporal progression that mourning had produced in me. I resolved to break open time with a journey through space to one of the most remote and least human places in this world, to the Antarctic Peninsula. So, I took the drawings of him with me and set them out, there amongst the penguins and ice and the white expanse of glaciers. Why?

In Modernism we encounter the presence of the past as fixed and necessary with the future opening out as the domain of freedom. Catastrophe is understood as a necessary event in the progression to a new and better future. The philosopher Agnes Heller in *The Theory of Modernity*, says that to modernists the past is seen as an unchangeable necessity and the future as total freedom.

> Modernism experiences 'being in the present' as living in a transitory state, stage, or world, compressed between past and the future. The past, which was normally seen as 'necessary' (for the reason that it cannot be changed) was supposed to have lead up to the present – the present on its part as a limit, a 'just now,' an insignificant moment which always transcends toward an infinite future, being conceived as the territory of freedom. (Heller 1999: 7)

I think I have a tendency to take the fantasy of travel as an escape from already determined past to totally free future. By train, boat or plane, this freedom of movement through space, to other places, comes to imitate the freedom from the restriction of our own histories, the limits of time and even the limits of our predetermined future. The ultimate pre-determined limit of time that our future faces us with is death; our own death and death of those that we love. So this fantasy of travel is a strategy for evading death.

In some muddle headed way, typical of the bereaved, I think I was pursuing two divergent strategies at once: the modernist dream of escapist travel, and dwelling with the past by turning history into geography. I had the impulse to go elsewhere, to step out of my sense of destiny, and to find a place in which to encounter the ongoing presence of the past that I was not yet willing to let go. I was also seeking a location for the afterlife in another place – an elsewhere. I was seeking somewhere in which the image of my dead father might have some hope of resurrection, rather than find myself locked into the irredeemably dead materiality of modernist art, held like a pinned moth on the surface of the paper, unable to take the imaginative leap into the next life, or even into a possible future.

It was with this in mind that the idea of a trip to Antarctica occurred to me. I wanted to go somewhere that I thought was nearly impossible. Antarctica is unpopulated and until recently had no human life on it. As a consequence it was outside human history. The human stories that relate to it are marginal and extraordinary. Time takes place on a geological scale, not on the scale of human biographies. In human terms it was without a past until the relatively recent history of Antarctic exploration.

So I sought out the sight of the sloping gradient of a glacier with its blue-ice strata of millennia-old frozen water. I wanted to see the way these masses of ice repeat the movement of a flowing river in slow motion. I wanted to see this sight, to try to perceive the passage of time in the scene that seems to have the stillness of a painting, that is, until a great piece breaks away and comes crashing down to the waters beneath in one catastrophic shift, a sudden acceleration.

It takes human effort to undertake the work of mourning. It takes a human effort for sense to be made of the forward movement of time. The forward movement is a quality that we impose upon our sense of time in order to understand it in terms of our mortal lives. Patterns form at a scale of the threads of human biographies weaving into communities of human histories on the time scale of human memory. It is not the memory of rock and stone, that takes the imprint of millions of years in their slow procession from earth core to surface, and which are the material of their own transmission, but the human scale of daily lives. These frozen mountains, icebergs and glaciers care nothing for us. Our memory is measured by the span of human lifetimes, and the chain of those lifetimes linked through personal memory to the collective memory of a shared historical past, and it relies upon the efforts of other human minds and lives to sustain them.

With this in mind, I booked myself upon a tourist trip and, after a considerable journey by air and sea, ended up on the shore of the Antarctic Peninsula. I carried the piece of watercolour paper with me that I had scrawled a few marks upon while teetering on the edge of that other shore, the 'just-before' crossing over to the 'afterwards' that the death of a dearly beloved person marks. I had kept it for the previous three years. The difference between that 'before' and 'after' feels like the space of separation between distant lands. These events structure our lives with the kind of seismic shifts that used to give rise to new continents. It is comparable to the earliest history of our planet, when continental shift broke one landmass into a set of many, sending Antarctica southwards to form the continent at the frozen antipodes. Before, while sitting by my father's hospital bed I had made those fragile drawings, and afterwards I had taken them, as a shadow of his former self, as a little scrap of paper and paint, and propped his image up on a pebbly beach on the Antarctic Peninsula. I set up a video camera resting on the beach and framed the watercolour resting on the stones (Figure 15.5).

The paper occupied the lower part of the frame. The left-hand edge of the paper, full of the broken perforations from the ring binder drawing pad, created a series of little cut-through windows to the scene behind. The top edge of the paper made a steep gradient descending from left to right, with the seashore and

Figure 15.5   Polly Gould, *Peninsular*, video still, 2005
Source: Image by the author

sea behind. The horizon divided the scene. The boat that had brought me there was at some distance at the far edge of the screen. Out in the far distance, the snowy glaciers and mountain peaks were softly coated. Streaks of light cloud were scattered across the sky and a pale blue showed through. In the mid-distance, a seal and a penguin moved away from one another in their funny lolloping gaits. They made a comical scene; from my position, it looked as if they were walking along the top edge of the paper upon which my father's deathbed portrait was scribbled. These creatures were oblivious to this. It was an illusion solely generated by my perspective.

I did not go there with the intention of making a work of art, but I did want to create the potential for a successful work of mourning that might also be the event of a creative act. It was a mixture of escapism and reparation, of going somewhere out of my own world, in which I might have some hope of letting go of the burden of inheritance. The Antarctica that I encountered was not a frozen waste, but a living place; not human maybe but certainly alive. In the same light as Horowitz characterises Freud's view of the work of art 'as the haunting of the present by the endless suffering of what is dead but still has claims over the living' (Horowitz 2001: back cover), I went to this other place as a way of stepping out of the dead past. Our own histories become the ground of ancestral burial sites, and 'art' is the noise of dead spirits, which cannot be put to rest.

There was no easy solution to my mourning. My motivations for going there were multiple and contradictory. There were no simple answers to the problem of

how to go on afterwards. I did not find transcendence, or escape, or a promise of a
next life but I did find a re-engagement with this one.

> ... modern historical self-consciousness can avoid falling back into myths of
> transcendence only by acknowledging that the past, while dead, is not gone, and
> that we coexist with it not as its afterlife but as its survivor. (Horowitz 2001:
> 22)

Antarctica, a land without ruins, is like a person without ghosts. This is perhaps
the appeal. What succour to the haunted, to encounter a place that has no past in
that sense, no historical ruins related to human lives, but which exists only as a
vivid tableau of nature regardless of human time. Antarctica is a continent outside
of history.

Antarctica had appealed to my Romantic notion of the artist's lonely journey
into awesome and sublime nature in the mode of Caspar David Friedrich, man
amongst the mountains. This resonates with the fantasised subject of artistic
individualistic self-expression, as well as the trope in Western philosophy of the
lone self, facing death in confronting the empty void. The mingling of mourning
with melancholia, in which the loss of the other becomes a loss of oneself, could
perhaps be disentangled in such a place; a place not riddled with memories and
histories but a space through which, and in which, I might be able to structure an
encounter with my authentic self. I imagined myself as Friedrich pictured himself
in the *Wanderer above the Sea Fog*; heroic, alone, facing the awesome beauty of
an un-peopled nature. In my imagination, Antarctica figured as an empty space, a
pure landscape of frozen wastes.

So it was that I found myself on a small Russian ship, sailing the waters of
the Antarctic Peninsula with thirty other tourists, a crew and a team of expedition
guides. Although there was no indigenous human population for thousands of
miles around, the immediate vicinity of the boat was densely peopled. I maintained
a modicum of aloneness by not talking much to my companions.

While on this boat, thinking about the claims of the dead and wondering how
I might inherit certain things, as I grappled with, for example, the question of
transmission, the dilemma of how to act in this world or the next, my existential
problems of how to *be* an artist, a fellow traveller asked me if I was an artist, to
which I replied, 'Yes'. Then she asked if I could do portraits. She did not ask if I *did*
portraits but if I *could* do them. I said, 'Yes I can.' I can draw. I *can* make a picture
of something look like its subject. I understand the principles of perspective. I can
make a representation of the reality I see with a pencil line and some colour. I can.
And she promptly pulled a sentimental photo of herself and her fluffy pet dog out
of her wallet and asked me to paint it for her.

And so I found myself, having flown to the other side of the world and taken
a ship across the notoriously torrid waters of the Drakes Passage, sailing between
the most astonishing icebergs and glacier landscapes in one of the most remote and

empty places on the planet, copying faithfully from a photo snapshot of a woman and her pet dog, to make a watercolour portrait as a souvenir (Figure 15.6).

Figure 15.6   Polly Gould, *untitled*, 14 × 12 cm, watercolour on paper, 2005
Source: Image by the author

Antarctica is outside of history. But I am not. Art is mnemonic. It is a salve against human forgetfulness, a memorial stone, a reparation for something lost, an *aide-mémoire*, a keep-sake, a witness, a souvenir. There is no art in Antarctica, no history, no humans, no ruins, no remnants of a lost civilisation. There is no rubbish on Antarctica. International protocol declares that you are not allowed to pee or shit on the continent either, as human excrement is not naturally indigenous to the environment. You can visit and look on with wonder, but you have to take all your crap, all your rubbish, all of your memories, back home with you.

References

Elkins, J. 2002. *Stories of Art*. London: Routledge.
Freud, S. 1917 [1915]. Mourning and melancholia, in *11: On Metapsychology*. London: Penguin.
Heller, A. 1999. *A Theory of Modernity*. Oxford and Massachusetts: Blackwell.
Horowitz, G.M. 2001. *Sustaining Loss: Art and Mournful Life*. Stanford, California: Stanford University Press.
Koster, M.L. 2003. The Arnolfini double portrait: a simple solution. *Apollo: The International Magazine for Collectors*, 499, 3–14.

Index